Principles of Analytical Chemistry

Springer-Verlag Berlin Heidelberg GmbH

Miguel Valcárcel

Principles of Analytical Chemistry

A Textbook

With 132 Figures

 Springer

Miguel Valcárcel
University of Córdoba
Department of Analytical Chemistry
Avda. San Alberto Magno, s/n
14004 Córdoba
Spain

ISBN 978-3-642-62959-4

Library of Congress Cataloging-in-Publication Data
Valcárcel Cases, Miguel.
  Principles of analytical chemistry : a textbook / Miguel Valcarcel.
    p. cm.
  Includes bibliographical references and index.
  ISBN 978-3-642-62959-4   ISBN 978-3-642-57157-2 (eBook)
  DOI 10.1007/978-3-642-57157-2
  1. Chemistry, Analytic. I. Title.

  QD75.2 .V35 2000
  543–dc21   00-033829

© Springer-Verlag Berlin Heidelberg 2000
Originally published by Springer-Verlag Berlin Heidelberg New York in 2000
Softcover reprint of the hardcover 1st edition 2000
The use of general descriptive names, registered names, trademarks, etc. in this publication does not imply, even in the absence of a specific statement, that such names are exempt from the relevant protective laws and regulations and therefore free for general use.

Product liability: The publisher cannot guarantee the accuracy of any information about dosage and application contained in this book. In every individual case the user must check such information by consulting the relevant literature.

Cover Design: design & production GmbH, Heidelberg
Typesetting: Fotosatz-Service Köhler GmbH, Würzburg
SPIN: 10575227   52/3020/M – 5 4 3 2 1 0 –

Foreword

In 1988, Professor Julian Tyson published an informative little book entitled "Analysis: What Analytical Chemists Do." Professor Valcarcel's text could well be called "What Analytical Chemistry Is, What Analytical Chemists Do, and How to Do It Right."

Thus, it is concerned with such things as what information is really needed, cost effectiveness considerations, tradeoffs in analyses, quality control and good analytical practice.

This book departs from traditional texts in concentrating on the structure of analytical chemistry, introducing students to analytical concepts, how to obtain reliable data, and how to interpret it. It does not teach the myriad of techniques and methodologies, which are reserved for the more conventional texts. It serves as an introductory text to be taught at the first year level or as a complement to conventional analytical chemistry texts. Practitioners of analytical chemistry will also find it useful for placing in context all that they do.

Professor Valcárcel takes a unique hierarchical approach to define and describe concepts and analysis in general. The text builds on and integrates a number of concepts he has introduced over the years in the analytical chemistry literature (see, for example, Suggested Readings 1, 2, 3, 7 and 9 in Chapter 1). The author makes generous use of relational figures throughout the text to aid the student in visualizing the concepts introduced in each chapter. The reader is recommended to look first at Figure 1.10, which nicely integrates and places in context the topics covered in the remainder of the text.

Each chapter is introduced with a list of learning objectives, a useful way to put the topics in perspective. There are boxed materials throughout that provide more detailed examples of the topics introduced. And at the ends of chapters are a series of "seminars." Many are discussion oriented analysis problems, which can serve for individual study or group discussions, led, for example, by the instructor, for student centered learning. Others provide worked example problems. There are also numerous questions at the ends of the chapters aimed at testing the student's understanding of the concepts introduced in the chapter. Finally, a list of suggested reading is provided for each.

Chapter 3 has a useful Annex in which actual standard analytical methods are reproduced to illustrate the analytical process and approach for some specific analyses. There is a useful glossary of terms at the end of the text. I recommend reading through this initially. It will give a quick flavor of topics to be covered, and provide a foundation for reading the text. Also, there is a helpful list of terms and symbols.

In order to enhance the flexibility of the text, and assist the instructor in adapting it for different learning environments or levels of instruction, Professor Valcárcel has provided a helpful list of topics to be selected at three levels, based on number of class hours.

A student who has mastered the material in this novel text will have an unusual understanding of what analytical chemistry is, how it helps us understand our world, and how it improves what we do. It will serve as an excellent basis for studying the analytical methodologies taught in quantitative analysis and instrumental analysis courses.

Gary D. Christian
University of Washington
Seattle, WA, USA

Preface

This book is intended to serve both as a generic introduction to Analytical Chemistry and as a body of essential analytical chemical knowledge. This grassroots approach aims to convey the underlying integral teaching message of this discipline in a proper manner in order to ensure effective, consistent learning leading to a solid analytical chemical background. As such, this textbook has been conceived as a tool of use to students undertaking Analytical Chemistry for the first time within the framework of a wide variety of undergraduate curricula. It aims to be a visiting card for this discipline. Because of their generic nature, the contents of this book must obviously be completed with the topics that constitute the typical core of other, high-quality textbooks of which this is by no means a competitor.

This is an atypical Analytical Chemistry textbook inasmuch as it provides no systematic description of analytical techniques (gravimetries, titrimetries, spectrometries, electroanalysis) or methods – not even of ionic equilibria, improperly deemed by some "the fundamentals of analysis". Rather, it deals systematically with the true intrinsic fundamentals of Analytical Chemistry, which are those that make it an independent, self-contained science and distinguish it unequivocally from other scientific and technical areas.

Few textbooks adopt such a generic approach to the subject, one that addresses topics such as analytical properties (Chapt. 2), standards and traceability (Chapt. 3), the analytical process (Chapt. 4), qualitative (Chapt. 5) and quantitative aspects (Chapt. 6), the analytical problem (Chapt. 7) and analytical quality (Chapter 8). These contents are intended to provide a solid foundation for subsequently acquired analytical chemical knowledge. Some of these topics are not easy to teach on an elementary level, so they are profusely illustrated with non-academic examples and consolidated with questions and seminars.

In a way, innovating involves leaping into the void, a risk that has consciously been assumed by the author. However, the pedagogical innovation implicit in this textbook has been tried and tested for six years in teaching Analytical Chemistry as a first-year subject of the Chemistry curriculum at the University of Córdoba, Spain. I am deeply grateful to my partners in this venture, Professors Manuel Silva and Angel Ríos, for their comments and expertise, which are reflected in this book in one way or another.

This novel conception is bound to meet with rejection by some Analytical Chemistry lecturers. It certainly departs considerably from the way they have traditionally approached first-time students of this subject. I just wish to ask those reluctant to adopt this new approach to reflect on it and let me know anything they may think fit about it. Their comments will no doubt help me enrich subsequent editions.

Because undergraduate curricula vary widely, the amount of time that can be devoted to this introduction to Analytical Chemistry also differs widely. By way of guidance, the table following this Preface establishes three different comprehensiveness levels and suggests the book sections lecturers might want to include in each in developing their curricula.

The effort expended on writing this book was funded by Spain's CICyT within the framework of its programme for the development of undergraduate teaching materials. I wish to thank my colleagues Professors Carmen Cámara, Miguel de la Guardia and José Manuel Pingarrón for their thorough revision of its draft, which greatly helped me improve the initial scheme, and also first-year Chemistry student Mercedes López Pastor. Also worthy of special acknowledgement here are Dr Marisol Cárdenas, a fellow teacher at the Department of Analytical Chemistry of the University of Córdoba, for her proofing assistance, José Manuel Membrives for preparing the typescript and artwork, and Antonio Losada for the English translation. This book would never have been published without the warm welcome and support of Peter Enders, Editor of Springer-Verlag.

Miguel Valcárcel   Córdoba, 1 July 2000

Recommendations for using the contents of this book at three different comprehensiveness levels

	First level 6–8 Credits (60–80 hours)	Second level 4–6 Credits (40–60 hours)	Third level 2–3 Credits (20–30 hours)
Chapter 1	Unabridged (the less generic concepts in each section may be omitted, if necessary)	1.1 Slightly abridged (last part) 1.2 Unabridged 1.3 Unabridged 1.4 Abridged (fundamental concepts in each section) 1.5 Abridged 1.6 Unabridged 1.7 Unabridged	1.1 Abridged 1.2 Abridged 1.3 Unabridged 1.4 Highly abridged (Exclude 1.4.4) 1.5 Exclude 1.6 Unabridged 1.7 Unabridged
Chapter 2	Unabridged	2.1 Unabridged 2.2 Abridged 2.3 Unabridged 2.4 Nearly unabridged 2.5 Nearly unabridged 2.6 Abridged 2.7 Abridged 2.8 Abridged	2.1 Unabridged 2.2 Highly abridged 2.3 Unabridged 2.4 Definitions only 2.5 Definitions only 2.6 Definitions only 2.7 Highly abridged 2.8 Exclude
Chapter 3	Unabridged	3.1 Unabridged 3.2 Nearly unabridged 3.3 Unabridged 3.4 Unabridged 3.5 Unabridged 3.6.1 Unabridged 3.6.2.6 Abridged	3.1 Unabridged 3.2 Abridged 3.3 Abridged 3.4 Unabridged 3.5 Nearly unabridged 3.6 Section 3.6.1 only
Chapter 4	Unabridged	4.1 Slightly abridged 4.2 Unabridged 4.3 Unabridged 4.4 Unabridged 4.5 Unabridged 4.6 Abridged	4.1 Slightly abridged 4.2 Unabridged 4.3 Slightly abridged 4.4 Abridged 4.5 Abridged 4.6 Exclude

Recommendations for using the contents of this book at three different comprehensiveness levels

	First level 6–8 Credits (60–80 hours)	Second level 4–6 Credits (40–60 hours)	Third level 2–3 Credits (20–30 hours)
Chapter 5	Unabridged	5.1 Unabridged 5.2 Unabridged 5.3 Unabridged 5.4 Unabridged 5.5 Slightly abridged 5.6 Slightly abridged	5.1 Slightly abridged 5.2 Abridged 5.3 Slightly abridged 5.4 Abridged 5.5 Abridged 5.6 Abridged
Chapter 6	Unabridged (Sections 6.2.1.2 and 6.2.2.1 may be ex- cluded or abridged)	6.1 Slightly abridged 6.1.5 Unabridged 6.1.6 Unabridged 6.2.1.1 Unabridged 6.2.1.2 Abridged 6.2.2.1 Abridged 6.2.2.2 Unabridged 6.3 Unabridged	6.1 Abridged (last two sections excepted) 6.2.1.1 Slightly abridged 6.2.1.2 Exclude 6.2.2.1 Exclude 6.2.2.2 Slightly abridged 6.3 Abridged
Chapter 7	Unabridged	7.1 Abridged 7.2 Abridged 7.3 Unabridged 7.4 Abridged 7.5 Abridged 7.6 Unabridged 7.7 Abridged	7.1 Exclude 7.2 Exclude 7.3 Highly abridged 7.4 Highly abridged 7.5 Highly abridged 7.6 Highly abridged 7.7 Exclude
Chapter 8	Unabridged (slightly abridged, focussing on the more relevant aspects of each section)	8.1 Abridged 8.2 Abridged 8.3 Unabridged 8.4 Abridged 8.5 Abridged 8.6 Abridged 8.7 Abridged	Exclude

Table of Contents

1 Introduction to Analytical Chemistry

Objectives

- To introduce the third basic component of Chemistry (Analysis).
- To define Analytical Chemistry from a modern perspective.
- To establish its essential landmarks, which are dealt with at length in subsequent chapters.
- To provide hierarchical definitions for some key aspects.
- To distinguish Analytical Chemistry from other scientific and technical areas, and to establish mutual relationships.

Table of Contents

1.1 Analytical Chemistry Today

Etymologically, "to analyse" involves to examine, study, learn, work out, weight, *etc.* The target of an analysis can be an object, system, fact, behaviour or attitude, named in increasing order of intangibility. The ultimate aim is to provide information about the target in order to facilitate diagnoses and the making of educated, effective decisions.

Unlike mathematical, grammar, economic and historical analysis, for example, Analytical Chemistry is concerned with the most tangible analytical targets, *viz.* objects and systems. Depending on the nature of the qualitative and quantitative parameters to be examined (Fig. 1.1), objects can be subjected to three types of analysis, namely: **physical** (*e.g.* the electrical resistance of a circuit), **chemical** (*e.g.* the concentration of pesticides in a plant), **biochemical** (*e.g.* the activity of an enzyme in plasma) and **biological** (*e.g.* the atmospheric concentration of pollen). Analytical Chemistry deals primarily – but not solely, as shown in this chapter – with (bio)chemical analyses.

Depending on the nature of the objects or systems involved, more specific designations such as "Clinical Analysis", "Environmental Analysis", "Food Analysis", "Pharmaceutical Analysis", *etc.*, are used. All are compatible with the distinction made in the previous paragraph according to the nature of the target parameters.

Analysing an object (*e.g.* an alloy, urine from an athlete, a doughnut, an electrical circuit, sea water) or system (*e.g.* an ecological or industrial entity) invariably entails measuring pertinent parameters. The science in charge of measurements is called Metrology (see Fig. 1.1). It is obviously concerned with all the above-described facets but, out of tradition, has so far been more extensively used in connection with physical parameters.

Based on the foregoing, Analytical Chemistry can be tentatively defined in broad terms as the *(Bio)Chemical Metrological Science*, the *Science of (Bio)Chemical Measurements* or the *Science of (Bio)Chemical Information*, all of which are mutually related. Such an elementary approach to Analytical Chemistry can be completed by a preliminary definition that takes account of its generic trends, namely: "the obtainment of more chemical information of a higher quality (augmentation goals) by use of increasingly less material, human, economic and time resources, with gradually decreasing hazards (diminution goals)."

The following is a more conventional and comprehensive definition of Analytical Chemistry that is also more precise: *Analytical Chemistry is a metrological science that develops, optimizes and applies measurement processes in-*

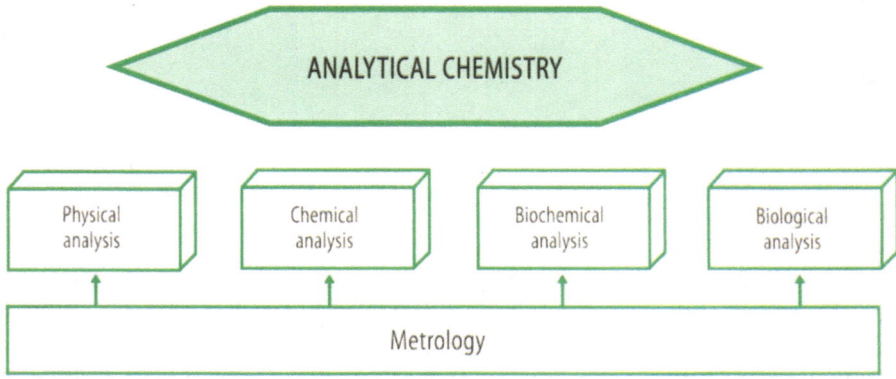

Fig. 1.1. Placement of Analytical Chemistry according to nature of the parameters measured in the analyses

tended to derive quality (bio)chemical information of global or partial type from natural or artificial objects or systems in order to solve analytical problems. This definition includes several key words mentioned earlier [*e.g.* "metrological", "(bio)chemical information"] and others such as "process", "problem" and "quality" that are considered in subsequent chapters.

However, this definition is too vague and necessitates more precise description of some concepts. A more comprehensive definition can be formulated by introducing the term "tools", specifying their nature and that of the information and objects involved, distinguishing between the "global" and "partial" attributes of the information, and including scientific, technical, economic and social problems. One such definition is as follows: *Analytical Chemistry is a metrological science that develops, optimizes and applies material, methodological and strategic tools of widely variable nature which materialize in measurement processes intended to derive quality (bio)chemical information of both partial [presence or concentration of bio(chemical) analyte species] and global nature on materials or systems of widely variable nature (chemical, biochemical and biological) in space and time in order to solve scientific, technical and social problems.*

The most salient features of this comprehensive definition are as follows:

(*a*) Analytical Chemistry is a metrological science;
(*b*) it possesses a dual character, *viz.* basic ("it develops") and practical ("it applies");
(*c*) it provides information about the material (or system) studied and distinguishes between global and partial information (components) that it places in space and time;
(*d*) it regards measurement processes as the universal vehicles for obtaining information;
(*e*) it considers "tools" of the strategic (planning, design, optimization, adaptation), methodological and material types (apparatuses, instruments, reagents), which obviously interrelate with one another when analytical measurement processes are performed (Fig. 1.2 *a*);
(*f*) it includes and distinguishes analytical, scientific, technical, economic and social problems;

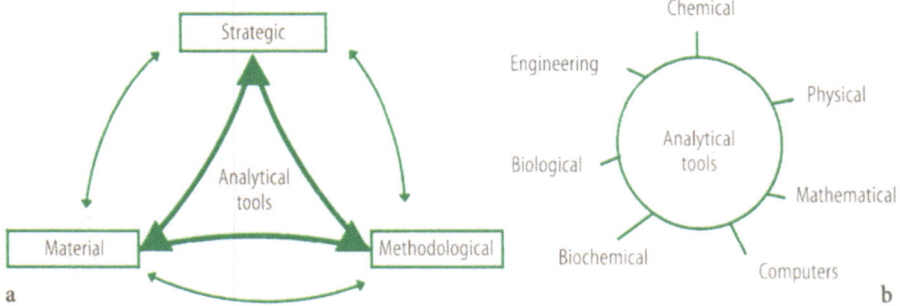

Fig. 1.2. Types (**a**) and nature (**b**) of the analytical tools included in the definition of Analytical Chemistry

(g) it considers quality, in a direct manner (referred to the information produced) and an indirect one (referred to tools and processes), as a strategic target with a view to achieving the external quality required to solve scientific, technical, economic and social problems.

(h) it defines the nature of the tools, information, analytes and materials (systems) involved, a frequent source of confusion in need of further specification.

As can be seen from Fig. 1.2 b, Analytical Chemistry uses a wide array of **tools** including, but not limited to, the following types:

- Chemical: reagents, chemical reactions.
- Biochemical: immobilized enzymes, immunoassay reagents.
- Physical: laser sources, apparatuses, instruments, balances.
- Mathematical: statistical treatment of primary data.
- Computers: software for data acquisition and processing.
- Biological: animal and plant tissues for building sensors.
- Engineering: large-scale processes adapted for micro scale implementation (*e. g.* supercritical fluid extraction, freeze drying).

On the other hand, the **information produced** by analytical chemical processes can be of the following types:

- Chemical: the presence or concentration of inorganic or organic species; the structure of a material; changes in the properties of the material surface (*e. g.* by effect of corrosion).
- Biochemical: the presence, concentration or activity of biochemical species such as enzymes or nucleic acids.
- Biological: the presence or concentration of microbes or allergens.

In fact, biological information does not strictly belong in the analytical chemical domain (analyte species are not molecules or ions but biological entities); there is, however, a growing trend to expanding the boundaries of Analytical Chemistry into the microbiological and allergological fields as a result of chemists being more experienced in measurements than are biologists and of the fact that characterizing a system (*e.g.* drinking water, the atmosphere) entails extracting integral information from it.

_____ Box 1.1

Nature of some elements included in the definition of Analytical Chemistry

	Chemical	Biochemical	Physical	Mathematical	Biological
Tools	×	×	×	×	×
Information	×	×			(×)
Analytes	×	×			(×)
Objects	×	×			×

The **analyte species** to be determined in analytical chemical measurement processes can be chemical (drugs, fat, inorganic ions), biochemical (aminoacids, proteins, enzymes) and, occasionally, also biological (microbes, pollen). The naturally occurring or artificial **material** to be analysed can be chemical (an alloy, a pharmaceutical preparation, a solvent), biochemical (serum, urine, sage) or biological (animal or plant tissue). Whether Analytical Chemistry is to deal with "biological" information and analytes may be a matter for debate; that it must deal with biological materials is beyond dispute.

As can be seen from Fig. 1.3, using the word **"chemical"** to designate analytical chemical measurements can be misleading. It is incorrect to restrict the scope of Analytical Chemistry to the following: (*a*) the use of chemical tools to derive information when, in fact, it uses other types of tools (see Fig. 1.2*b*); (*b*) the analysis of chemical materials when biochemical and biological materials are also frequent analytical objects; and (*c*) the determination of chemical analyte species as analyses can also include biochemical and, less often, biological determinations.

For years, Analytical Chemistry has been defined in mystifying, confusing terms. The original definition of "Chemical Analysis" was appropriate in the past as the tools, samples and analytes involved were almost exclusively chemical in nature. Based on the previous definition, speaking of chemical analysis at present makes no sense; however, dividing Analytical Chemistry into "Chemical Analysis" and "Physical Analysis" according to the nature of the principal tool

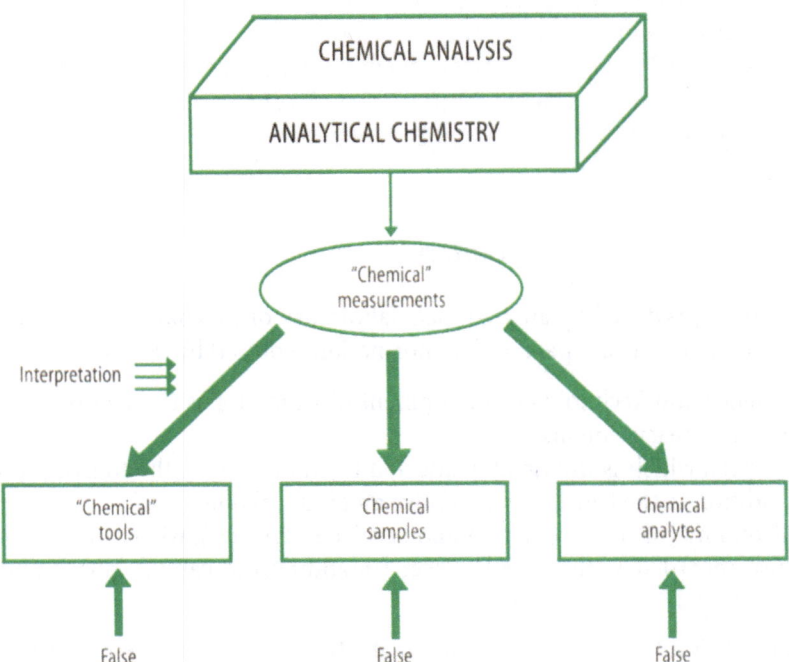

Fig. 1.3. Incorrect interpretations of Analytical Chemistry. All are incomplete as they deem tools, objects and analytes exclusively chemical

used is rather a common mistake propitiated by those who place all emphasis on the use of chemical reactions and (physical) measuring instruments, respectively. These terms tend to be misleading and should be avoided in every context, but particularly in textbooks.

Recently, the designation **Analytical Science** has been used instead of Analytical Chemistry in some geographic areas in an attempt at emphasizing its basic side; however, this term is meaningless as it can be applied to any facet of the word "analysis". Occasionally, the term "Instrumental Analysis" is used to emphasize the fact that Analytical Chemistry uses mostly "instruments" for its purposes; however, this designation is also pointless since the tendency to distinguishing it from Classical Analysis, which is discussed in Sect. 1.7, is now a thing of the past. Other designations such as *Bioanalytical Chemistry*, *Clinical Analysis*, *Environmental Analysis* and *Industrial Analysis* are but specific tasks of Analytical Chemistry and can be used in their respective contexts.

_____ Box 1.2

Using the designation "Physical Analysis" to refer to analytical processes based on physical measurements (*e.g.* X-ray techniques, luminescence spectroscopy) may be misleading. In fact, this term can also be applied to measurements such as the thickness of a steel sheet, the temperature of a water tank or the compressive strength of a plastic block, for example. All these are obviously outside the scope of Analytical Chemistry.

However, physical and physico-chemical parameters are often used for analytical chemical purposes. The identification of analyte species in chromatography and electrophoresis relies on time or distance measurements. Also, the ethanol content in a wine is routinely determined from density measurements.

So-called "Thermal Analysis" can be performed in various ways. In one (thermal differential analysis or TDA), a typically physical parameter such as temperature is used to extract analytical information; in another (thermogravimetric analysis or TGA), it is used as a "tool".

1.2 Analytical Chemical Information

The results produced by an analytical laboratory or by analytical systems that deliver data on-site are part of the information required by (Fig. 1.4):

(*a*) Science and Technology, development of which relies on data provided by quality measurements;
(*b*) Society, which is unarguably affected by information (the fourth power in addition to the legislative, executive and judicial ones); and
(*c*) Economy (of the production and service industries), where information is also regarded as the fourth power (in addition to capital, labour and raw materials).

Hence the significance of Analytical Chemistry, which should deliver as truthful information as possible in order to facilitate educated, efficient, timely decisions in the above-described domains.

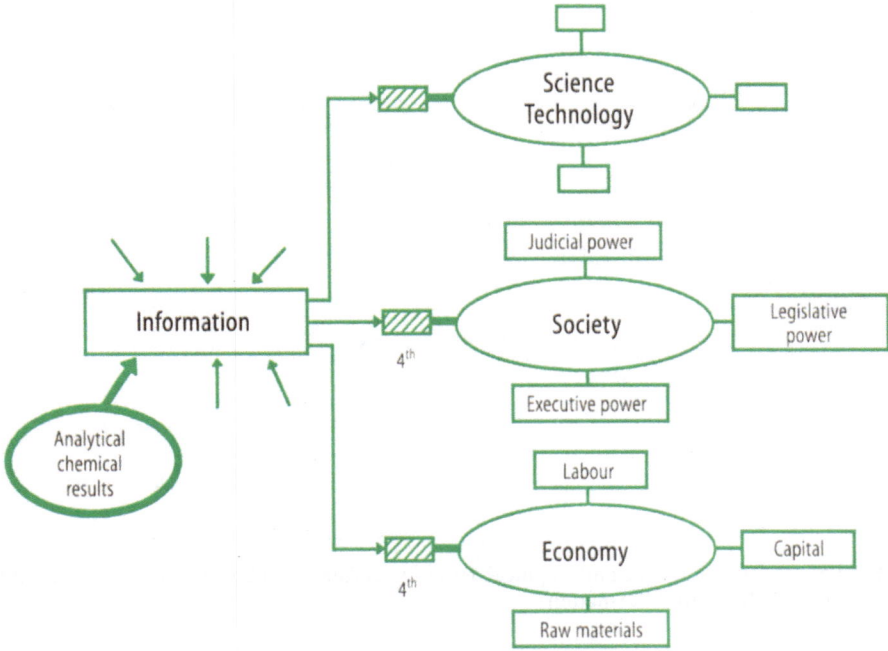

Fig. 1.4. Impact of analytical chemical information

From its definition it immediately follows that Analytical Chemistry is an **information science** which provides (bio)chemical – and, occasionally, also biological – information on materials of scientific, technical, industrial or social interest.

_____ Box 1.3

In strictly economic terms, analytical information is crucial.

On the one hand, one must consider the expenses incurred in making chemical measurements. An estimated 3 – 5% of the gross domestic product (GDP) of a typical developed country is used to obtain the chemical information required to sustain living standards, trade, industry and R&D activities. This proportion can be used as a reliable indicator of a country's social welfare and economic competitiveness.

On the other hand, one must take into account that analytical information of inadequate quality can lead to incorrect decisions that may result in severe economic losses. According to Professor Tölg, such losses might have amounted to DM 6,000 million in Germany in 1986.

As in any information science, the primary aim of Analytical Chemistry is to minimize uncertainty in the qualitative, quantitative and structural information extracted from the target material. The word "uncertainty" tends to be used in a more specific sense to characterize an analytical result and is equivalent to, though not identical with as shown in Chap. 2, precision – or rather, the lack of it.

According to Professor Malissa, _the general aim of Analytical Chemistry (Analytiks) is to make the intrinsic but latent information about an object_

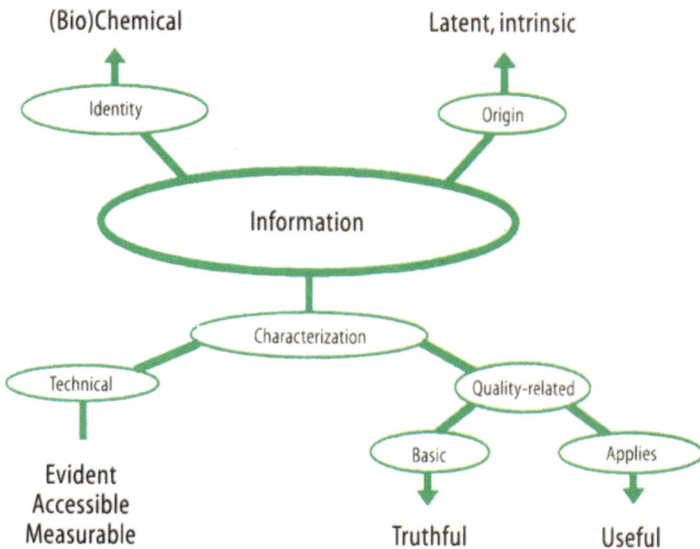

Fig. 1.5. Professor Malissa's philosophical definition of Analytical Chemistry revolves around attributes of the word "information"

evident, accessible, measurable, truthful and useful. Such a basic approach to Analytical Chemistry, which is illustrated schematically in Fig. 1.5, relies on several attributes of the word "information" that identify it ("chemical"), specify its origin ("latent", "intrinsic"), assign it technical features ("evident", "accessible", "measurable") and delimit its quality in both basic ("truthful") and practical terms ("useful").

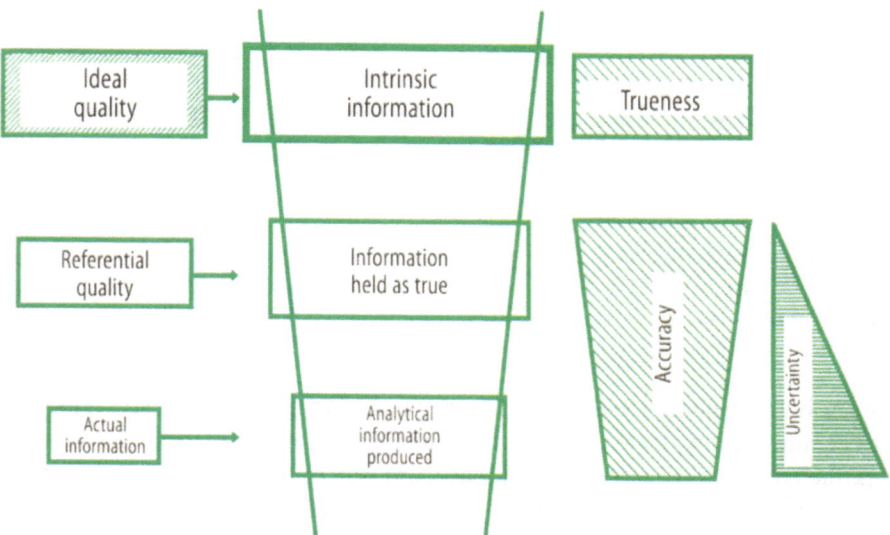

Fig. 1.6. Analytical chemical information levels ranked according to quality and to their relationship to trueness, accuracy and uncertainty

_____ Box 1.4

On arrival at the laboratory, a rock may have a silica (SiO_2) content from 0 to 100% (the highest possible uncertainty). By analysis, the laboratory obtains a silica content of, say, 78.3 ± 0.1%; as a result, the rock can be said to contain from 78.2 to 78.4% silica and the uncertainty will have been substantially diminished. The specific uncertainty for the sample (i. e. the range over which the true result can be expected to vary) will be ±0.1% .

Not knowing whether bay water is polluted with mercury species also represents the highest possible uncertainty, which can be reduced by using a qualitative test to obtain a yes/no response and, to an even higher degree, by quantifying the total concentration of mercury. If, in addition to confirming the presence of mercury and determining its concentration, one distinguishes among specific mercury species (e. g. the metal ion in free form and as a part of chelates or organometallic compounds), the initial uncertainty will be further decreased.

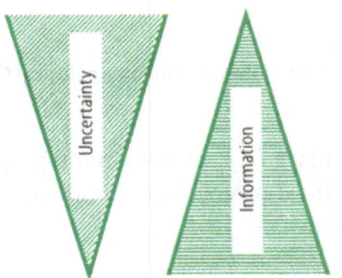
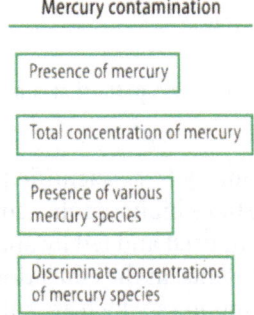

Materials possess intrinsic information that persists or changes in space and/or time; ideally, this is the type of information to be supplied by the analytical chemist. This **intrinsic information** is subject to no uncertainty, represents the top level of analytical quality and is characterized via an ideal, unattainable property: trueness (see Fig. 1.6). At a lower level is **referential information**, which is factual, reasonably accurate – nearly as much as intrinsic information – and scarcely uncertain. This type of information is obtained under uncommon conditions (e. g. in intercomparison tests involving many participating laboratories using different analytical processes) and is basically used as a reference to extract **ordinary analytical information** or assess its goodness. This last type of information is on the lowest step of the quality hierarchy; it is also based on facts but is less accurate and more uncertain than referential information.

The way analytical information is conveyed strongly affects its quality, which may diminish or even be lost altogether in the way. Analytical chemists can deliver analytical information by establishing one- or two-way links with the following (Fig. 1.7):

(1) Society, other scientific and technical areas, and the departments of the body to which they are answerable (the one that requested the analytical information in the first place);

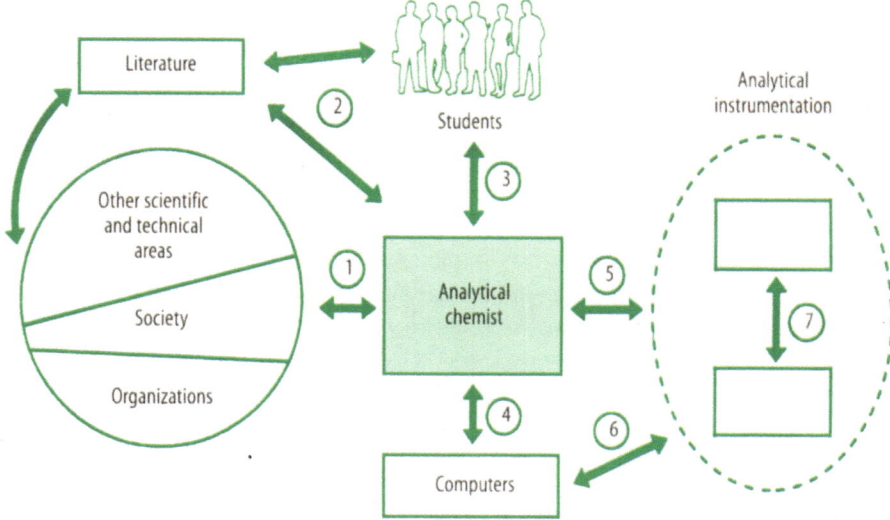

Fig. 1.7. Two-way paths that can be followed to transmit information in Analytical Chemistry

(2) primary (*e.g.* scientific papers), secondary (*e.g.* reviews, monographs) and tertiary bibliographic sources (*e.g.* official protocols, textbooks), which are both used and fed by analytical chemists;

(3) students, in the educational context;

(4) computers, which are supplied with software and data in order to extract the information required; and

(5) instruments, which are programmed to supply the data needed. The link between the analytical chemist and instruments is currently established via a two-way computer–instrument connection (see the 5–6–4 cycle in Fig. 1.7).

Other ways of conveying analytical information (*e.g.* those that link instruments, literature sources or students and society) are not directly related to the analytical chemist. While each link involves a specific type of information, all links share the fundamentals of Analytical Chemistry outlined in its definition. The scheme of Fig. 1.7 illustrates the variety and complexity of Analytical Chemistry relationships and reveals the need to develop effective systems to ensure fluent, reliable, efficient communication between links.

1.3 Analytical Chemical References

As a metrological science, Analytical Chemistry relies on the measurement of parameters [specifically, (bio)chemical parameters]. To measure is to compare. In order to be able to compare, one needs references that are called "standards". Just as measuring length requires using the metre, making analytical chemical measurements entails using reference landmarks, on the basis of which analytical results are expressed. When, for example, the concentration of a given pesticide in drinking water is to be determined, a specimen of the pure pesticide (or a sample standard containing it) is subjected to an appropriate analytical process in order to obtain a qualitative and quantitative response that can be used as a reference in identifying and quantifying the pesticide in the target object (water). If the "instrument" used is a human sense (*e.g.* in classical qualitative analysis), then the analyst must previously memorize the response of a suitable standard.

Box 1.5

The typical odour of hydrogen sulphide (H_2S) must be known if the presence of sulphide in a sample is to be identified. For this purpose, the sample is supplied with hydrogen chloride (HCl), which produces H_2S, a volatile compound with a distinctive smell:

$$\underset{\text{sample}}{S^{2-}} + 2H^+ \rightarrow H_2S\uparrow$$

In order to recognize the odour, the analytical chemist must previously have smelt it. The odour can be experienced by adding HCl to a solution of sodium sulphide (a standard).

The concentration of copper in an alloy can be quantified by dissolving the sample and placing and aliquot of the resulting solution in an atomic absorption spectrometer to obtain an absorbance signal. This piece of information only allows one to confirm the presence of copper. In order to determine its concentration, one must establish a relation between absorbance and concentration by using standards containing variable amounts of copper in order to construct a "calibration curve" that is then fitted to an equation of the type

[absorbance] = a + b × [copper concentration]

The absorbance value obtained allows one to calculate the concentration of copper in the solution and hence its proportion, by weight, in the original sample (alloy).

Consequently, Analytical Chemistry is meaningless unless one or more standards suited to the type of information required are available. One essential goal of Analytical Chemistry is thus to find and properly use reliable references. Poor or improperly employed references produce analytical chemical results of inadequate quality, however suitable the technical means employed and high the laboratory performance achieved may be.

Analytical chemical standards are discussed in detail in Chap. 3, which is devoted to traceability (a metrological property to which the standards are closely related).

1.4 Essential Features of Analytical Chemistry

While definitions must inevitably be strict and austere, they should be completed with a description of all those aspects required to delineate their contents precisely. Such is the case with the definition of Analytical Chemistry, the current and foreseeable connotations of which are discussed in the following sections.

1.4.1 Basic Elements

For Analytical Chemistry to conform to the above definition, it should embody the following three essential components (Fig. 1.8):

(a) Efficient **research and development** (R&D), so that analytical tools can be of use in implementing novel analytical processes, improving existing ones and solving new analytical problems. This element makes Analytical Chemistry an independent science. Section 1.4.4 discusses developments that promote innovation in Analytical Chemistry.

(b) The **body of existing analytical tools and processes**, which can be found in the literature (scientific papers, encyclopaedias, official methods, protocols) and used to solve analytical problems resembling to a greater or lesser extent previously addressed situations with little or no R&D effort. It suffices to follow the recommended procedure in order to implement the ensuing analytical process in the laboratory. This analytical arsenal is becoming increasingly massive – but also increasingly accessible thanks to the widespread availability of computerized information management systems. Occasionally, this component is incorrectly referred to as "Chemical Analysis". Note that this element involves two distinct components of the definition of Analytical Chemistry: tools and processes.

(c) **Education** is the third essential component of Analytical Chemistry. It guarantees consolidation of a modern, proper conception of this science by transferring its true landmarks to future professionals and fostering interest in the R&D component.

Obviously, these three elements are related to one another. Thus, R&D enriches tools and processes, both of which in turn provide technical support for it. These two essential components of Analytical Chemistry are the primary sources of

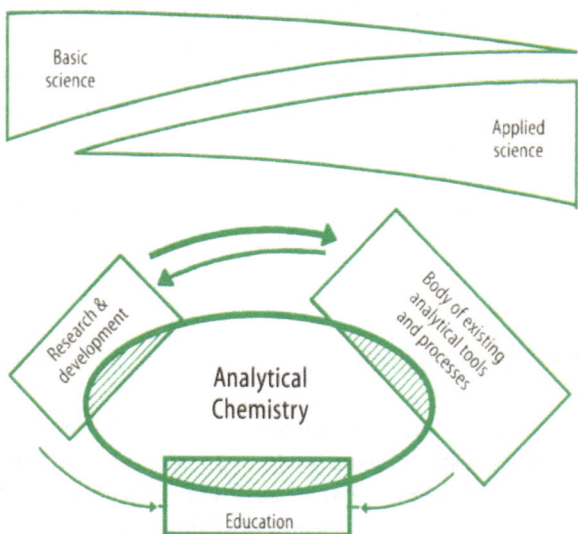

Fig. 1.8. Essential elements of Analytical Chemistry and their mutual relationships

the teaching materials used by the third element. Both should feed this element and be the driving forces of pedagogical renovation. Confining the teaching of Analytical Chemistry to the mere description of the second component is a glaring mistake.

1.4.2 Fundamentals

A science must possess some distinct fundamentals if it is to possess a structure and character of its own and be distinguishable from other sciences. In Analytical Chemistry, the two groups of fundamentals or principles shown in Fig. 1.9 are frequently confused.

Analytical Chemistry has some **shared fundamentals** in common with other areas of Chemistry, Physics, Mathematics, Biology and Engineering, among others. Such fundamentals make it no self-contained entity; also, they encompass its applied facet only. However, they play a crucial role in the development of analytical tools, which provide theoretical and practical support for instruments, apparatuses, methodologies, reagents, approaches, *etc.*

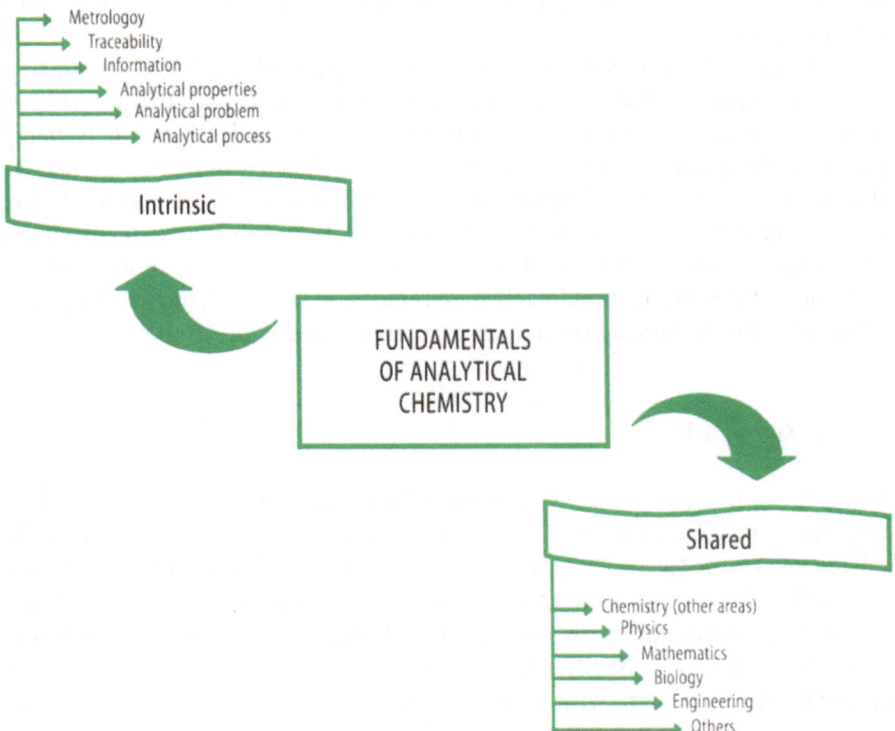

Fig. 1.9. Intrinsic and shared fundamentals of Analytical Chemistry

Here are some typical examples of shared fundamentals of Analytical Chemistry:

- Spectroscopy lies on the boundary between Physics and Physical Chemistry. Its adaptation and application to (bio)chemical analysis is the responsibility of Analytical Chemistry, which in turn feeds on spectroscopic developments.
- Chemometrics is a subdiscipline that possesses and uses mathematical tools to increase the amount and quality of the information that can be derived from primary data provided by analytical instruments.
- Enzymology and immunology, two substantial components of biochemistry, provide the analytical chemist with highly useful tools to develop modern analytical processes under highly favourable conditions as regards selectivity.
- Modern (bio)chemical engineering processes are quite applicable on the laboratory scale provided they are properly adapted and miniaturized to ensure efficient sample treatment. Such is the case with freeze-drying and supercritical fluid extraction.

Analytical Chemistry also possesses **intrinsic fundamentals** that make it a distinct scientific and technical entity. These fundamentals arise from the above tentative and final definitions of Analytical Chemistry and materialize in several key words that are commented on in the following section and should be its primary goals.

Both types of fundamentals are compulsory parts of Analytical Chemistry. The problem arises from the relative prominence that is assigned to each. At present, emphasis in Analytical Chemistry research and education is being placed on the shared fundamentals; this supports the view of those who believe this science possesses no contents of its own. Equally detrimental is placing undue emphasis on its intrinsic fundamentals and refusing to incorporate, utilize and adapt scientific and technical developments from other areas. A position in between the previous two extremes and balanced, flexible emphasis adjusted to the circumstances in each case, are obviously preferable.

1.4.3 Key Words

The definition of Analytical Chemistry, and its shared and intrinsic fundamentals, encompass several basic factors that materialize in different, mutually related key words (Fig. 1.0). All revolve around **information**, which relies on the **traceability** inherent in a **metrological** science. A scientific, technical, economic or social problem raises an **analytical problem** that calls for an **analytical process** in order to produce results that are interpreted in the light of the particular analytical problem with a view to extracting the information required to solve the problem addressed in the first place.

Figure 1.10 relates the above-mentioned key words in terms of attributes and to one another (dashed lines). **Quality** is primarily related to information and information relies on traceability (*i.e.* on the quality of standards and their proper use). Quality can also be assigned to the analytical process and assessed

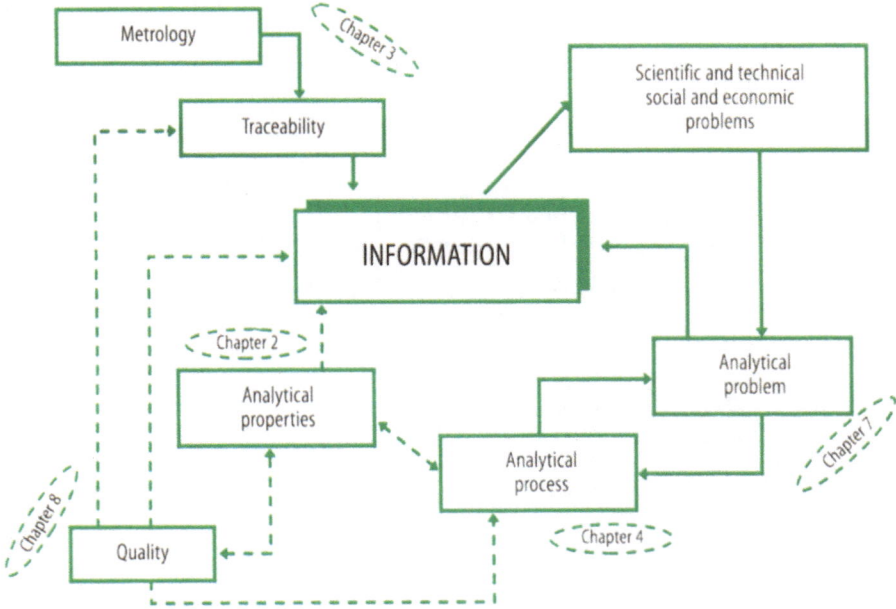

Fig. 1.10. Key words of Analytical Chemistry and their mutual relationships

through the ability to solve the analytical problem. **Analytical properties** are the materialization of analytical quality inasmuch as they are attributes of both the information produced and the analytical process used.

Analytical Chemistry key words are discussed in detail in different parts of this book, so this section simply outlines them and their mutual relationships. Thus, Chap. 2 is devoted to analytical properties and Chap. 3 to traceability. The analytical process, the analytical problem and analytical quality are the subject matters of Chaps. 4, 7 and 8, respectively.

1.4.4 Developments

Every science relies on developments that differ markedly in quality and accessibility as they depend on the particular (external or internal) stimuli involved and objectives pursued. Based on the essential elements of Analytical Chemistry stated in Sect. 1.4.1, a variety of developments are possible. The most representative are shown schematically in Fig. 1.11 and briefly discussed below in two groups according to the type of stimulus [external (A, B, C) or internal (D, E, F)] involved.

Scientific, technical, economic and social needs (problems) can be the driving forces for various types of analytical chemical developments. One [links 1-2-3-4 (A) in Fig. 1.11] is clearly innovative: an external problem gives rise to a new analytical problem that requires both basic and applied research, research that leads to new analytical tools and processes which, properly used, produce the results needed to solve the analytical problem and hence the original, external

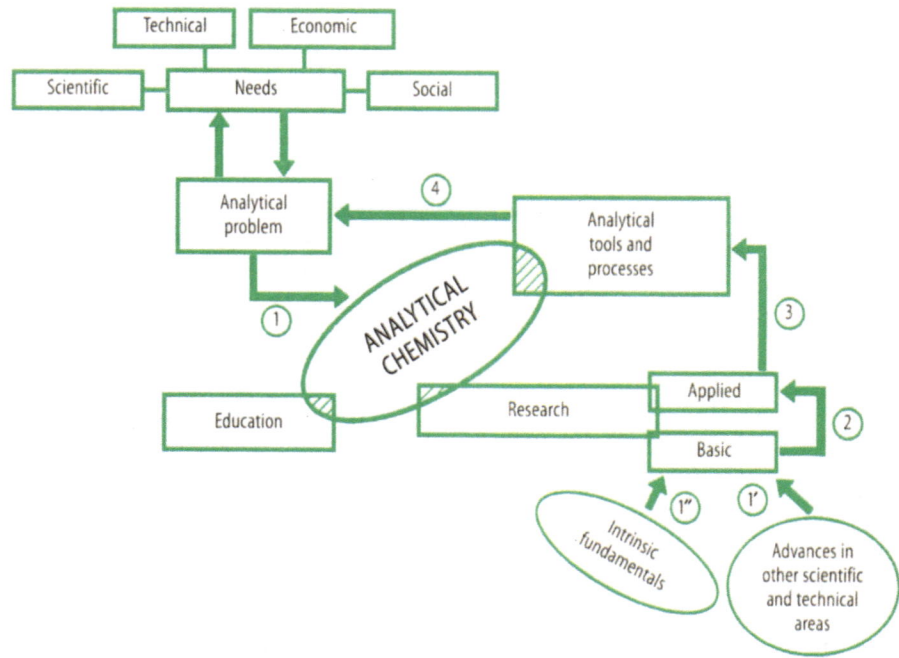

Fig. 1.11. Types of developments possible in Analytical Chemistry. (A) 1-2-3-4. (B) 1-3-4. (C) 1-4. (D) 1′-1-3. (E) 1″-2-3. (F) 3

problem. This type of development is consistent with the definition of Analytical Chemistry; two typical examples are ultratrace analyses in semiconductors – the properties of which are dictated by metals present at ultra-low concentrations – and the determination of enantiomer proportions in the active principles of pharmaceutical formulations. These developments are representative of new problems external to Analytical Chemistry that have fostered developments of the A type in Fig. 1.11. Less innovative, but equally interesting and pertinent, are those that require no basic research [links 1–3-4 (B) in Fig. 1.11]; the mere adaptation of existing means enables the development of modified tools and processes to solve the analytical problem. The routine use of available tools and processes (links 1 to 4 C in Fig. 1.11) is appropriate when the analytical problem in question has already been addressed previously, so only one of the basic elements of Analytical Chemistry need be involved.

Other types of development (links D, E and F in Fig. 1.11) do not arise in response to external stimuli from analytical problems. One (links 1′-2-3 D in Fig. 1.11) relies on developments in other scientific and technical areas, incorporation of which into Analytical Chemistry requires both basic and applied research with a view to developing new analytical tools or resources, or to improving existing ones, in order to potentiate this basic element. Such is the case with laser technology, which has revolutionized some spectroscopies; the design of new materials for sensors; or microelectronic technology, which has provided strong support for the miniaturization of analytical equipment. The intrinsic fundamentals of Analytical Chemistry can propitiate developments in response to internal stimuli, which provide the strongest support for its existence as an independent entity [see links 1″-2-3 (E) in Fig. 1.11]. Such is the case with new standards, and calibration and standardization procedures, among others. Finally, one must deem incorrect those developments that simply involve routine research (link 3 F in Fig. 1.11), with disregard of fundamentals and advances in other areas, as they simply introduce worthless variations that expand but never enrich the second basic element of Analytical Chemistry (tools and processes).

1.4.5 Placement

Speaking of independent sciences is gradually becoming obsolete. To be consequent, Analytical Chemistry should maintain two-way, symbiotic links with other scientific and technical areas. Placing it in this context should pose no problem provided its intrinsic fundamentals are established, maintained and fostered. Otherwise, an irreversible shift to its shared fundamentals will deprive it of its natural connotations and possibly, also of its identity and independence.

Placing Analytical Chemistry in the Chemistry domain is especially crucial since chemical areas are the strongest mutual competitors for a place of their own in Science and Technology. Chemistry rests on the four essential, interrelated cornerstones shown in Fig. 1.12, namely: Theory, Synthesis, Analysis and Applications (industrial, environmental, pharmaceutical). Thus, Analysis is a substantial part of Chemistry and is closely related to other components. For example, no synthesis can be conceived without analysing the ingredients, intermediates and end products; nor is it reasonable to analyse without solid

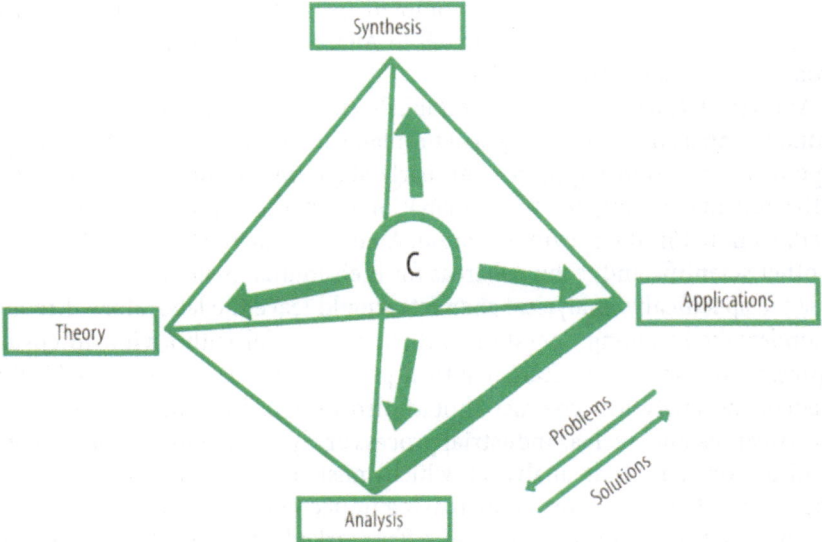

Fig. 1.12. Placement of Analytical Chemistry in the Chemistry domain

_____ Box 1.7

The regular tetrahedron of Fig. 1.12 can be distorted by pulling one, two or three of its vertices.

Placing Analytical Chemistry and Physical Chemistry in this geometric model is very easy (at the theory and analysis vertices, respectively). Organic and Inorganic Chemistry share the synthesis vertex and other branches such as Chemical Engineering, Agricultural Chemistry, Pharmaceutical Chemistry, etc., are at the applications vertex. However, such a stiff distribution is not factual. The chemistry of today and tomorrow fits in the centres of the tetrahedron edges, along which it can move in both directions. Thus, for a synthesis to be properly approached, the underlying theoretical principles must be known and the results of the analysis of the raw materials, intermediates and end products available. Chemical Engineering relies on synthetic developments and on analyses providing quality results.

Analysis rests on physico-chemical principles and on a wide variety of synthetic products that are used as reagents or standards. However, it also provides solid support for synthesis and applications, as well as for experimental testing of basic theories.

theoretical chemical support. The link to the application component is a frequent source of confusion since, if Analytical Chemistry does not rely on its intrinsic fundamentals, its applied facet prevails and the tetrahedron of Fig. 1.12 becomes a triangle: improperly approached, Analytical Chemistry is a mere application of Chemistry.

1.4.6 Boundaries

Analytical Chemistry cannot be confined to the laboratory or library. To be consistent with its definition, it must expand its traditional boundaries. This anomalous situation has been promoted by a departure from its intrinsic fundamentals. One modern approach to Analytical Chemistry is based on the broad horizon depicted in Fig. 1.13.

Analytical Chemistry should establish effective links with the social and economic areas that raise analytical problems in order to accurately identify the type of information required at an early stage; in addition, it should take an active role in ensuring that the information produced is properly and efficiently used. As noted in the previous section, Analytical Chemistry should also relate to other scientific and technical areas for their mutual benefit.

More specifically, analytical chemists should leave the laboratory if required to undertake sampling operations, which are crucial with a view to ensuring representativeness (a capital analytical property as shown in Chap. 2). When needed, the analytical chemist should also design, implement and maintain analytical systems such as industrial process analysers, environmental pollution monitors or spacecraft analysers, which must inevitably operate outside the laboratory. These are issues of increasing impact and interest which provide *in situ* information via procedures that differ markedly from traditional analytical laboratory approaches.

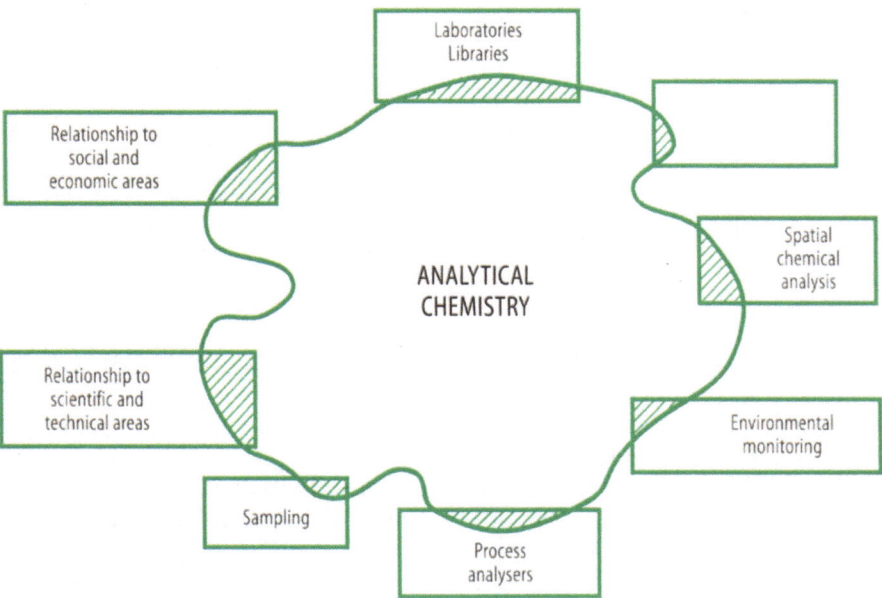

Fig. 1.13. Boundaries of Analytical Chemistry

1.5 Evolution of Analytical Chemistry

The history of Analytical Chemistry can be divided into three periods the boundaries between which are difficult to establish precisely but can be inferred from the types of tools used in each. The classical period was marked by chemical reactions, the balance and the burette; the modern period by the systematic use of a wide array of instruments (photometers, polarographs, fluorimeters, potentiometers, chromatographs, mass spectrometers, X-ray spectrometers); and the contemporary period by the use of computers.

The key words of Analytical Chemistry obviously changed from one period to the next. Thus, in the classical period, the words were qualitative analysis, analytical schemes, equilibrium constants, inorganic and organic reagents, reactions in solution, gravimetry and titrimetry. In the modern period, terms such as molecular absorption and emission spectroscopy, gas chromatography, high performance liquid chromatography, mass spectrometry, nuclear magnetic resonance spectroscopy, atomic emission and atomic absorption spectroscopies, voltammetry, *etc.*, took over. Finally, the key words of contemporary Analytical Chemistry are automation, miniaturization, robots, interfaces, immunoassay, process control, sensors, screening techniques, chemometrics, expert systems, neural networks and micellar media, among others.

The history of Analytical Chemistry can also be viewed from a different, but complementary, conceptual perspective. In its eight schemes, Fig. 1.14 depicts the evolution of Analytical Chemistry from eight different points of view that are briefly commented on below.

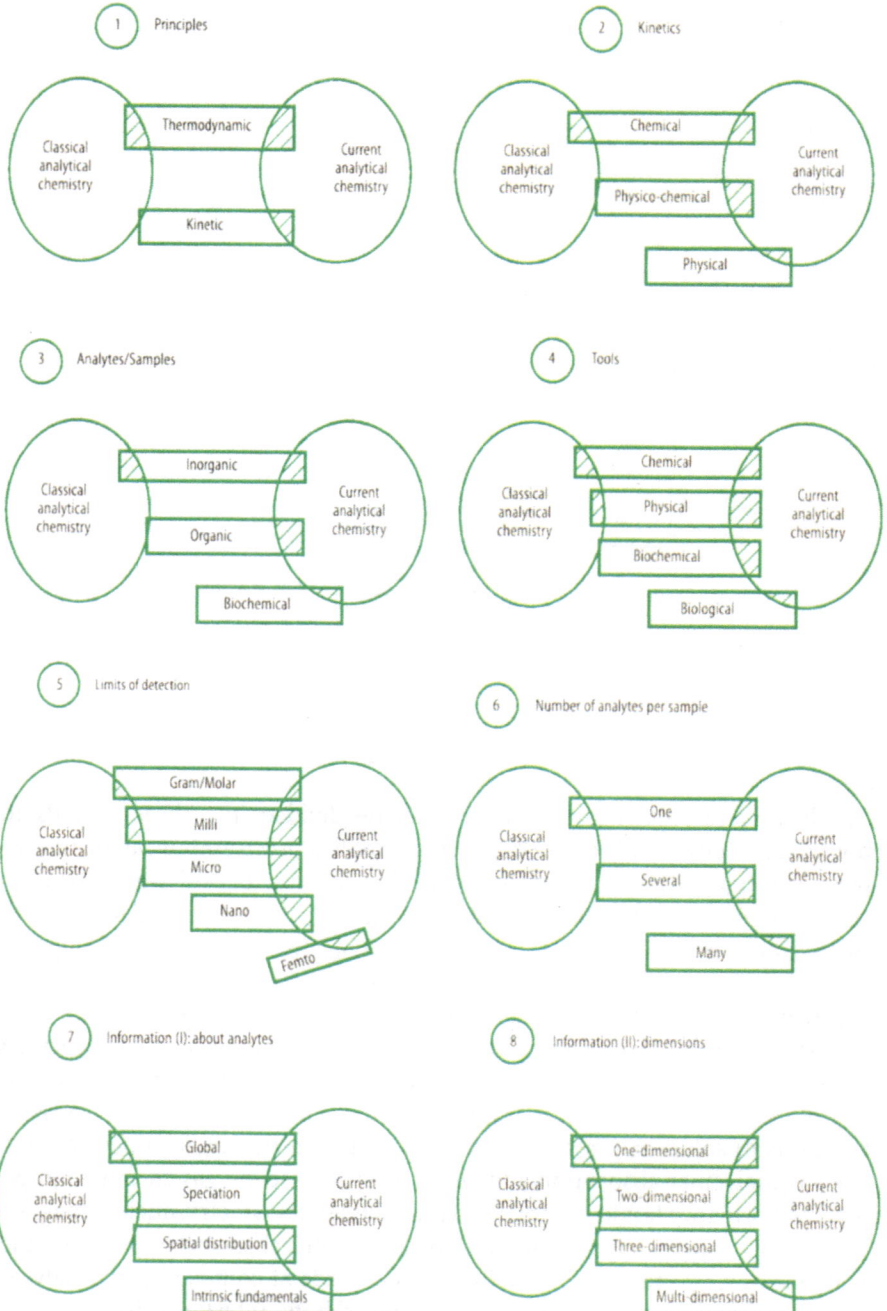

Fig. 1.14. Conceptual and technical evolution of Analytical Chemistry as seen from eight different standpoints

The **basic principles** of classical Analytical Chemistry were essentially thermodynamic, with some kinetic connotations; at present, however, kinetics is akin to thermodynamics in importance (Fig. 1.14.2). In classical Analytical Chemistry, chemical **kinetics** was virtually the sole type of kinetics considered, either to circumvent its adverse effects in slow reactions or to exploit them (*e. g.* to develop kinetic methods of analysis). Currently, physico-chemical and purely physical kinetics also play prominent roles: introducing the time dimension has enabled the development of major analytical chemical techniques such as picosecond spectroscopy and time-domain electroanalytical modes, among others (Fig. 1.14.2).

Materials, reactions and **analytes,** which used to be mostly inorganic in the classical period, are now varied in nature, as seen in analysing the definition of Analytical Chemistry (Fig. 1.14.3). Tools have also evolved from primarily chemical to widely variable in nature (Fig. 1.14.4).

The minimum amounts or concentrations required for analytes to be identified and quantified (*i. e.* limits of detection) have also changed dramatically: from the gram/mole level to others several orders of magnitude lower (Fig. 1.14.5). As the potential of Analytical Chemistry has been expanded with new tools and processes, the **number of analytes** that can be determined in a straightforward, convenient way in a single sample has also risen dramatically (Fig. 1.14.6).

The **aims** of analytical determinations have also changed markedly in response to the ever increasing demand for information (Fig. 1.14.7). Thus, the overall content in a given analyte can easily be inadequate for the intended purpose; often, one needs to know not only the forms in which the analyte may be present in the unknown sample or system, but also their relative proportions. So-called "speciation" is a key to Environmental Analytical Chemistry, but is also significant to determinations in other fields such as Clinical Chemistry (*e. g.* ionic and total calcium, free and HDL cholesterol), Food Chemistry (*e. g.* free and bound sulphur dioxide in wine) and Pharmaceutical Chemistry (*e. g.* enantiomer proportions in an active principle). For solid materials, one must also know the spatial distribution of the composition and the structure. The **dimensions of analytical information** have also expanded dramatically (Fig. 1.14.8). Thus, gravimetry and titrimetry provide one-dimensional information $[F = f(x)]$, whereas conventional analytical techniques (*e. g.* UV-visible spectroscopy, chromatography) produce two-dimensional information of two types, *viz.* $F = f(x, y)$ and $F = f(x, t)$, where x is the measured signal, y an instrumental variable and t time. Three-dimensional information is becoming increasingly affordable and multi-dimensional data is bound to become available in the very near future.

The evolution of each facet of Analytical Chemistry can be used as the starting point to assess the global change undergone by this science and anticipate its future.

1.6 Conceptual and Technical Hierarchies

The growing variety of relations that the analytical chemist may establish (see Fig. 1.7), and changes in Analytical Chemistry (see Sect. 1.4), have created some gaps and confusion that usually arise from a lack of time for conceptual and technical consolidation. It is utterly indispensable to develop reliable communication systems based on the use of proper, universally accepted words to describe objects, facts and systems in unequivocal terms. One way of making this goal more reachable is by establishing conceptual and technical rankings in the form of significance and scope hierarchies. This section discusses some; others, are dealt with in subsequent chapters. This approach allows one to define some aspects of great interest in the context of Analytical Chemistry.

Figure 1.15 shows two hierarchies related to analytical information. At the top of the ranking on the right are reports, which, in addition to results, contain their interpretation in the light of the analytical problem addressed. At the next information level down the ranking are qualitative and quantitative results; these are obtained by mathematical (chemometric) manipulation of primary data (absorbance, mass, intensity, potential, *etc.*) from samples and standards, which are at the third information level. Secondary data (fourth level) are referred to the technical parameters of analytical processes and tools (*e.g.* the pressure and temperature of a chromatograph, the rotation speed of a centrifuge, the ambient relative humidity) and provide support for the goodness of primary data. On the other hand, secondary data are not directly related to results, so they cannot be regarded as analytical information proper.

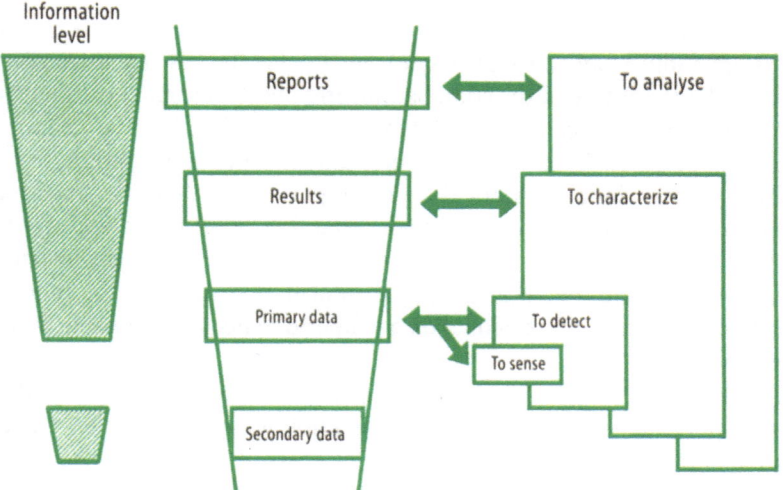

Fig. 1.15. Significance and scope hierarchies (and relationships between them) for chemical information

The rank of the analytical information produced also increases from the term "to sense" to the verb "to analyse" as follows (Fig. 1.15):

- **To sense** involves using a device responsive to the presence (and concentration, if applicable) of a (bio)chemical species in a sample.
- **To detect** involves the direct use of an instrument to acquire or produce a signal and transduce it into a readily measured physical (usually electrical) quantity. This function is served by instruments (*e.g.* photometers, potentiometers, balances), which produce primary data (*e.g.* absorbance, electrical potential or mass units). Indirectly, instruments are the basis for sensors.
- **To characterize** entails establishing distinct features of an object or system in order to define it on the basis of results obtained by mathematical processing of primary data produced by an instrument.
- **To analyse** involves interpreting – in addition to characterizing – in qualitative, quantitative and structural terms, which entails converting results into reports.

_____ Box 1.8

pH paper, the pH electrode and a fibre-optic probe containing an immobilized reagent on its tip for immersion in the sample are typical examples of sensors. A sensor can only be used if connected, in a direct or indirect manner, to an instrument for detection. Thus, the "instrument" for pH paper is the human body, which compares the resulting colour with those on a scale (the "standard"). The pH electrode is a sensor connected to a potentiometer to measure the potential difference between it and a reference electrode. A fluorimetric fibre-optic sensor must unavoidably be connected to a spectrofluorimeter, which is the actual instrument that measures the fluorescence intensity.

The two rankings of Fig. 1.15 are significance and scope hierarchies; as a result, the upper levels rely on the lower ones. Thus, a report cannot be made without results and a system cannot be characterized without detection. Also, the levels in the two hierarchies are mutually related (reports → to analyse, results → to characterize, primary data → to detect).

A different conceptual and technical hierarchy (Fig. 1.16) is needed to clarify the meaning of words such as "process", "technique", "method" and "procedure", which are often used indifferently. A **technique** is a generic principle that is used to extract information; it uses an instrument and comes into play in the second step of the analytical process. The **analytical process** is the body of operations that separates the uncollected, unmeasured, untreated sample from the results, expressed as required; Chapter 4 is devoted to it. A **method** is the specific implementation of an analytical technique to determine one or more analytes in a given sample; it is thus the materialization of the analytical process. A **procedure** is the detailed sequence of actions performed in carrying out an analytical method. Technique and method should never be confused; techniques (polarography, titrimetry, X-ray spectrometry) are generic in nature whereas methods are specific to given types of samples and analytes.

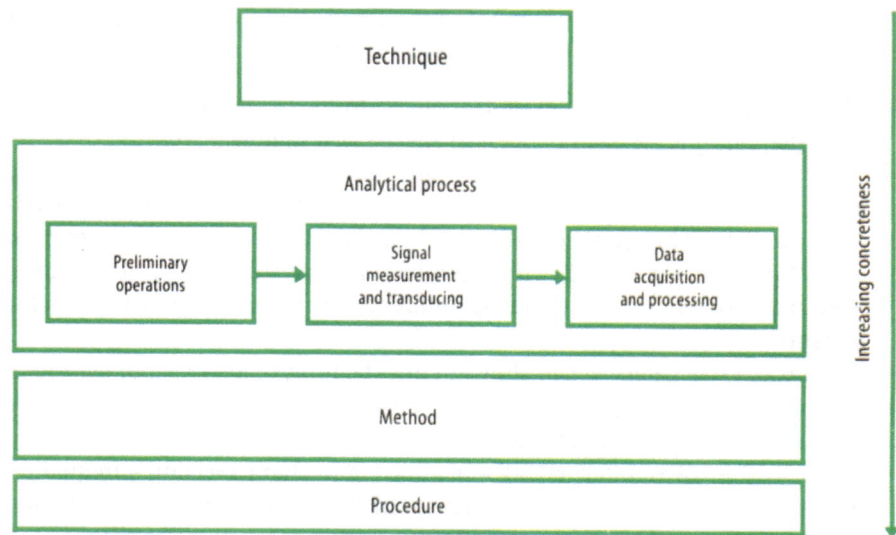

Fig. 1.16. Conceptual hierarchy of the words "process", "technique", "method" and "procedure" in the analytical chemical domain

_____ Box 1.9

The hierarchy for technique, process, method and procedure can be illustrated with two different examples where the technique and analyte are the same but the nature of the sample calls for rather different processes, methods and procedures.

The determination of the concentration of cadmium (the analyte) by atomic absorption spectroscopy (technique) can be performed by using one of different analytical processes (and hence different methods and procedures); the choice will be dictated by whether the sample is a mineral, a yellow-painted toy or a biological material, for example. As a result, the preliminary operations to be carried out will also differ markedly.

The determination of traces of sulphur dioxide (SO_2) by UV – visible molecular absorption spectroscopy (the technique) is based on the formation of a dye that absorbs maximally at 520 nm by reacting the analyte with p-rosaniline and formaldehyde. The process, method and procedure of choice will vary depending on whether the sample is red wine (to which the analyte is added as a preservative) or the atmosphere (SO_2 is one source of so-called "acid rain").

Figure 1.17 shows a scope hierarchy for analytical tools and their relationship to the previous ranking. An **analyser** performs (almost) the whole analytical process – and hence the method and procedure involved –; an **instrument** is the materialization of an analytical technique that provides data directly related to the analytes. An **apparatus** performs a function but produces no analytical information and a **device** is a part of an apparatus. Thus, an analyser comprises one or several instruments, apparatuses and devices, and an instrument consists of several apparatuses and devices. Only analysers and instruments provide analytical information that is directly related to the analytes; all four types of tool, however, can produce non-analytical information on their operating conditions (_e.g._ the rotation speed of a centrifuge, the pressure and temperature of a microwave digester, the

Fig. 1.17. Hierarchical distinction among analyser, instrument, apparatus and device, and relationship to the hierarchy of Fig. 1.16

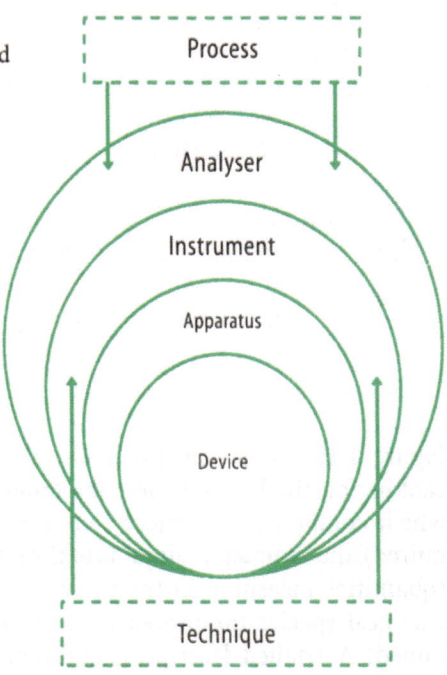

temperature program of a gas chromatograph). Special emphasis in this context should be placed on the difference between instrument and apparatus, two words that are often used indifferently – and misleadingly – in practice. The difference lies in whether or not they can provide analytical information.

_____ Box 1.10

The hierarchy for analyser, instrument, apparatus and device can be illustrated with the following examples:

Type of information	Element ranked	Examples
Analytical	Analyser	• Analyser for gases (O_2, CO_2) in blood • Autoanalyser for C and H in steel
	Instrument	• Balance • Polarograph • Mass spectrometer • Chromatograph (gas, liquid, supercritical fluid)
Operational	Apparatus	• Microwave digester • Extractor (liquid–liquid, solid–liquid) • Centrifuge
	Device	• Pressure or temperature sensor • Electronic interface

A supercritical fluid extractor (SFE) is an apparatus because it performs a function (leaching analytes from a solid sample) but produces no analytical information; it uses devices (sensors) that provide non-analytical information (pressure, temperature) about its operation.

A gas chromatograph (GC) is at the very least an instrument because it accommodates an on-line detector (of the flame ionization or electron capture type, for example) which produces a chromatogram that is used to extract analytical information. However, it also uses a column to separate analytes from interferents, as well as an injector. If the sample can be inserted as such, a gas chromatograph can be considered an analyser; if, on the other hand, it requires some pre-treatment – which is usually the case –, the chromatograph can be deemed an instrument.

An instrumental combination of supercritical fluid extraction (SFE) and gas chromatography (GC) via an appropriate interface can be viewed as an analyser since it can accept untreated solid samples and produce chromatographic data related to the analytes.

Figure 1.18 relates the previous conceptual and technical hierarchies. The ranking on the left considers the tangible contents of the **analytical problem**, which comprises the **object** (*viz.* the system from which information is required), the **sample(s)** (aliquots of the object in space and time), the **measurands** (quantities measured in the samples) and the **analytes** (measurands that are chemical species the presence and/or concentration of which must be determined). A detailed description of this hierarchy is provided in Chap. 7. The word "object" has a more specific meaning here than in Sect. 1.1. One should also distinguish analysis, determination and measurement. **Analysis** refers to the sample, **determination** to the measurands or analytes, and **measurement** to a property of the measurands (or their reaction products, for example). Consequently, a sample is analysed, an analyte is determined and one or more of their physico-chemical properties are measured.

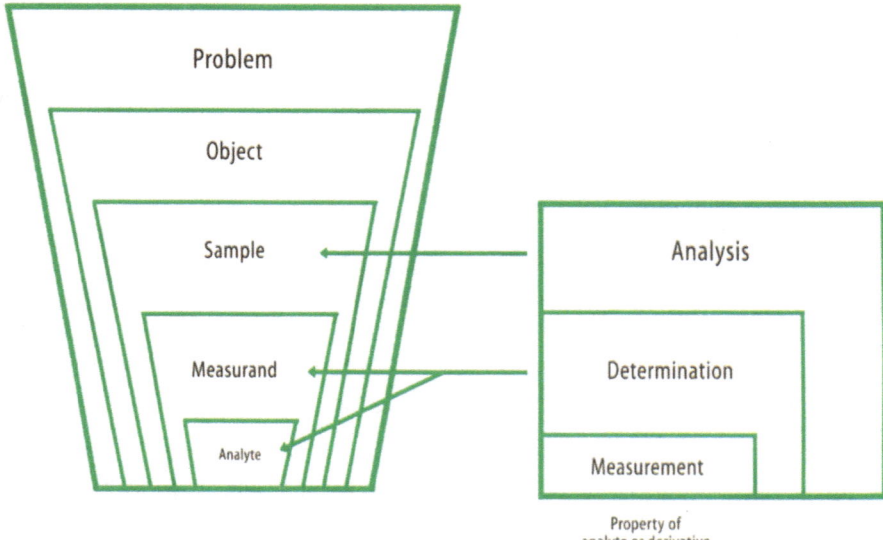

Fig. 1.18. Hierarchies topped by the words "problem" and "analysis", and relationships between them

Box 1.11

Quantity	Measurand	Analyte

It is interesting to distinguish the following three terms, which can be ranked with respect to one another:

Quantity: An attribute of a phenomenon, object or substance that can be distinguished qualitatively and determined quantitatively.

Measurand: A quantity that is measured via a comparison that provides the required information. It is equivalent to **quantity** in the physical domain.

Analyte: A chemical species (element, ion, molecule, radical) that can be identified (and its amount or concentration determined) via a chemical measurement process. It is a special type of measurand in chemical metrology.

Quantity and measurand not always coincide in the analytical chemical domain. Thus, the signal (quantity) provided by most instruments is of the electrical type (volts or amperes); what one actually wants to measure, however, is the measurand (*e.g.* the absorbance, in a UV-visible molecular absorption spectrophotometer).

All analytes can be considered measurands; however, most measurands, but not all, are analytes. One case in point is the pH (a measurand) provided by a glass electrode, which is sensitive to the concentration of protons (analytes); the potentiometer measures the potential difference between the glass electrode and a reference electrode.

Box 1.12

The ranking on the left of Fig. 1.18 is materialized here in various representative examples.

	Problem	Object	Sample(s)	Analytes
Example 1	Contamination of a river	The river, with its geographic and temporal features	Aliquots of the object collected at different places and times	Organic and inorganic contaminating species
Example 2	Drug abuse at the olympic games	Athletes	Urine	Amphetamines, hormones, β-blockers, etc.
Example 3	Adulteration of olive oil with extraneous fat	Factory output	Aliquots representative of the output	Vegetable and animal fat
Example 4	Toxicity of yellow-painted toys (cadmium paint)	Toys from an imported batch	Surface scrapings from various toys selected according to sampling plan	Cadmium
Example 5	Economic feasibility of gold recovery from mining waste	The waste dump as a whole	Samples of the object collected at different depths at different places	Gold

Note: The heading "Problem" here encompasses the two types of problem (*viz.* the information needed and the analytical problem involved) included in the definition of Analytical Chemistry given in Sect. 1.1.

_____ Box 1.13

One should avoid using the words analysis, determination and measurement indifferently as they refer to different types of actions.

Thus, it is correct to speak of the following:

- Determination of cadmium in toys.
- Determination of hydrides in waste water.
- Analysis of a lunar rock.
- Measurement of the absorbance at 540 nm of the copper-cuproin complex.

However, it is incorrect to refer to the following:

- Analysis of cadmium in toys.
- Measurement of copper in sea water.
- Analysis of pesticides in milk.

1.7 Classifications

The intrinsic contents and scope of Analytical Chemistry have promoted the development of classifications according to the different criteria outlined in the previous sections. The most salient of such classifications are schematically depicted in Fig. 1.19.

According to **purpose** and **type of information required**, Analytical Chemistry can be classified as follows:

- **Qualitative analysis,** which is concerned with the identification of analytes (ions, molecules, atoms or chemical groups) present in a sample being subjected to an analytical process. It produces a yes/no binary response (*e.g.* to the presence or absence of aromatic hydrocarbons in the atmosphere, pesticides in foods, drugs and their metabolites in blood) that is of a high analytical value. Uncertainty is substantially diminished when the qualitative response is obtained. However, one should be aware of the inability to detect very low concentrations of some analytes – below their limits of detection. As a result, a theoretically ideal, categorical yes/no response is in fact an actual reply to a more specific question: is the analyte sought present at a concentration above the limit of detection of the analytical process used? Qualitative analysis therefore possesses quantitative connotations. A more detailed description is provided in Chap. 5.
- **Quantitative analysis** aims to determine the amounts or proportions of analytes, atoms or chemical groups in the materials studied. It produces a numerical response suited to the particular purpose. This information is subject to the uncertainty inherent in experimental measurements (see first example in Box 1.4). Chapter 6 deals more extensively with this analytical concept.

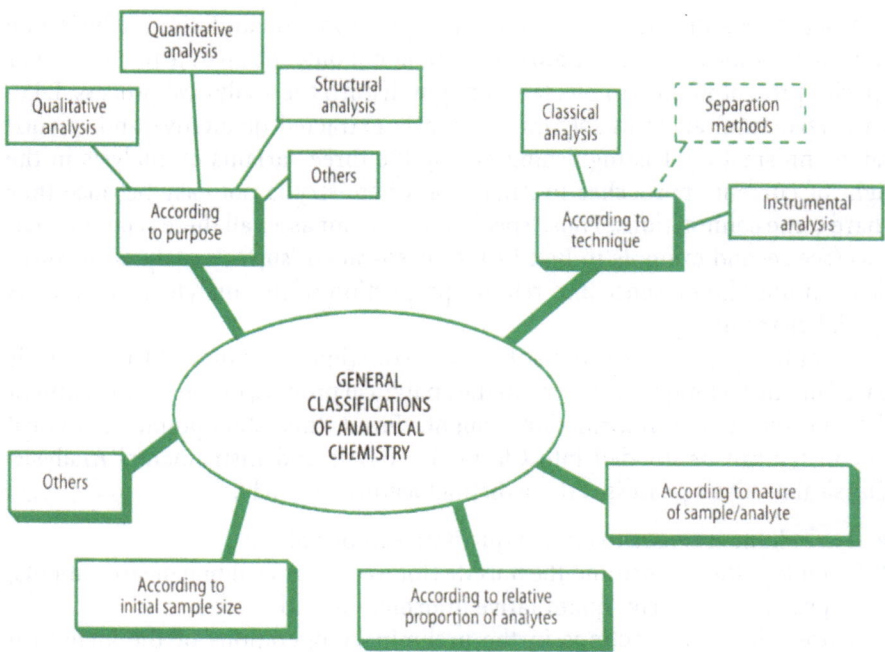

Fig. 1.19. Typical classifications of Analytical Chemistry

• **Structural analysis,** as implied by its name, aims to elucidate (bio)chemical structures. The information it produces can pertain to an individual analyte (a pure compound such as a protein) or a whole sample (*e. g.* the spatial distribution of its components).

Qualitative Analysis identifies a property of the analyte (or its reaction product), Quantitative Analysis measures it in numerical units and Structural Analysis interprets it. As can be seen in Fig. 1.20, the three concepts can be related to one

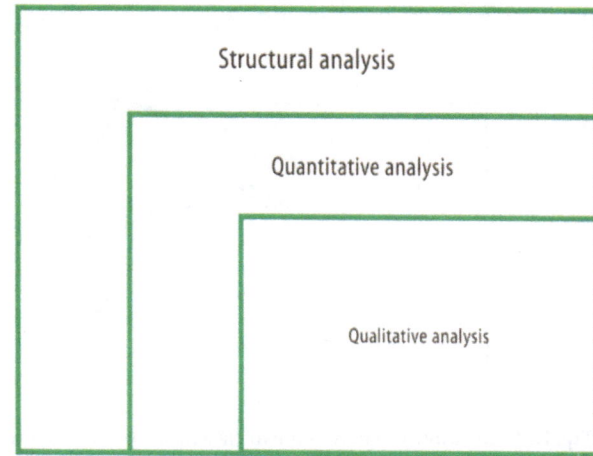

Fig. 1.20. Scope hierarchy for the three types of analysis according to purpose

another via a scope hierarchy. In order to perform a quantitative analysis, one must have some qualitative knowledge about not only the analyte but also other species present in the sample, which might disturb the analytical process. Likewise, structural analyses rely on previously extracted qualitative and quantitative information. Distinguishing among the three variants of analysis in the light of current approaches to Analytical Chemistry is not easy because they share some connotations. Thus, "speciation" encompasses all three types of analysis (see second example in Box 1.4). Also, so-called "surface analysis" involves determining the presence and relative proportion of the analyte, as well as its spatial distribution.

A gradually abandoned but still used distinction in Analytical Chemistry is based on the technique employed in the analytical process, *i. e.* on the instrument that produces the information sought. From this standpoint, Analytical Chemistry can be divided into Classical Analysis and Instrumental Analysis. Classical Analysis possesses three distinct features, namely:

- it uses human senses to extract qualitative information;
- it employs the balance and the burette (for weight and volume measurements, respectively) to derive quantitative information; and
- it uses chemical reactions in the preliminary operations of the analytical process.

As can be seen from Fig. 1.21, Qualitative Analysis has a qualitative purpose: the "human" identification of ions and molecules by use of (bio)chemical reactions, whether in isolation or as parts of an established analytical separation scheme. Its quantitative connotations materialize in Titrimetries (burette) and Gravimetries (balance). Therefore, it also uses "instruments", which reveals the artificial nature of this classification – the persistence of which is only justified on historical grounds.

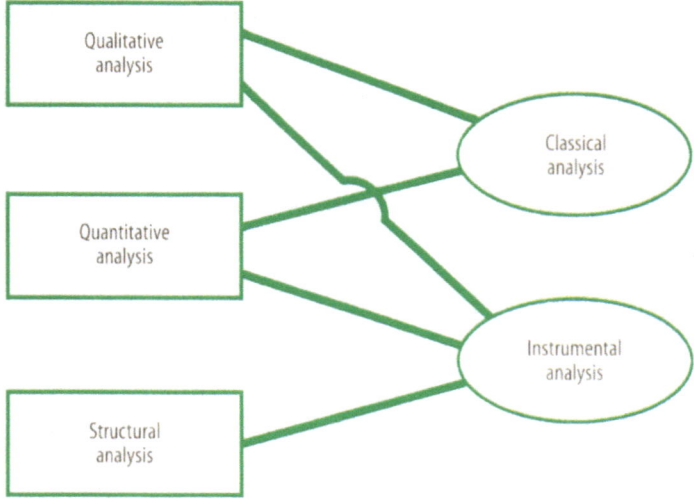

Fig. 1.21. Relationship between two classifications of Analytical Chemistry

By exclusion, Instrumental Analysis thus involves obtaining information in the second step of the analytical process by using instruments other than a burette (volume), a balance (weight) or human senses. Its purpose can be qualitative, quantitative or structural. According to the type of matter-energy interaction involved, instrumental analytical techniques can be classified as follows:

- *Optical* (*e.g.* atomic spectroscopy and molecular spectroscopy, both of which can measure absorption and emission of light);
- *Electroanalytical* (*e.g.* conductimetry, potentiometry, polarography, coulometry);
- *Magnetic* (*e.g.* mass spectrometry, nuclear magnetic resonance spectroscopy); and
- *Radiochemical* (*e.g.* isotope dilution).

Even a brief description of each group is well beyond the scope of this section. Their discussion is thus deferred to Chaps. 5 and 6.

Strictly speaking, **separation methods** are not techniques proper but warrant some comment in respect of this classification by virtue of their extensive use in current analytical processes and their impact on analytical properties. Separations involve mass transfer between two phases, which can take place in widely variable forms during the preliminary operations of the analytical process. Separations can be classified according to various criteria. The most simple divides them into the following categories according to the state of aggregation of the phases involved:

- *Gas-liquid* (gas diffusion, distillation, chromatography).
- *Liquid-liquid* (dialysis, extraction, chromatography).
- *Solid-liquid* (precipitation, ion exchange, chromatography).
- *Solid-supercritical fluid* (extraction, chromatography).

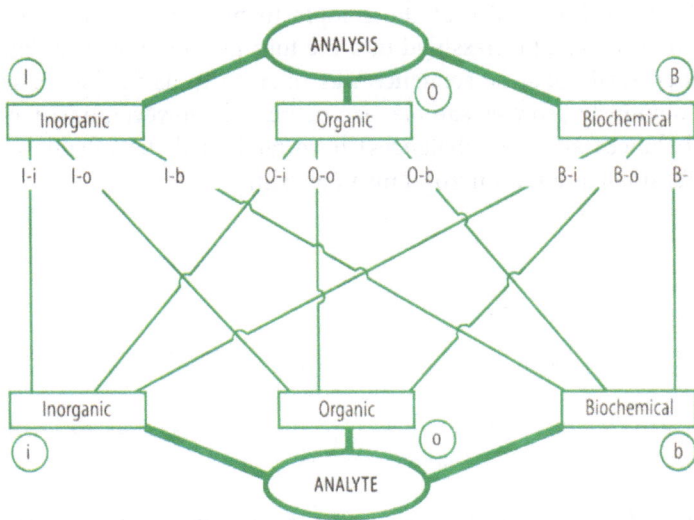

Fig. 1.22. Types of analysis according to nature of samples and analytes

_____ Box 1.14

Here are some practical examples of the different possibilities illustrated in Fig. 1.22 regarding the nature of the sample and analytes:

Sample	Analyte(s)	Example
I	i	• Determination of gold in a mineral
I	o	• Determination of a pesticide in soil
I	b	• Determination of traces of biochemical molecules in a meteorite (*e.g.* in search of evidence of life on other planets)
O	i	• Determination of metal traces in organic pharmaceutical preparations
O	o	• Determination of nitrogen-containing organic compounds in petroleum crudes
O	b	• Determination of enzyme activity in an organic solvent
B	i	• Determination of calcium in biological fluids
B	o	• Determination of drugs and their metabolites in human urine
B	b	• Determination of protein content in milk

Two other possible classifications of Analytical Chemistry are based on the **nature of the sample analysed and analytes determined** and were outlined in its definition. Figure 1.22 shows both classifications and their multiple mutual relationships. It should be noted that assignations in this context can rarely be categorical. Thus, a metal may occur as such or in free ionic form as a part of a complex or an organometallic compound; this last is difficult to place in the classification because it lies on the organic-inorganic boundary.

Based on the **initial size of the sample** to be subjected to the analytical process, analyses can be classified into the four groups shown in Fig. 1.23 – in some, older, textbooks the sequence was referred to as the "work scale." Difficulties obviously grow as sample size decreases. However, new technologies (*e.g.* in balances and microbalances) have facilitated the reliable analysis of samples as small as a few micrograms in weight.

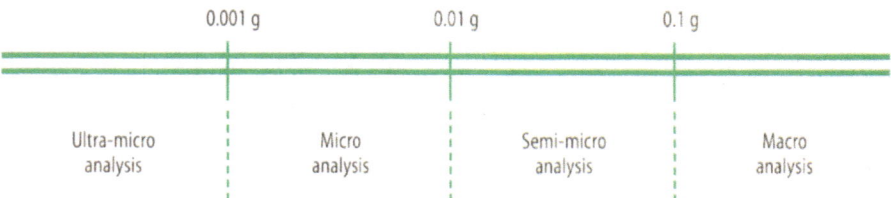

Fig. 1.23. Classification of (bio)chemical analyses according to initial sample size (in grams, g)

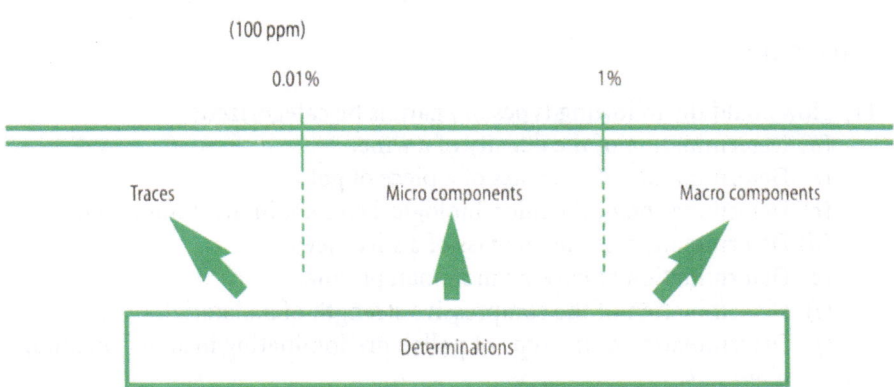

Fig. 1.24. Types of analytical chemical determinations according to relative proportion (percentage) of analyte mass in the sample

Depending on the **relative proportions** (concentrations) of the analytes in the sample, the three types of determinations depicted in Fig. 1.24 – frequently, and improperly, referred to as "analyses"– are possible. Difficulties here also increase with increase in concentration of the target analytes (see Fig. 1.14.5).

A combination of the classifications in Figs 1.23 and 1.24 results in a variety of situations of increasing complexity that range from determinations of macro compounds (present in proportions exceeding 1%) in macro samples (weighing over 0.1 g) to determinations of traces (contents below 0.01%) in ultramicro samples (weighing less than 0.001 g). The widespread designation "Trace Analysis" is justifiable because it designates a specially adapted analytical process; however, it is incorrect in the light of the above hierarchies. In any case, one should never confuse Microanalysis and Trace Analysis.

Questions

(1) How could the following types of analysis be categorized?
    (a) Determination of the density of a wine.
    (b) Determination of the mass of a piece of gold.
    (c) Determination of the microbiological content in drinking water.
    (d) Determination of the fineness of a silver jewel.
    (e) Determination of cocaine in human plasma.
    (f) Determination of the compressive strength of a material.
    (g) Determination of the type of pollen predominating in an urban atmosphere.

(2) Give some examples of analyses that combine the nature of the measured parameters and examined objects (e.g. determination of the protein content in milk).

(3) Give several examples of analyses that are outside the scope of Analytical Chemistry.

(4) Comment on the key words and tentative definitions of Analytical Chemistry.

(5) Define analytical tools and give several examples.

(6) Why are "Analytical Science" and "Chemical Analysis" not synonyms for "Analytical Chemistry"?

(7) Distinguish the analytical information inherent in an object from referential information and from that routinely produced by laboratories.

(8) State the most salient features of ideal analytical information.

(9) Give some examples where analytical results allow one to make economic and social decisions.

(10) Distinguish generic uncertainty and specific uncertainty in the analytical chemical domain.

(11) Give some everyday examples where references are used to obtain information.

(12) On what should teaching materials for Analytical Chemistry feed?

(13) Discuss several examples of fundamentals shared by Analytical Chemistry and areas such as Physics, Physical Chemistry, Organic Chemistry, Biology, Biochemistry, Engineering and Mathematics.

(14) To which other key words of Analytical Chemistry is quality related?

(15) Discuss specific examples of Analytical Chemistry developments that have arisen in response to external stimuli.

(16) Same, with regard to internal stimuli.

(17) Relate those aspects of Figs. 1.7, 1.8, 1.10 and 1.11 that make Analytical Chemistry an independent, self-contained scientific entity.

(18) Give some examples of two-way links between Analysis and other Analytical Chemistry landmarks.

(19) What instrumental landmarks have signalled the three historical periods of Analytical Chemistry?

(20) What is speciation? Where does it fit in Analytical Chemistry classifications?

(21) Why were the fundamentals of Classical Analytical Chemistry essentially thermodynamic in nature?

(22) Give some examples of analytical techniques that provide three-dimensional information.

(23) How would you define "hierarchy"? Look it up in a dictionary, if necessary. Discuss the definition in the light of the analytical chemical hierarchies.

(24) Why is the operating pressure of a chromatograph not analytical information proper? Why is the distance travelled by an organic cation to the cathode in an electrophoretic system related to analytical information?

(25) Distinguish ordinary analytical information from that held as true.

(26) Relate generic uncertainty to schemes 5–8 in Fig. 1.14.

(27) Distinguish the two meanings of "object" used in this chapter.

(28) Use some examples to illustrate the difference between analytical and non-analytical information.

(29) Why does Qualitative Analysis possess quantitative connotations?

(30) Discuss the qualification of a chromatograph.

(31) Distinguish between "Microanalysis" and "Trace Analysis".

(32) Based on the two classifications of Fig. 1.22, give some examples other than those of Box 1.14.

(33) Discuss the difference between Classical Analysis and Instrumental Analysis.

(34) What would the most complex situation in Analytical Chemistry be regarding initial sample size and relative proportion of analytes?

(35) Place separation methods in the analytical chemical domain.

(36) Relate Analytical Chemistry to other applied chemical areas (Industrial, Environmental, Agricultural Food, Pharmaceutical and Clinical Chemistry).

(37) Discuss the relationship between the hierarchies of Fig. 1.15.

(38) Same, with regard to Figs. 1.17 and 1.18.

(39) State several incorrect uses of the words "technique" and "method".

(40) Same, with regard to "analysis" and "determination".

(41) Same, with reference to "instrument" and "apparatus".

(42) Summarize and discuss suggested reading no. 6.

(43) Based on suggested reading no. 7, expand on the concept "analytical quality".

(44) Based on Fig. 1.17 and suggested reading no. 9, discuss the significance of communication in Analytical Chemistry.

(45) Distinguish "measurand" and "analyte"; "object" and "sample".

(46) Distinguish "accuracy" and "trueness" in the analytical chemical domain.

(47) What does expanding the traditional boundaries of Analytical Chemistry involve?

(48) Describe different types of analytical information in the light of the hierarchies, classifications and other concepts discussed in this chapter.

(49) Place the examples given in reply to question (32) (and in Box 1.14) in different contexts: Clinical, Forensic, Industrial, Environmental, Agricultural Food Analysis.

Seminar 1.1

A Negative View of Analytical Chemistry

- **Learning objective**
  To identify what Analytical Chemistry is not in order to provide a final definition for it.
- **Questions for debate**
  - Is Analytical Chemistry an applied science or discipline?
  - Is it a body of methods and techniques for obtaining results?
  - Does it merely provide technical support for society, industry, trade and science?
  - Is it solely concerned with samples and analytes?
  - Does it only encompass inorganic samples and analytes?
  - Is it confined to the laboratory?
- **Misconceptions and wrong approaches among analytical chemists**
  - With reference to the fundamentals of Analytical Chemistry.
  - With regard to developments.
  - With reference to boundaries.
- **Misconceptions of other professionals**
  - Social (politicians, legislators).
  - Physicians, engineers.
  - Physicists.
  - Chemists other than analytical ones.

Seminar 1.2

A Positive View of Analytical Chemistry

- **Learning objective**
  To establish, via representative examples, an integral conception of Analytical Chemistry that can hold at the turn of the third millennium.
- **Questions for debate**
  - Is Analytical Chemistry independent of other scientific and technical areas? Why?
  - In what way should it relate to other professional areas?
  - Where does it fit in the context of Chemistry?
  - What developments make it a self-contained science?
  - Why does expanding its boundaries help consolidate it?
  - What does it imply to include "problems" in its definition?
  - Which key words make it especially independent?
  - What aspects of its conceptual and technical evolution are the most consistent with its definition and key words?

- Are any of its intrinsic and shared fundamentals related? In what way? Give some examples.
- What would a correct research balance between its intrinsic and shared fundamentals be?

Seminar 1.3

Analytical Chemistry in the Context of Chemistry

- **Learning objective**
  To place Analytical Chemistry in the context of today's and tomorrow's Chemistry in the light of the positive and negative approaches discussed in the previous two seminars.
- **Topics for discussion**
  - Relate (graphically, if possible) the cornerstones of Chemistry (Fig. 1.12) and the traditional classification reflected in current university curricula: Organic Chemistry, Inorganic Chemistry, Analytical Chemistry and Physical Chemistry.
  - Discuss the consistency of the interface between each of the two classifications of Chemistry.
  - Starting from the tetrahedron of Fig. 1.12, draw different distorted bodies by pulling one vertex, one edge, two edges and three edges (a triangle). Of the many possible distortions, outline the more significant via specific examples. Bear in mind the contents of Box 1.7.
- **Conclusions**
  In the light of Chemistry evolution so far, will the traditional classifications (*e.g.* that of Organic, Inorganic, Physical, Analytical, Industrial Chemistry) survive? Will an alternative one take over? What type of classification is it bound to be? Why? Discuss the potential role of Analytical Chemistry in the new scene.

Suggested Readings

(1) "A Modern Definition of Analytical Chemistry", M. Valcárcel, *Trends Anal. Chem.*, 1997, 16, 24.
    Discusses at length tentative and final definitions for Analytical Chemistry, and the facets that complement them. The backbone for this chapter.
(2) "Analytical Chemistry: Today's Definition and Interpretation", M. Valcárcel, *Fresenius J. Anal. Chem.*, 1992, 343, 814.
    Second prize at an international competition on the definition of Analytical Chemistry. A foretaste of the previous paper, which it completes in some respects.
(3) "Analytical Chemistry, Today", M. Valcárcel, *Quím. Anal.*, 1990, 3, 215.
    Summarizes the state of the art and current trends in Analytical Chemistry. Contains some sections of this chapter in expanded form.

(4) "Theory of Analytical Chemistry", K. Booksh and B. Kowalski, *Anal. Chem.*, 1994, 66, 782A.
Emphasizes the information side of Analytical Chemistry and establishes its landmarks as an information science.

(5) "Two Sides of Analytical Chemistry", G. Hieftje, *Anal. Chem.*, 1985, 57, 257A.
Discusses in simple, yet categorical, terms the basic and applied facets of Analytical Chemistry and their mutual relations.

(6) "Analytical Chemistry; Feeding the Environmental Revolution", J. G. Graselli, *Anal. Chem.*, 1992, 64, 677A.
Envisages the present and future of Analytical Chemistry in the light of changes in various contexts.

(7) "Analytical Chemistry and Quality", M. Valcárcel and A. Ríos, *Trends Anal. Chem.*, 1994, 13, 17.
Establishes the analytical connotations of quality and also landmarks intended to avoid the frequent confusions in this context.

(8) "Analysis: What Analytical Chemists Do", J. F. Tyson, *Royal Society of Chemistry*, Cambridge, 1988.
A straightforward book that departs from traditional textbooks. A singular approach that is complementary to the one adopted in this book.

(9) "A Hierarchical Approach to Analytical Chemistry", M. Valcárcel and M. D. Luque de Castro, *Trends Anal. Chem.*, 1995, 16, 242.
Proposes, for the first time, the use of conceptual and technical hierarchies as alternatives to the traditional definitions of facts, objects and systems of significance to Analytical Chemistry in order to facilitate internal and external communication.

Objectives

- To define analytical properties in a global, consistent manner, with emphasis on their theoretical and practical significance.
- To assign analytical properties unequivocally to specific aspects of Analytical Chemistry.
- To establish foundation, complementary and contradictory relationships among analytical properties.
- To relate analytical properties to different facets of analytical quality.

Table of Contents

2.1 Introduction

In broad terms, "quality" is defined as the totality of features (properties, attributes, capabilities) of an entity that make it equal to, better or worse than others of the same kind. Analytical quality materializes in various properties that are ultimately used as quality indicators. Such indicators allow one to assess, compare and validate various basic and applied facets of Analytical Chemistry in a well-founded, efficient manner.

Analytical properties can be defined in the light of the following considerations:

(*a*) They constitute a part of the intrinsic fundamentals of Analytical Chemistry (see Fig. 1.9);
(*b*) they are key aspects of Analytical Chemistry (see Fig. 1.10);
(*c*) they are the rational vehicles for Qualimetrics (the statistical, computer-based support for analytical quality);
(*d*) they characterize, in an orderly manner, different analytical facets;
(*e*) they vary in rank and significance, the ranking being flexible and adjustable to the particular analytical problem;
(*f*) they are dependent on one another; and
(*g*) they are related to classical metrological properties.

Figure 2.1 provides an overview of analytical properties and classifies them into three hierarchical categories, namely; *capital properties* (accuracy and representativeness); *basic properties* (precision, sensitivity, selectivity and proper sampling); and *accessory properties* (expeditiousness, cost-effectiveness and personnel-related factors such as safety and comfort). Capital analytical properties are typical of results (analytical information). The quality of the analytical process is obviously related to that of its results; consequently, basic

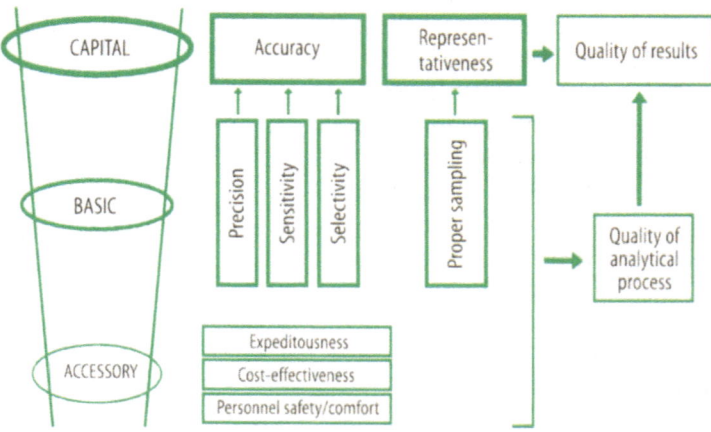

Fig. 2.1. Types of analytical properties and relationships among them and with analytical quality (results and processes)

properties support capital properties. In addition, analytical properties bear complementary and contradictory relationships that are discussed later on in this chapter, following a concise description of each analytical property. A prior, brief definition of errors in Analytical Chemistry may be of help to better define the properties. Thus, analytical chemical errors arise from the so-called "chemical metrological hierarchy", which allows one to define such analytical features as uncertainty and trueness, which are not included in the scheme of Fig. 2.1 but are obviously related to classical analytical properties.

2.2 The Chemical Metrological Hierarchy: Uncertainty and Trueness

Subjecting n aliquots of the same sample to an analytical process provides n results. Obviously, the quality of the information obtained will increase with increase in n. As can be seen from the central portion of Fig. 2.2, the lowest information level corresponds to an individual result (x_i) provided by a single sample aliquot. Next in the rank is the arithmetic mean $\bar{X}$ of a small number of results or sample aliquots ($n < 30$). The mean obtained with $n > 30$, μ', will be of

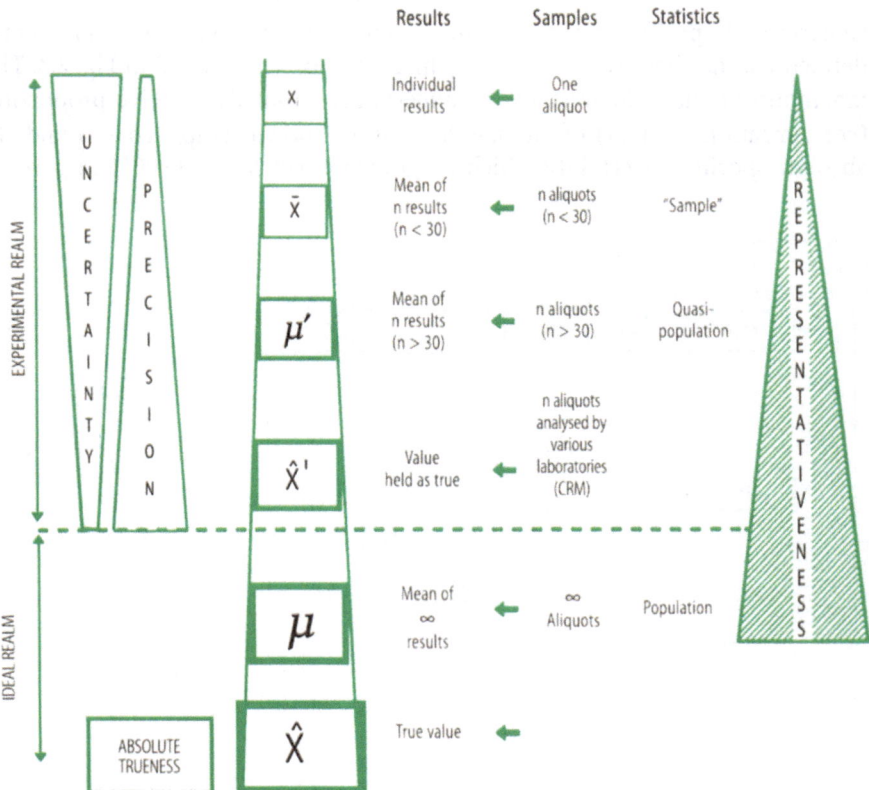

Fig. 2.2. The chemical metrological hierarchy and its relationships to analytical properties

higher quality than the previous one, the theoretical upper quality limit for the series being the mean for a statistical population ($n = \infty$), which is denoted by μ.

When n sample aliquots are analysed by different laboratories using various analytical processes and a consensus is reached from a thorough technical study, a result $\hat{X}'$ is obtained that is held as true and represents referential quality in Fig. 1.6; this kind of quality is typical of certified reference materials (CRMs), which are discussed in Chap. 3. Total quality or theoretical ideal quality corresponds to the true value $\hat{X}$ associated to intrinsic information.

The former four landmarks (*viz.* x_i, $\bar{X}$, μ' and $\hat{X}'$) are obtained through testing; by contrast, the latter two (μ and $\hat{X}$) are only ideal, theoretical data that can never be obtained in practice.

As the chemical metrological hierarchy is climbed from x_i to $\hat{X}$:

(*a*) Generic and specific uncertainty (Box 1.4) both decrease to 0, which corresponds to absolute trueness ($\hat{X}$);

(*b*) closeness to the true value (accuracy), which is defined below, increases;

(*c*) representativeness in the information obtained increases with increasing number of sample portions used from 1 to ∞; and

(*d*) the number of data (results) that may accompany the result on each step of the hierarchy decreases (thus, x_i may be accompanied by ∞ results on the first step while no datum other than $\hat{X}$ may exist on the last).

Generic and specific **uncertainty** in Analytical Chemistry can be efficiently defined via the chemical metrological hierarchy as schematized in Fig. 2.3. The maximum possible dubiousness or uncertainty about the relative proportion (concentration, content) of an analyte in an unknown sample corresponds to absolute specific uncertainty, which ranges from 0.00 % to 100.00 % in percen-

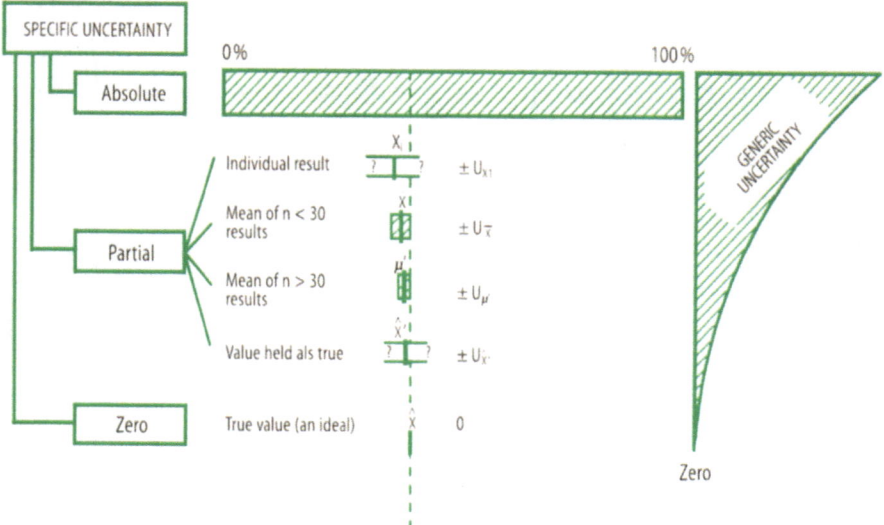

Fig. 2.3. Graphical depiction of generic and specific uncertainty, and of their relationships to the chemical metrological hierarchy

tage terms. The true value, $\hat{X}$, is subject to zero specific uncertainty, which coincides with the absolute absence of dubiousness in the analyte proportion in the sample. This is intrinsic information that corresponds to ideal quality (see Fig. 1.6).

Partial specific uncertainty exists over an interval $\pm U$ around a specific datum; the interval includes the analyte proportion in the sample. The ultimate expression of partial specific uncertainty is that consisting of a single result (x_i), which expresses no interval by itself; however, it is subject to less generic uncertainty as regards the maximum dubiousness since it bounds an approximate zone or interval where the information sought may lie. The mean of $n < 30$ results, $\bar{X}$, obtained by analysing multiple aliquots of the same sample, is subject to a statistical uncertainty $\pm U_{\bar{x}}$, as is the mean of $n > 30$ results $(\pm \mu')$. As n increases, the uncertainty decreases $(|U_{\bar{x}}| > |U_{\mu'}|)$, and $\bar{X}$ and μ' approach the true value – from which x_i can be more or less distant.

_____   Box 2.1

Let us illustrate the uncertainty concepts described in the text and Fig. 2.3 with a specific example.

In the determination of an amphetamine in a urine sample, the true value (the ideal of quality), which is experimentally inaccessible, was $\hat{X} = 3.2510$ mg/L. The following four series of experiments were conducted on aliquots of the same sample:

(a) a routine determination that gave $x_i = 3.6$ mg/L;
(b) a routine determination of $n = 8$ aliquots that yielded $\bar{X} \pm U_{\bar{x}} = 3.4 \pm 0.3$ mg/L:

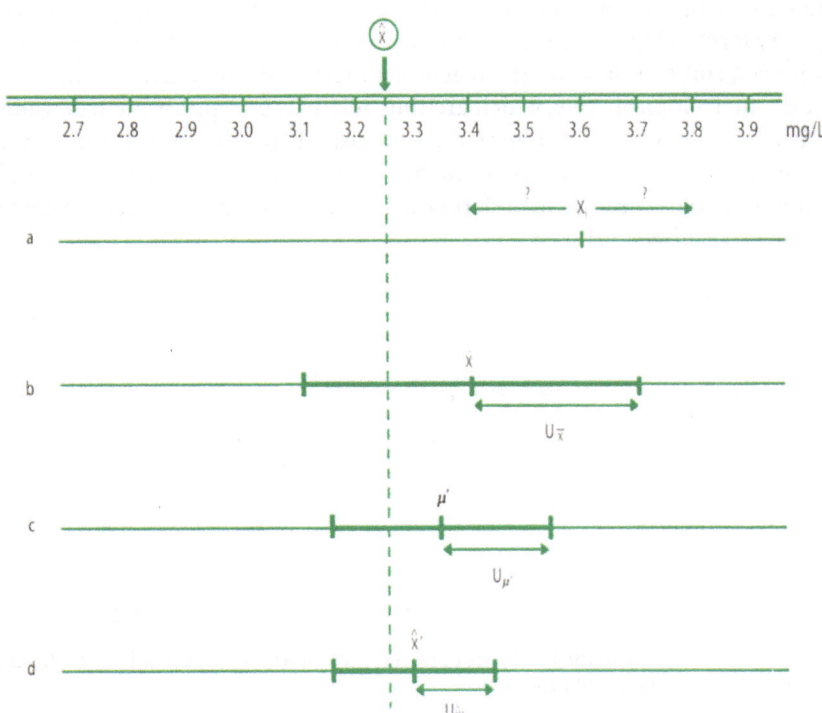

(c) a routine determination of $n = 40$ aliquots that yielded $\mu' \pm U_{\mu'} = 3.35 \pm 0.20$ mg/L; and

(d) an interlaboratory exercise involving 14 laboratories that analysed 8 samples each to provide 14 means $(\bar{X}_i)$ from which a global mean $\hat{X}' = 3.30 \pm 0.15$ mg/L for the exercise was calculated.

As can be seen, specific uncertainties decrease from U_{x_i} – which cannot be determined – to $U_{\mu'}$. In this example, special interlaboratory testing led to lower uncertainty.

The generic uncertainty in x_i is very large but that in $\hat{X}'$ is very small because the latter is very close to the true value $(\hat{X})$. Also, the specific uncertainty $(U_{\hat{X}'})$ is quite small.

Even though the referential value ($\hat{X}'$, considered to be the true value) is subject to little generic uncertainty, its specific uncertainty, $U_{\hat{X}'}$, can be greater or less than $U_{\mu'}$ inasmuch as it is calculated from n' means $(\bar{X}_{n'})$ obtained by n' laboratories analysing different aliquots of the same sample, which will have introduced much more significant sources or variability. The specific uncertainty in the mean of ∞ results, μ, is theoretically zero.

As shown graphically in Fig. 2.4, specific uncertainty is unequivocally related to the analytical property precision; on the other hand, generic uncertainty is related to both accuracy and precision since closeness of experimental means $(\bar{X}, \mu', \hat{X}')$ to the true value $(\hat{X})$ decreases dubiousness in the sample composition; also, the smaller the respective uncertainties $(U_2, U_m, U_{\hat{X}'})$ are, the narrower will be the dubious zone.

The chemical metrological hierarchy can also be used to clarify the meaning of **trueness** in Analytical Chemistry. Figure 2.5 illustrates different approaches to this concept. Absolute trueness coincides with the sample's intrinsic information, *i.e.* with the true value $(\hat{X})$, and represents the maximum possible (ideal) quality level in Fig. 1.6 and the top level in the metrological hierarchy of Fig. 2.2. However, some authors define trueness in terms of differences between values on the metrological scale, which are thus relative concepts. If the true value $(\hat{X})$ is used as reference, differences are established from the mean for an infinite number of samples (μ) or the value held as true $(\hat{X}')$; if, on the other hand, the last parameter is used as the reference, then differences can be established from $\mu (n = \infty)$ or $\mu' (n > 30)$.

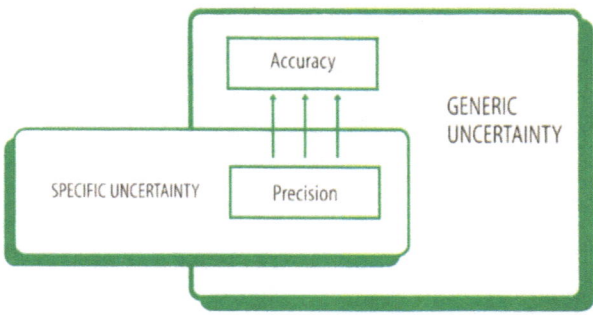

Fig. 2.4. Relationships of the two types of uncertainty in Analytical Chemistry to the analytical properties accuracy and precision

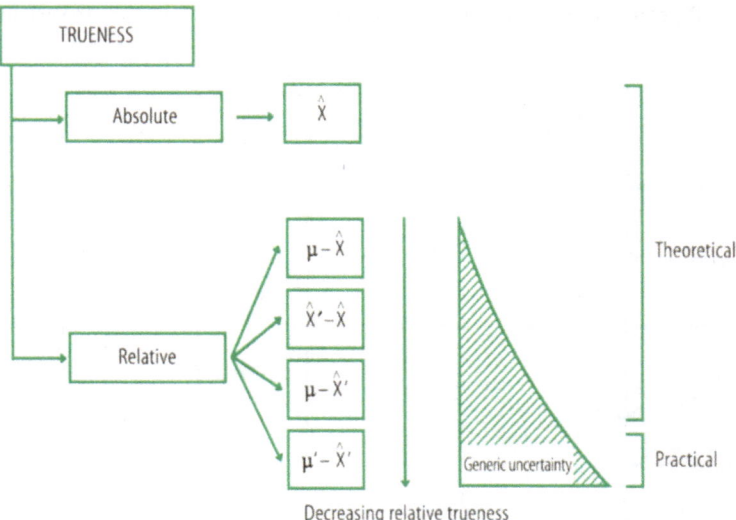

Fig. 2.5. Connotations of the word "trueness" in Analytical Chemistry

From Fig. 2.5 it follows that relative trueness decreases from the difference $|\hat{X} - \mu|$ to $|\hat{X}' - \mu'|$ and that uncertainty increases in the same direction; as noted above, absolute trueness is equivalent to the absence of uncertainty. Note that only the last step in the relative trueness ranking is experimentally accessible; all others rely on ideal data ($\hat{X}, \mu$) as references. The trueness concept is unequivocally related to accuracy in a direct manner and to precision in an indirect one – as shown in Sect. 2.8, precision is one of the cornerstones of trueness.

2.3 Errors in Analytical Chemistry

The word "error" means, mistake, falsehood, departure from something that is held as correct or true. In Analytical Chemistry, "error" is used in broad terms to refer to alterations in the information supplied and, ultimately, to describe differences between the true value ($\hat{X}$), or the value held as true ($\hat{X}'$), and an individual result (x_i), the means of multiple results ($\bar{x}, \mu'$) or differences between results. An error can thus be ascribed to a result or to an analytical process. It can be expressed in absolute (a difference and its sign) or relative terms (a fraction of unity or a percentage obtained by dividing the difference into the reference value).

There are three general types of error (random, systematic and gross) according to magnitude, sign, source and the reference used to establish the ensuing differences. Their features are summarized in Table 2.1.

Random or **indeterminate errors** stem from typical experimental fluctuations. They arise when the same analyte is subjected to several determinations in

Table 2.1. Types of errors in Analytical Chemistry, characterized according to various criteria

TYPES OF ERROS	DIFFERENTIAL				
	SOURCE	Reference used for definition	SIGN	Relative magnitude (as a rule)	Analytical property affected
RANDOM (indeterminate)	Random	Mean of a set	variable $\oplus$ $\ominus$	small	PRECISION (specific uncertainty)
SYSTEMATIC (determinate)	Well-defined alterations	$\hat{x}'(\hat{x})$	unique $\oplus$ positive or $\ominus$ negative	small	ACCURACY (traceability)
GROSS				large	

the same sample or, simply, when the same measurement is made several times. The reference for calculation of this type of error is the arithmetic mean ($\bar{X}, \mu'$) for a series of determinations. Random errors vary in magnitude, which is rarely very large. They result in overestimation or underestimation, depending on whether they lead to data above or below the mean. Consequently, this type of error is meaningless for an individual datum (result) unless a data set is available for comparison. It can be ascribed to individual results or to a method (means) and is characterized by using statistical principles based on the normal or Gaussian distribution (see Box 2.2). Random errors are the basis for the analytical property precision and hence for specific uncertainty.

Box 2.2

Principal statistical landmarks for handling analytical chemical data

- The statistics applied to analytical chemical data is a part of the multidisciplinary entity called CHEMOMETRICS, which aims to expand and enhance analytical chemical information by using less material, time and human resources (with the aid of computers).
- Statistics is a key support for analytical properties in general and the following in particular:
  - Representativeness (sample collection rules)
  - Accuracy (traceability)
  - Precision (uncertainty)
  - Sensitivity (definition of "limit of detection" and "limit of quantitation" via statistical parameters).

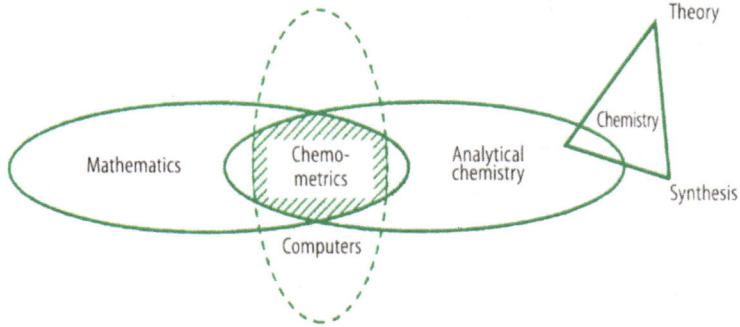

- The "sample" and "population" concepts as referred both to a set of aliquots (samples) of an object and to the results obtained by subjecting each aliquot to an individual analytical process.

	Aliquots (n)	Arithmetic mean	Standard deviation
Population	∞ ($n > 30$)	μ'	σ
Sample	$n < 30$		s

- Statistical processing of data (results)
  - n measurements $x_1, x_2, \ldots x_n$
  - absolute frequency with which they occur $f_1, f_2, \ldots f_n$
  - normal (Gaussian) distribution   bell-shaped plot of relative frequency (f_n/n) as a function of the data.

  (*a*) Characteristic of dispersion among data.
  (*b*) Characterized by some mathematical terms

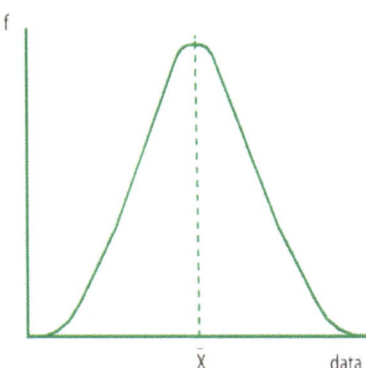

- Parameters that characterize a normal distribution of data
  - Arithmetic mean ($\bar{X}$ and μ').
  - Standard deviation: the distance (to the right or left) from the mean to the first inflection point in the curve. The most elementary measure of dispersion (expressed in the same units as the data).

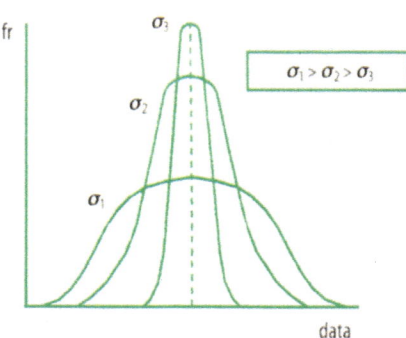

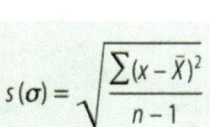

$$s(\sigma) = \sqrt{\frac{\sum(x - \bar{X})^2}{n - 1}}$$

The standard deviation is a measure of the proportion of data that can exist within a given interval around the mean.

Interval	$\mu \pm \sigma$	$\mu \pm 1.96\,\sigma$	$\mu \pm 2.57\,\sigma$	$\mu \pm 3\,\sigma$	$\mu \pm 3.29\,\sigma$
% DATA	68	95	99	99.7	99.9

- Variance: the standard deviation squared (S^2, σ^2)
- Relative standard deviation

$$rsd = \frac{s}{\bar{x}} \left(\frac{\sigma}{\mu} \right)$$

- Coefficient of variation (a percentage)

$$CV = rsd \cdot 100$$

- Standard deviation of the means
  - Means ($\bar{X}_1, \bar{X}_2, \ldots$) resulting from grouping data for samples of the same population.
  - The mean of these means is $\bar{X}_m$; their standard deviation (sdm) does not coincide with that of the population (σ, s).

$$sdm = \frac{\sigma(s)}{\sqrt{n}}$$

  The greater n, the lower the dispersion among the means
  - When $n \to \infty$, sdm $\to 0$ (all measurements tend to provide the same result)
  - sdm is a measure of the uncertainty incurred in estimating μ from $\bar{x}$
- Limits of confidence for the mean
  - The interval within which a given measurement can be assumed to fall with a preset likelihood or the value held as true, μ, to lie when an analytical process is repeated.

- For a sample ($n < 30$), which is the most usual case,

$$\mu = \bar{x} \pm t \cdot \frac{s}{\sqrt{n}}$$

- Student's t depends on (a) the number of degrees of freedom ($n-1$); (b) the confidence level required, expressed as a percentage or as a probability [$P = 1 - (\%/100)$].

Confidence interval	t values			
	90%	95%	98%	99%
Likelihood	0.10	0.05	0.02	0.01
Degrees of freedom $V = n - 1$				
1	6.31	12.71	31.82	63.66
2	2.92	4.30	3.96	9.92
3	2.35	3.18	4.54	5.84
4	2.13	2.78	3.75	4.60
5	2.02	2.57	3.36	4.03
6	1.94	2.45	3.14	3.70
7	1.89	2.36	3.00	3.50
8	1.86	2.31	2.90	3.36
9	1.83	2.26	2.82	3.25
10	1.81	2.23	2.76	3.17
12	1.78	2.18	2.68	3.05
14	1.76	2.14	2.62	2.98
16	1.75	2.12	2.58	2.92
18	1.73	2.10	2.55	2.88
20	1.72	2.09	2.53	2.85
30	1.70	2.04	2.46	2.75
50	1.68	2.01	2.40	2.68
∞	1.64	1.96	2.33	2.58

- The confidence interval depends on:
  (A) Sample size (n). As n increases, the interval decreases because $\sqrt{n}$ is in the denominator [and t decreases as a result (there are fewer degrees of freedom)].
  (B) The confidence level (probability).

_____ Box 2.3

Here is an example of random errors in an analytical process based on gravimetry: the determination of iron in a sample. Each measurement operation involved in this methodology produces a random error that is usually established previously from the characteristics of the measuring system (instrument) used.

The process is started by weighing a solid sample or measuring the volume of a liquid sample; the measurement will contain a random error of ± 0.0002 g and ± 0.03 mL, respectively. Thus, if the sample weight is 0.5280 g, then the datum will be subject to a fixed random error, so it should be expressed as 0.5280 ± 0.0002 g.

Since the analytical process involves dissolution and oxidation of iron to Fe^{3+} ion, precipitation as $Fe(OH)_3$ with NH_3 in the presence of an ammonium salt, filtration through paper, drying and calcination of the precipitate in a crucible to obtain a weighable form of iron (Fe_2O_3), the clean, dry crucible to be used must be "tared" (weighed) before it receives the filter paper containing the precipitate. This produces a weighing error of ± 0.0002 g.

After the hydroxide is thermally converted into the oxide (and the paper burnt), the crucible is cooled down and weighed again, which introduces a new random error of ± 0.0002 g.

Consequently, the analytical process is subject to at least three sources of random errors that coincide with the three measurements made. These indeterminate errors obviously affect the final result. Box 2.7 shows the way the random error in the final result is calculated.

Systematic or **determinate errors** are due to well-defined operational alterations in the analytical process (*e.g.* the presence of interferents, incomplete filtration, carry-over and adsorption losses in trace analyses, deteriorated reagent or standard solutions). They are referred to the true value ($\hat{X}$) or that held as true ($\hat{X}'$) and materialize in differences (deviations) of the results from them. Systematic errors affect the analytical property accuracy; a result that is subject to a small systematic error may be an accurate result. Consequently, deviations are of a definite sign: positive (overestimation) or negative (underestimation). Errors that are very large are called "gross errors". Systematic errors are constant or proportional depending on whether or not they depend on the analyte concentration. Also, they can be ascribed to an individual result (x_i) or to a method characterized by the mean ($\bar{X}$, μ') of the results produced by its repeated application to the same sample. This gives rise to three different designations, namely (Fig. 2.2):

(a) **Accuracy** proper, when the error refers to a result (in which case it coincides with the difference $[x_i - \hat{X}']$);
(b) **bias**, when the error refers to a method that was used to perform $n < 30$ determinations (the difference sought being $[\bar{X} - \hat{X}']$); and
(c) **relative trueness**, when the error refers to a method that was used to carry out $n > 30$ determinations (the difference being $[\mu' - \hat{X}']$) (see Fig. 2.5).

Gross or **spurious errors** are essentially similar to systematic errors (see Table 2.1) but considerably larger in magnitude. They are positive or negative errors that introduce severe alterations in the results (*e.g.* between -60% and $+300\%$ in relative terms). Gross errors can be easily detected and avoided. Also, they require no special treatment – the results concerned are simply rejected, even though statistical rules can be used for this purpose (see Box 2.10).

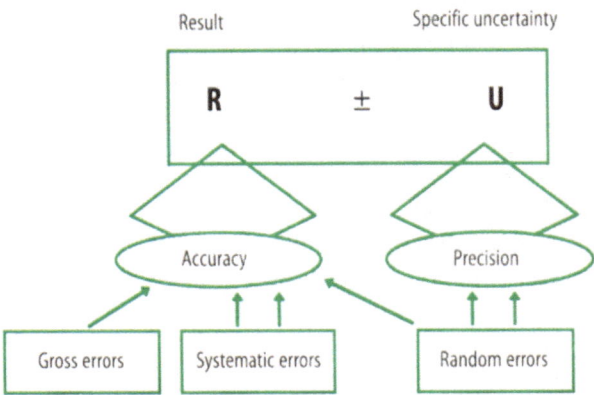

Fig. 2.6. Expression of a result (and its uncertainty) and relationships to the different types of error and to the analytical properties accuracy and precision

This brief introduction to errors in Analytical Chemistry is the basis for defining accuracy (a capital analytical property) and precision (a basic analytical property). Both were mentioned earlier in this section. Figure 2.6 shows the relationships of gross and systematic errors to the analytical result, R, via accuracy, and that of non-specific uncertainty, U, to random errors via precision. Such a categorical difference is not factual: as shown in forthcoming sections, accuracy is impossible in the absence of precision.

Box 2.4

Let us illustrate the previous definitions of errors in Analytical Chemistry with the following example: four different methods (A, B, C and D) are applied n times each to aliquots of the same sample in order to determine the concentration of an analyte for which the value held

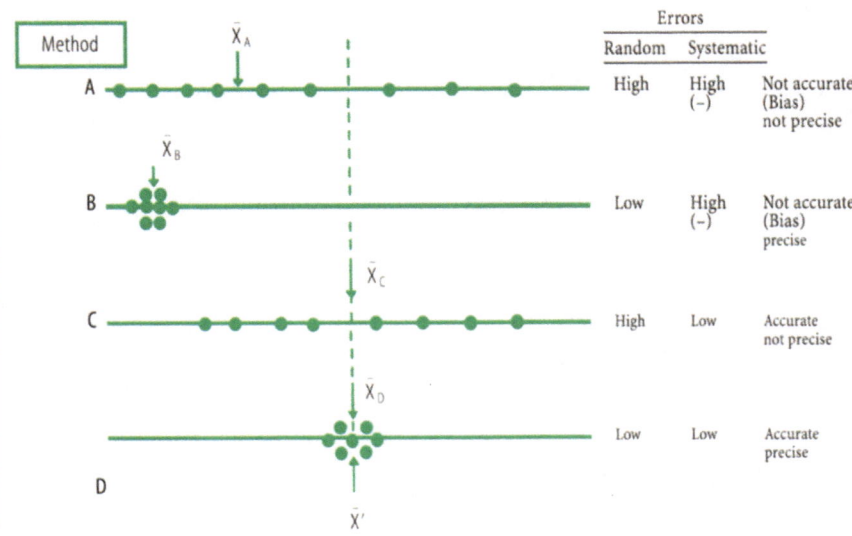

as true, $\hat{X}'$, is known beforehand. Each method produces n results the respective means of which are X_A, X_B, X_C and X_D. A graphical depiction of the results and means allows one to envisage the different types of errors involved and provides a qualitative description of two analytical properties: accuracy and precision.

As far as systematic errors are concerned, methods A and B are subject to bias, $|\bar{X}_A - \hat{X}'| > |\bar{X}_B - \hat{X}'|$, whereas methods C and D are not. Both means, $\bar{X}_A$ and $\bar{X}_B$, are subject to a negative error. Each datum possesses an intrinsic accuracy that is its specific difference from $\hat{X}'$; such a difference coincides with the respective means for methods C and D ($\bar{X}_C$ and $\bar{X}_D$).

Regarding random errors, the results of methods A and C are highly disperse (*i.e.* the differences from the respective means are significant), whereas those of methods B and D are tightly clustered around their means. Therefore, methods B and C are more precise than are A and C.

2.4 Capital Analytical Properties

The adjective "capital" is synonymous with "foremost" and "topmost", and is used here to qualify the top-rank properties in the hierarchy of Fig. 2.1 (accuracy and representativeness). These two are directly ascribed to results and hence to analytical information, the Analytical Chemistry key word around which all others revolve (see Fig. 1.10).

It is very important to realize that both capital properties are indispensable with a view to ensuring proper quality in the results (see the scheme in Fig. 2.7). Thus, an accurate (and scarcely uncertain) result that is not representative of the problem addressed will be a poor result (for example, operating under excellence laboratory conditions will be worthless if the sample used is unfit for the particular purpose).

Below are summarized the more relevant features of capital analytical properties.

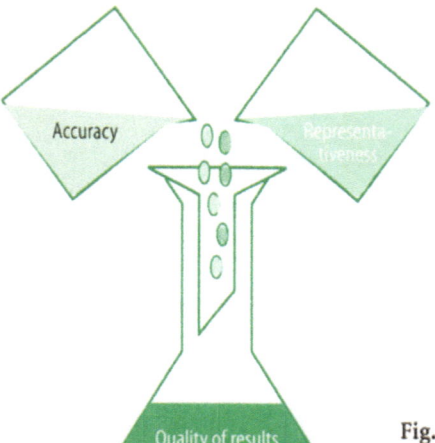

Fig. 2.7. Capital analytical properties are the essential ingredients of quality of analytical results

2.4.1 Accuracy

Accuracy is defined in broad terms as the **"degree of consistency between the result of a determination (x_i) or the mean of n results $(\bar{X}, \mu')$ and the true value $(\hat{X}$, intrinsic information) of the analyte, measurand or determinand in a sample."** This definition is purely theoretical; to endow it with practical significance, the true value $(\hat{X})$, which is purely ideal, must be replaced with the value held as true $(\hat{X}')$.

Mathematically, accuracy is characterized by a systematic (or gross) error, which is a difference with a fixed sign (positive or negative). It can be expressed in absolute (a difference and its sign) or relative terms (a fraction of unity or percentage).

Accuracy can be referred to a result; such is the case with accuracy proper, which is expressed as the difference

$$e_{x_i} = \pm \left| x_i - \hat{X}' \right| \qquad (e_{x_i})_{rel} = \frac{\pm e_{x_i}}{\hat{X}'} \cdot 100$$

Accuracy can also be referred to a method, via the mean $(\bar{X}, \mu')$ of n results produced by the same method as applied to aliquots of the same sample. In this case, accuracy is designated differently depending on the magnitude of n:

Accuracy of a method	Bias	$n < 30$	$e_{\bar{X}} = \pm \left| \hat{X} - \hat{X}' \right|$	$(e_{\bar{X}})_{rel} = \dfrac{\pm e_{\bar{X}}}{\hat{X}'} \cdot 100$
	Relative trueness	$n > 30$	$e'_{\mu} = \pm \left| \mu' - \hat{X}' \right|$	$(e'_{\mu})_{rel} = \dfrac{\pm e'_{\mu}}{\hat{X}'} \cdot 100$

Also, it is not uncommon to refer to "accuracy of the mean".

In theoretical terms, accuracy (systematic or gross errors) is not the same as precision (random errors). In practice, however, accuracy cannot be correctly defined without considering the precision (uncertainty) involved, as shown in Box 2.5. In the example of Box 2.4, the two concepts are dealt with separately. Case C shows that a method can be accurate without being precise, but this is merely a coincidence. For a method to be accurate, individual results must be highly likely to fall close to the true value $(\hat{X}')$, which requires that the method be highly precise. On these grounds, only method D in Box 2.4 will be accurate since, in addition to $\bar{X}$ being close to $\hat{X}'$, the precision is quite high. Also, the comparison of $\bar{X}$ (or μ') with $\hat{X}'$ should be extended to their respective uncertainties: $U_{\bar{X}}$ (or U'_{μ}) and $U_{\hat{X}'}$.

_____ Box 2.5

In order to define the accuracy of a method, one must previously know its precision and compare its uncertainties (ranges) and that in the value held as true ($\hat{X}'$). The graph below shows six different situations; the first four lead to the same result, which is different from that for the fifth and sixth (mutually identical but subject to a different uncertainty each).

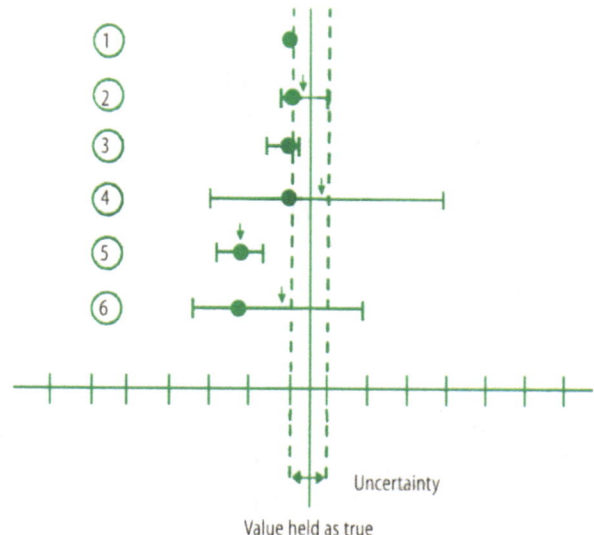

1. The accuracy of the method cannot be defined because the uncertainty in the result is unavailable.
2. The result is accurate as it falls within the uncertainty range for the value held as true and both ranges are virtually coincident – the mean for the set of results is closer to the true value.
3. The result is less accurate than the previous one because its respective ranges are less consistent, even though its absolute difference from $\hat{X}'$ is the same.
4. The result can be considered inaccurate – despite its closeness to $\hat{X}'$ – because it is subject to a high uncertainty relative to $\hat{X}'$.
5. The result is not accurate – however precise – because it is very distant from $\hat{X}'$ and the respective uncertainties are different.
6. The result is neither accurate nor precise. The uncertainty range for $\hat{X}'$ is only a small fraction of that for the result.

2.4.2 Representativeness

This capital property is also ascribed to results. It relies on proper sampling (see Fig. 2.1) and requires expanding the traditional boundaries of the analytical chemical laboratory (see Sect. 1.4.6 and Fig. 1.13). As shown in Chap. 7, representativeness is an essential element of the analytical process, where the purpose of the analytical information to be derived must be clearly established.

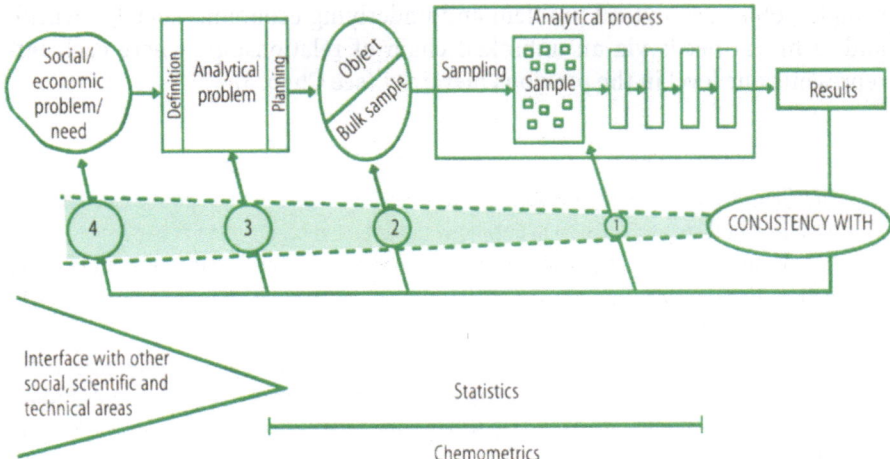

Fig. 2.8. Hierarchical ordering of the different levels of representativeness of analytical results

In the context of analytical information, the word representativeness is synonymous with consistency in the results. To ensure consistency, one must use some type of reference such as those depicted in Fig. 2.8.

Level 1. The results are consistent with the samples (or aliquots) that are received by the laboratory and subjected to the analytical process. This is the lowest level of representativeness that can be achieved via the so-called "sample custody chain". It is intended to correlate samples and results; where applicable, it assigns results to the object that raises the analytical problem [*e. g.* a global mean for the samples received or discriminate results (means) for each type of sample].

Level 2. The results are consistent with the object or system being studied (*e. g.* a river, a tile factory, a dairy factory, a bioreactor). This entails proper sampling and the ability to extract additional information from the results (*e. g.* detecting the presence of contaminants or establishing the kinetics of a bioreactor). This level relies on the representativeness of the previous one.

Level 3. Representativeness here also involves consistency with the analytical problem addressed (see Chap. 7), *i. e.* considering the purpose of the analytical information to be produced.

Level 4. The results are consistent with established social, industrial, scientific and technological information needs. This level relies on the representativeness of the previous three.

These four levels can be ranked according to significance and scope. If a result is representative at a given level, then it must also inevitably be representative at lower levels. There are some traceability connotations in this assertion; the ultimate aim is to relate, in a documented, rigorous manner, the results to the samples, object, analytical problem and underlying economic, social, scientific and technical needs via an unbroken chain of relationships where each link represents one level in the previous hierarchy (see Chap. 3).

_____ Box 2.6

The analytical property representativeness is illustrated here with a specific example as an advance of the many discussed in dealing with the analytical process in Chap. 7.

Gold found in a mining area! In fact, a rock was analysed at the laboratory and found to contain 12 mg/kg of the precious metal. This result is consistent with the sample analysed but any further conclusions are unwarranted. In fact, whether mining for gold in the area would be profitable is an industrial problem that must be solved by addressing an analytical problem – subject to the technical profitability constraints imposed by the mining company –; this latter will entail sampling a wide zone (the object) to an appropriate depth. The laboratory will receive many samples allegedly representative of the object that will be analysed in accordance with the analytical problem. The results thus obtained should be consistent with (a) the samples received, (b) the zone explored, (c) the analytical problem and (d) the mining company's problem. If the information produced solves the last, then absolute representativeness will have been achieved.

The analytical property representativeness encompasses qualitative and quantitative connotations. At the upper consistency levels, intangible aspects become relevant. At the lowest two, Chemometrics (Statistics) provides indispensable technical support (e. g. statistical rules for sampling). The analytical problem can be approached via Statistics, with assistance from Society, Industry, Science and Technology inputs in stating their information needs. Obviously, for Analytical Chemistry to ensure the maximum possible representativeness in its actions, it must establish effective two-way links with other areas.

2.5 Basic Analytical Properties

There are three basic analytical properties: precision, sensitivity and selectivity. "Proper sampling", one other basic analytical goal, is strictly not an analytical property.

As can be seen in the scheme of Fig. 2.1, these properties are at the second level in the significance hierarchy for analytical properties and provide the strongest support (foundation relationships) for capital analytical properties, the magnitude of which depends on them. No accuracy is possible without adequate precision, sensitivity and selectivity. Also, no representativeness can be claimed in the absence of proper sampling.

Basic analytical properties, in addition to accessory properties, characterize the analytical process. Section 2.7 discusses the complementary and contradictory relationships among the former and with the other types of properties. There follows a non-exhaustive description of the more relevant aspects of each property that is consistent with the approach adopted throughout this book.

2.5.1 Precision

The precision concept was previously described in dealing with errors in Analytical Chemistry (Sect. 2.3). In formal terms, precision can be defined as the **"degree of consistency among results obtained by applying the same analytical method separately to individual aliquots of the same sample"** or as the **"dispersion of the results with respect to one another and their mean."** Dispersion is the opposite of precision: the higher the precision, the lower the dispersion. On the other hand, precision is synonymous with "clustering". As noted earlier, precision is one cornerstone for accuracy.

Precision materializes in random or indeterminate errors. The reference from which differences or deviations are established is internal, *viz.* the set of results, characterized by its mean ($\bar{X}, \mu'$). By contrast, accuracy is referred to an external landmark ($\hat{X}'$). The mathematical support for precision is provided by statistics (Chemometrics).

Precision is not a unique, isolated concept; rather, it depends on the object to which it is applied – which gives rise to a variety of statistical parameters – and on the way the set of results is obtained. Both notions are commented on below.

Precision can be referred to an individual result, all the results or their mean. The **precision of an individual result** is defined as the difference between the result and the arithmetic mean of the set of results. Such a difference coincides with the systematic error, which can be positive or negative.

$$d_{x_i} = \pm \, |x_i - \bar{X}| \qquad d_{x_i} = \pm \, (x_i - \mu')$$

In relative terms, the random deviation from a result can be expressed as a fraction of unity or a percentage:

$$(d_{x_i})_{\text{rel}} = \frac{\pm d_{x_i}}{\bar{X}} \cdot 100 \qquad (d_{x_i})_{\text{rel}} = \frac{\pm d_{x_i}}{\mu'} \cdot 100$$

$$n < 30 \qquad\qquad n > 30$$

The quantities used to characterize the **precision of a set of results** rely on statistical parameters based on normal (Gaussian) distributions. The higher the precision is, the smaller the values of such parameters will be as they are directly related to dispersion. The best known parameter of this type is the **standard deviation**, which is described in qualitative terms as the distance (to the left or right) of the mean to the inflection point in the bell-shaped (Gaussian) curve, and in quantitative terms as

$$s = \sqrt{\frac{\sum (x_i - \bar{X})^2}{n-1}} = \sqrt{\frac{\sum d_{x_i}^2}{n-1}} \qquad \sigma = \sqrt{\frac{\sum (x_i - \mu')^2}{n-1}} = \sqrt{\frac{\sum d_{x_i}^2}{n-1}}$$

$$n < 30 \qquad\qquad n > 30$$

The standard deviation (s, σ) has the same dimensions (e.g. mg/mL) as the result to which it applies.

The **variance**, v, which is defined as the standard deviation squared,

$$v = s^2 \qquad v = \sigma^2$$

$$n < 30 \qquad n > 30$$

is of great practical interest on account of its additive nature. In fact, it allows one to calculate cumulative random errors when an analytical result is the combination of several previous measurements, all subject to individual random errors. The standard deviation can also be expressed in relative form, either as a fraction of unity (**relative standard deviation**, rsd) or a percentage (**coefficient of variation**, CV).

$$\text{rsd} = \frac{s}{\bar{X}} \qquad \text{rsd} = \frac{\sigma}{\mu'}$$

$$n < 30 \qquad n > 30$$

$$\text{CV} = \frac{s}{\bar{X}} \cdot 100 \qquad \text{CV} = \frac{\sigma}{\mu'} \cdot 100$$

$$n < 30 \qquad n > 30$$

Precision can also be **referred to the mean** $(\bar{X}, \mu')$ of the set of results since ∞ results (a population with a mean μ) can be split into groups with means that will differ from one another. The differences will materialize in the **standard deviation of the mean** (sdm), which is often also referred to as the "standard error of the mean."

$$\text{sdm} = \frac{s}{\sqrt{n}} \qquad \text{sdm} = \frac{\sigma}{\sqrt{n}}$$

The sdm for $n = \infty$ (a population) is zero.

As a rule, increasing the number of results (n) increases precision and decreases standard deviations (s and σ); also, $s > \sigma$ since n is in the denominator of the previous expressions.

_____ Box 2.7

Calculating specific uncertainty from precision-related parameters

Specific uncertainty has been defined as a symmetric interval around a result $(R \pm U_R)$ that describes the range of values where the result can fall – if the method is replicated – at a given probability level.

U is calculated from the standard deviations (s, σ) for a set of results, using the following expressions

$$\bar{X} \pm \frac{s}{\sqrt{n}} \qquad U_R = t \cdot \frac{s_R}{\sqrt{n}}$$

on the assumption that the result will be the arithmetic mean of the data set (see Box 2.2). The specific uncertainty can be calculated more easily from

$$U_R = K \cdot s_R$$

where K depends on the probability imposed (2 and 3 at the 95% and 99.6% confidence levels, respectively).

However, calculating the specific uncertainty, U, is usually a more complex process as it also involves knowing and accumulating or integrating the random errors made throughout the analytical process via the additivity of variances. When the result R is obtained as an algebraic combination (additions and subtractions, $e.g.$ $R = a + b - c$) of intermediate measurements (a, b and c) subject to individual uncertainties (U_a, U_b and U_c), the standard deviation, s_R, will be

$$s_R = \sqrt{s_a^2 + s_b^2 + s_c^2}$$

irrespective of sign. When, on the other hand, K is obtained by multiplying or dividing measurements ($e.g.$ $K = a \cdot b/c$), the relative standard deviation is preferred.

$$\frac{s_R}{K} \sqrt{\left(\frac{s_a}{a}\right)^2 + \left(\frac{s_b}{b}\right)^2 + \left(\frac{s_c}{c}\right)^2}$$

The relative standard deviation, s_R, can be calculated in two different ways from the uncertainty (U_R). When the probability of the confidence limits is unknown, s_R can be obtained from the expression $U_R = t \cdot s_R/\sqrt{n}$, where t is a tabulated value. When neither the probability nor n is known, S_R is obtained from the general expression.

$$s_R = \frac{U_R}{\sqrt{3}}$$

These quantitative connotations of precision are complemented by qualitative implications of the way the set of results is obtained in each case. Comprehensive information about the experiments conducted and whether the operator, instruments, apparatuses, reagents, standards and times used where the same or different is essential. The more dissimilar the experimental conditions are, the more varied will be the sources of variability, the higher will be the dispersion of the results and the lower the precision. Obviously, all experiments should use the

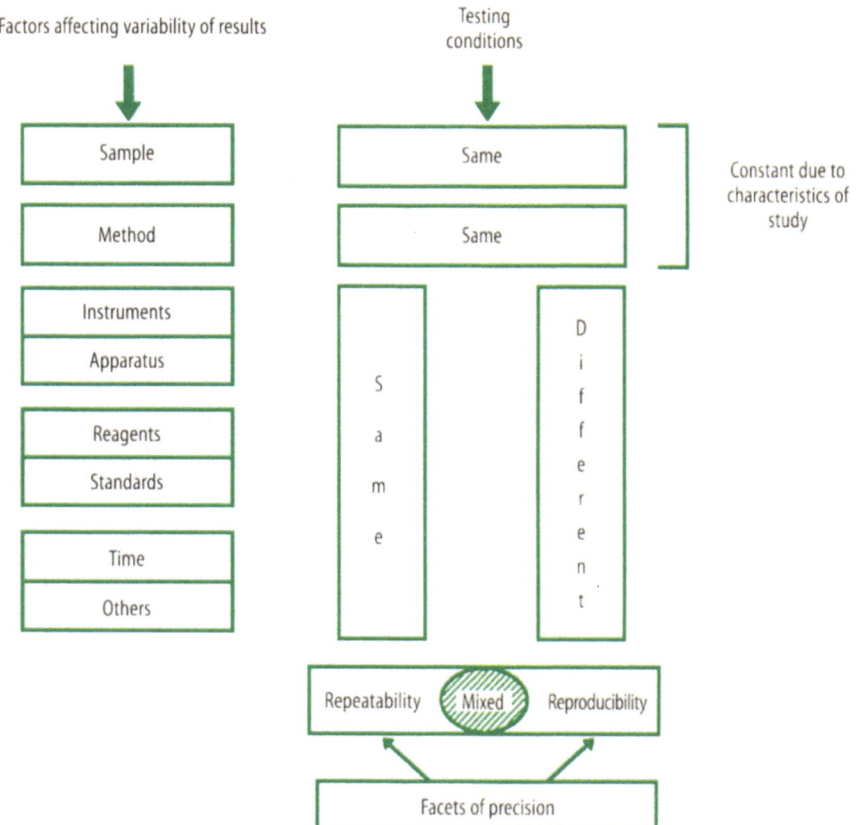

Fig. 2.9. The repeatability and reproducibility concepts in relation to precision

same analytical process and aliquots of the same sample. These qualitative connotations introduce two new nuances of meaning for precision, *viz.* repeatability and reproducibility, which are illustrated in Fig. 2.9.

The lowest level of rigour as regards precision is provided by **repeatability**, which is defined by the International Standards Organization (ISO) as the "dispersion of the results for mutually independent tests using the same method as applied to aliquots of the same sample, at the same laboratory, by the same operator, using the same equipment over a short interval of time." Repeatability is a measure of internal variability (variance) and reflects the highest precision a given method can be expected to reach.

According to ISO, **reproducibility** is the "dispersion of the results for mutually independent tests performed by applying the same method to aliquots of the same sample under different conditions: different operators, different equipment or different laboratories." Thus, reproducibility reflects variable rigour that depends on whether one, several or all experimental conditions (*e.g.* in interlaboratory exercises) are different. As a result, reproducibility must always be expressed with reference to the particular conditions. Most often, reports refer to reproducibility between days, operators or laboratories.

In summary, repeatability reflects the minimum possible dispersion (highest precision) and reproducibility the maximum possible dispersion (lowest precision) of a given analytical process. These two extremes represent the upper and lower limits for the parameters on the basis of which dispersion is expressed. As can be seen in Fig. 2.9, there exists a diffuse zone in between that is rather difficult to assign categorically.

It should be noted that the general definition of precision and the practical approaches to it include the concept of "independence" of the tests. Figure 2.10 illustrates this concept. The precision is correctly determined when *n* complete analytical processes are applied to *n* sample aliquots which provide the *n* results that are used to calculate it. Relatively frequently, statistics is incorrectly used to evaluate the precision of a set of *n* results obtained from a single sample aliquot subjected to a single manipulation leading to a treated sample that was simply split as required to make *n* measurements. There is no test-independence in this approach; also, the greatest source of variability in this example is preliminary operations. This incorrect approach leads to reporting conspicuously favourable precision values (unrealistically low uncertainties) resulting from use of the technically most simple – but false – strategy.

_____ Box 2.8

This example shows how precision varies with variability in the experimental conditions.

The precision of an analytical method for determining the total concentration of copper in sea water was assessed. Preliminary operations involved, among others, preconcentrating the analyte on a chelating exchange resin by passing 1 L of sea water through it; eluting the resin with 10 mL of 2 M HCl and inserting an aliquot of the eluate into an atomic absorption

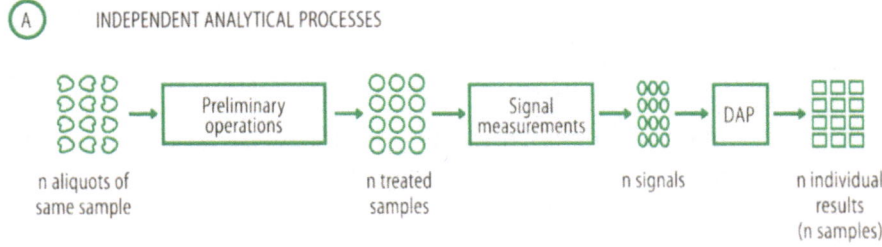

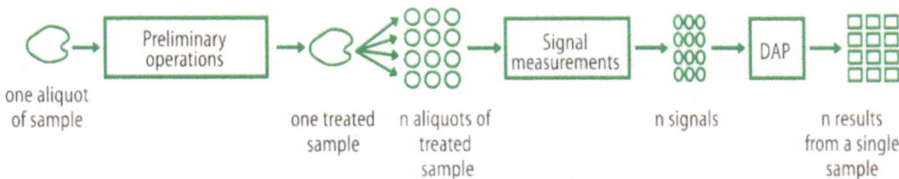

Fig. 2.10. Independence of analytical processes in determining precision. (A) Correct approach. (B) Incorrect approach. DAP = Data Acquisition and Processing

spectrophotometer to make an absorbance reading that was interpolated into a previously run calibration curve to obtain the copper concentration, corrected for the volumes used in the analytical process.

The precision of the method was assessed by determining the analyte six times ($n = 6$) under five different types of experimental conditions, namely:

(1) Using 1 L of sample, and preconcentrating and splitting the eluate into 6 aliquots that were inserted into the instrument to obtain 6 data and 6 results.
(2) Using 6 aliquots of 1 L each, and preconcentrating and inserting each eluate into the instrument, with the same equipment, reagents, ion exchanger, instrument, calibration curve, *etc.*, in the course of the same morning.
(3) Same as in (2), but with the processes conducted on different days.
(4) Same as in (3), but using different operators, reagents and AAS instrument.
(5) Six laboratories analysed 6 aliquots of 1 L of sample, using the same method.

The results obtained in the determinations, as well as some statistical parameters for each data set – calculated as described in the text –, are given in the following table.

Study	Results (mg/L)	$\bar{X}$ (mg/L)	s (mg/L)	CV (%)
1	1.32 1.31 1.32 1.33 1.30 1.31	1.31	0.0105	0.80
2	1.34 1.30 1.28 1.31 1.33 1.29	1.31	0.0232	1.77
3	1.28 1.36 1.30 1.27 1.31 1.33	1.31	0.0331	2.53
4	1.30 1.27 1.40 1.37 1.26 1.30	1.32	0.0560	4.24
5	1.35 1.45 1.21 1.37 1.30 1.28	1.33	0.0826	6.21

The best (apparent) precision was obtained in (1), which was incorrectly approached because the results were not test-independent. Study (2) expressed the precision in terms of repeatability. Reproducibility was assessed in studies (3) and (4) – the latter involved more extensive changes, so it was less precise than the former. The maximum possible rigour in relation to precision was achieved in study (5), the interlaboratory exercise; in fact s and CV were the highest of all.

_____   Box 2.9

Significant figures in a result

The significant figures of a datum (result) are all its relevant numbers that are reliable plus the first that is subject to some uncertainty. Consider a calculation (*e. g.* the mean of a data set) that produces a result containing a large number of digits such as

$$\bar{X} = 10.23442136$$

How many will be significant? The **rule of thumb** here is to use the standard deviation to estimate the uncertainty of the digits. If $S = 0.01324$, then the second decimal place will be uncertain and the mean should be expressed as

$$\bar{X} = 10.23 \quad or \quad \bar{X} = 10.23_4$$

The subscript is used to avoid losing potentially important information – whether to keep it or leave it out can be decided upon later.

Rounding is a frequent need when the uncertain digit in a number is followed by others. If the digit to the right is greater than 5, then the uncertain digit is increased by one (*e.g.* 3.248 is rounded to 3.25); if it is less than 5, the digit is left unchanged (*e.g.* 3.242 is rounded to 3.24); finally, if it is exactly 5, the digit is rounded to the nearest even number (*e.g.* 9.65 becomes 9.6, and 4.75 is rounded to 4.8).

The position of zeros alters the number of significant figures in a series of non-zero digits. Their distribution in decimal numbers does not change, however.

Example:		
	0.01234	
	0.1234	*4 significant figures*
	1.234	
	1234	
	1234.05	*5 significant figures*

When data containing different numbers of significant figures are added, subtracted, multiplied or divided, the following two rules apply:

(*a*) The final result should never contain more significant figures than the initial datum with the smallest number of significant digits; and

(*b*) Initial data should not be rounded (rounding should be delayed until the final result is calculated).

―――― Box 2.10

Rejection of outliers

An outlier is a datum not belonging to a data set (for a sample or population) or one the probability of which belonging to the set concerned is below a preset value, which, however, has found its way into the set as a result of an isolated methodological mistake (a systematic or accidental error).

In order to statistically distinguish outliers from the extreme values in a data set, an **acceptance/rejection test** is used to ensure that the set will have a normal or Gaussian distribution. Whether or not potential outliers are rejected has considerable effects on both the mean and the standard deviation –particularly when n is small. Let us comment on two different types of test.

(A) DIXON'S CRITERION is based on the "span", *viz.* the difference between the largest and smallest value in the set – the suspected outlier included. The procedure used to apply it is as follows:

(1) Data are arranged from largest to smallest.

(2) The suspected outlier (x_q) is identified.

(3) Q_{cal} is calculated from the expression $Q_{cal} = |x_q -$ nearest value $|/$span.

(4) The Q_{cal} value thus obtained is compared with a tabulated value Q_t at a given significance level and number of data in the set.

Data	n	4	5	6	7	8	9	10
90% conf.	Q_t	0.76	0.64	0.56	0.51	0.47	0.44	0.41
95% conf.	Q_t	0.831	0.717	0.621	0.570	0.524	0.492	0.464

(B) THE LIMIT OF CONFIDENCE CRITERION uses the following procedure:

(1) $\bar{X}$ and S are calculated from all available data and the limits of confidence, $\bar{X} \pm t \dfrac{S}{\sqrt{n}}$, at a given probability level are established.

(2) If the suspected outlier, x_q, does not fall within the interval, then it should be rejected and statistics recalculated.

This criterion has the disadvantage that, because it uses the outlier for calculation, it expands the limits artificially – particularly if n is small.

Example: In the determination of the ferric oxide content (% Fe_2O_3) in a mineral, the following data were obtained: 32.15%, 32.10%, 32.13%, 31.96% and 32.18%. Should any be rejected?

DIXON'S CRITERION

(a) Data are arranged in decreasing order and the potential outlier is identified.

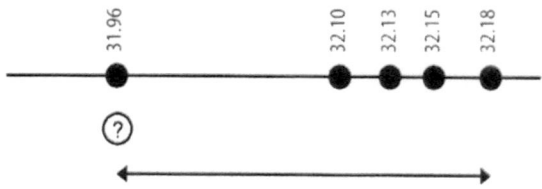

(b) Q_{cal} is calculated.

$$Q_{cal} = \frac{32.10 - 31.96}{32.18 - 31.96} = 0.64 \quad \text{(two significant figures)}$$

(c) Q_{cal} is compared with Q_t.

$n = 5$ | 90% conf. $Q_t = 0.64$ $Q_t = Q_{cal} \Rightarrow$ at the limit, so should be rejected
         | 95% conf. $Q_t = 0.717$ $Q_t > Q_{cal} \Rightarrow$ can be accepted

LIMIT OF CONFIDENCE CRITERION

The data set is subjected to statistical treatment, which provides the following results:

$\bar{X} = 32.10\%$ $s = 0.086\%$ $n = 5$

Limits of confidence are established at two different probability levels:

$$P = 0.10 \ (90\% \text{ confidence}), \quad t = 2.02 \quad 32.10 \pm 2.02 \cdot \frac{0.086}{\sqrt{5}} \quad 32.10 \pm 0.08\%$$

$$P = 0.05 \ (95\% \text{ confidence}), \quad t = 2.57 \quad 32.10 \pm 2.57 \cdot \frac{0.086}{\sqrt{5}} \quad 32.10 \pm 0.10\%$$

Based on this criterion, the datum 31.96% (x_q) will fall outside the limits at both $P = 0.10$ and $P = 0.05$, so it should be rejected.

If the outlier is rejected and the remaining four data are used to recalculate statistics, then the following results are obtained:

$$\bar{X} = 32.14\% \quad s = 0.034\% \quad n = 4$$

As can be seen, discarding the outlier results in a larger mean and a significantly smaller relative standard deviation.

2.5.2 Sensitivity

This analytical property can be ascribed to analytical methods (processes) and defined as **"the ability to discriminate among similar concentrations of the analyte"** or **"the ability to detect (qualitative analysis) or determine (quantitative analysis) small amounts of an analyte in a sample."** The more similar the analyte concentrations that can be distinguished or the lower those that can be detected or determined, the higher the sensitivity.

Sensitivity is a basic property that sustains accuracy; in fact, no method can be accurate unless it allows one to discriminate low concentrations in the detection or determination of an analyte. Mathematically, sensitivity can be expressed via various parameters that are described below.

_____ Box 2.11

Let us illustrate a qualitative approach to sensitivity with a specific example: the task was to determine the relative sensitivity levels of three different methods A, B and C that were used to discriminate the total hydrocarbon index (THI) of water, a relevant parameter with a view to assessing water pollution.

The three methods were individually applied to two standard samples of known concentrations: (1) 0.12 and (2) 0.11 µg/L.

Method A: Neither sample contained hydrocarbons.
Method B: Both samples contained hydrocarbons and had an identical global index: 0.1 µg/L.
Method C: Both samples contained hydrocarbons but possessed a different global index:
(1) 0.125 and (2) 0.116 µg/L.

Which method is the most reliable? No doubt, the sensitivity of method A is inadequate to determine THI. Method B can only detect the presence of hydrocarbons, whereas method C can additionally discriminate between samples of similar composition. Consequently, method C possesses the highest sensitivity and method A the lowest. Also, method C will give slightly overestimated values.

The most widely accepted form of expressing sensitivity, S, is the **"variation of the analytical signal, x, with the analyte concentration, C"**, which can be expressed in absolute, differential or incremental terms:

$$S = \frac{x}{C} \qquad S = \frac{\delta x}{\delta C} = \frac{\Delta x}{\Delta C}$$

The first part of the definition only holds if the signal is exclusively due to the presence of the analyte, *i. e.* in the absence of spurious signals called "analytical blanks" ($C = 0$). The more marked the change in the signal produced by a small change in the analyte concentration is, the higher will be the sensitivity. This idea can be expressed graphically via the calibration curve, a plot of the form $x = \varphi(C)$ constructed from data points obtained in a test series where the analytical process is applied to n samples (usually artificial samples) containing increasing concentrations of the analyte.

Signal (x) = intercept + sensitivity $(S) \cdot$ concentration (C)

Figure 2.11 shows a graph of this type. Its central portion fits the previous algebraic expression, so

Signal (x) = intercept + $S \cdot C$

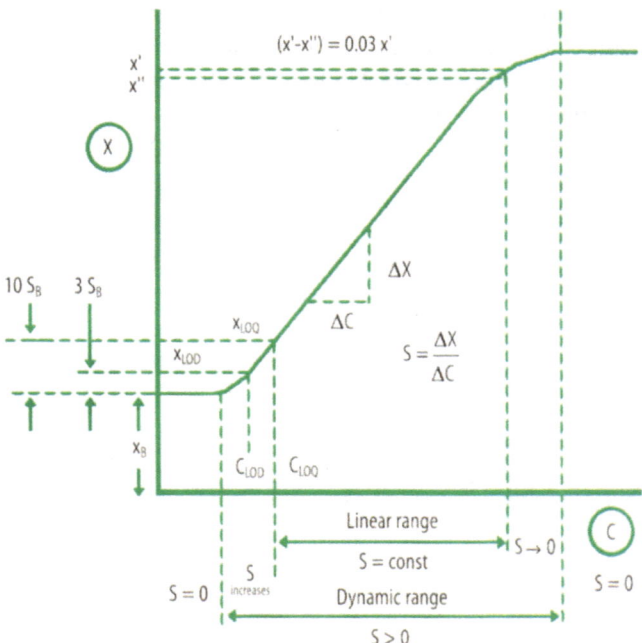

Fig. 2.11. Graphical definition of sensitivity parameters based on the calibration curve. For details, see text

The sensitivity is the slope of the central linear portion of the curve. A sample that contains no analyte (an analytical blank) produces an analytical signal x_B due to both the sample matrix and the instrument's background noise that is called a "blank signal".

The calibration curve consists of three distinct portions as regards sensitivity, namely:

(a) At low analyte concentrations ($S = 0$), the analyte cannot be detected or determined. The signal, $\bar{x}_B$, is produced by the blank and coincides with the intercept. If the blank gives no signal, then the curve intersects the origin ($x = 0, C = 0$).

(b) At intermediate analyte concentrations, the sensitivity is non-zero ($S > 0$). In this case, a **dynamic range** is bounded the lower limit of which is the limit of detection. This range comprises three sub-zones. In the central one, S remains constant; this is the so-called "**linear range**", which is bounded by the other two regions (where S decreases and increases, respectively). The lower limit of the linear range defines the coordinates for the limit of quantitation. The upper limit of the linear range is reached when the deviation from linearity ($x' - x''$) equals or exceeds 0.03 times x' (the value where linearity starts); if the deviation is greater, then the curve falls outside the linear range.

(c) At very high analyte concentrations, the signal no longer varies with the concentration ($S = 0$) and the curve is parallel to the initial portion – at a higher signal level, however.

The quantitative definition of sensitivity is completed by several parameters that establish the minimum concentrations that can be detected or determined. Thus, the **limit of detection** is the analyte concentration C_{LOD} which produces a signal x_{LOD} that can be statistically distinguished from the blank signal. This parameter is quantified in preliminary experiments involving measuring n blanks to obtain a mean and its standard deviation (σ_B). By convention, the limit of detection is defined as

$$x_{LOD} = \bar{X}_B + 3\sigma_B$$

A signal x_{LOD} is thus produced by a concentration C_{LOD} that can be calculated by graphical extrapolation.

If the blank signal is zero (*i. e.* if the intercept $\bar{X}_B = 0$), then

$$x_{LOD} = \bar{X}_B + 3\sigma_B = 0 + 3\sigma_B$$

Interpolation of this signal into the calibration curve gives

$$x_{LOD} = (= 3\sigma_B) = 0 + SC_{LOD}$$

from which the relation between the two forms of expressing sensitivity,

$$C_{LOD} = \frac{\bar{X}_B + 3\sigma_B}{S}$$

is derived.

If $\bar{x}_B > 0$, then the signal for the limit of detection will be

$$x_{LOD} = \bar{X}_B + 3\sigma_B$$

Interpolation of this signal into the calibration curve provides the relation between sensitivity and the limit of detection

$$x_{LOD}(= \bar{X}_B + 3\sigma_B) = \bar{X}_B + SC_{LOD}$$

$$C_{LOD} = \frac{3\sigma_B}{S}$$

Because S is not constant along the initial portion of the curve, this expression is only approximate. The relationship between C_{LOD} and S depends on the value of the blank.

The **limit of quantitation** is defined as the analyte concentration C_{LOQ} which produces a signal x_{LOQ} that can be considered the lower limit of the linear range. Its mathematical expression is also based on statistical processing of blanks:

$$x_{LOQ} = \bar{X}_B + 10\sigma_B$$

It is therefore a value above the limit of detection ($C_{LOQ} > C_{LOD}$). Its relationship to sensitivity can be established similarly to that of the limit of detection:

$$C_{LOQ} = \frac{10\sigma_B}{S}$$

A comprehensive study of sensitivity relies on statistical hypothesis testing, a detailed description of which is beyond the scope of this book. Interested readers are referred to the book by Kateman and Buydens (suggested reading no. 4) for an extensive justification of the previous definitions.

Some authors use the so-called **limit of decision**, C_{LD}, the value of which lies in between those of the previous two parameters.

$$x_{LD} = \bar{X} + 6\sigma_B$$

Figure 2.12 illustrates in graphical form the concepts of blind, decision, detection and quantitation region, which arise in placing the three types of limit on an analyte concentration scale that is related to the analyte signal scale and the blank signal scale via the previous expressions.

- In the **blind region** ($0 < C < C_{LOD}$), the analyte can be neither detected nor determined as the signal obtained is not statistically different from the blank signal or background noise from the measuring instrument.
- In the **detection region** ($C_{LOD} < C > C_{LOQ}$), the analyte can be detected but not quantified with accuracy since C falls outside the linear range.

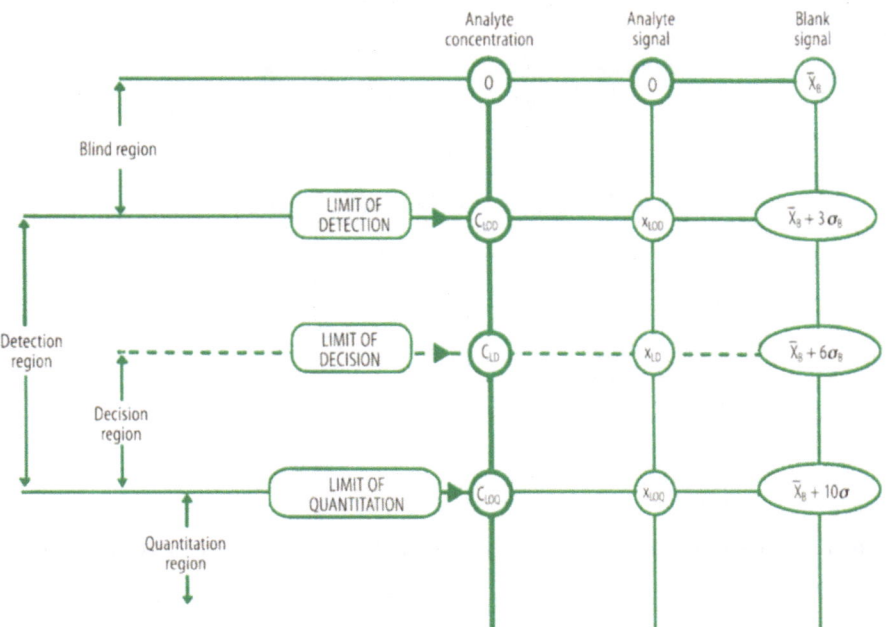

Fig. 2.12. Analyte concentration zones bounded by the limits of detection, decision and quantitation

- The **quantitation region** $(C > C_{LOQ})$ is the zone spanning the linear range $(\leq = \text{const})$ of the calibration curve.
- Finally, the **decision region** $(C_{LOD} > C > C_{LD})$ is a zone that encompasses concentrations in between the limits of detection and determination – but closer to the latter – and hence provides less reliable qualitative detection of the analyte.

2.5.3 Selectivity

Selectivity is one other basic analytical property that can be ascribed to analytical methods (processes). It is defined as the **"ability to produce results exclusively dependent on the analyte(s) for its/their identification or quantitation in the sample."** Selectivity (or rather, the lack of it) is dictated by interferences, which are perturbations that alter some or all of the steps of the analytical process.

Selectivity provides direct support for accuracy; in fact, interferences produce systematic errors that increase or decrease the result relative to the value held as true.

There are many types of analytical interferences. Figure 2.13 shows four different – though not mutually exclusive– classifications based on origin, sign, effect on the calibration curve and mechanism via which the perturbation is exerted.

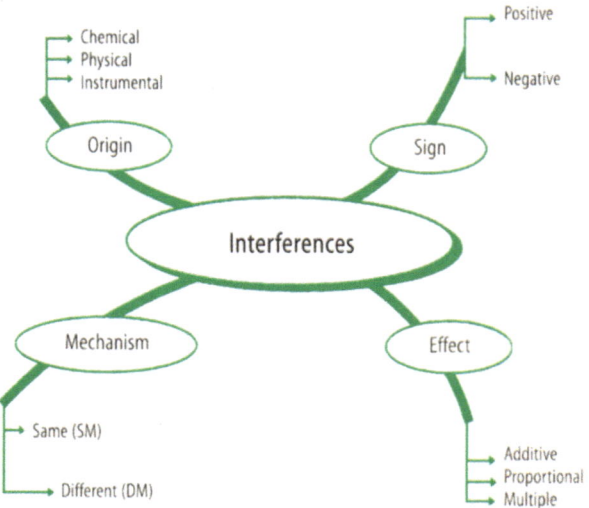

Fig. 2.13. Classification of interferences according to four non-mutually exclusive criteria

According to origin, interferences can be **chemical** [*e.g.* if the interfering chemical species gives a colour-forming reaction similar to that produced by the analyte on addition of a reagent (ligand) to the sample], **physical** (*e.g.* caused by suspended particles in a sample that scatter light during measurements of the fluorescence intensity of the analyte or its reaction product) and **instrumental** (*e.g.* when the operating flow-rate and/or temperature are not carefully controlled in gas chromatography).

Interferences can spuriously increase the analytical signal and lead to results subject to **positive** systematic errors (see the example for chemical interferences above) or decrease the signal, thus giving rise to **negative errors** (*e.g.* in the presence of a masking species reacting competitively with a reagent that should interact with the analyte only). Accordingly, a positive or negative interference, respectively, arises.

The **mechanism** by which the interference is produced can be **similar** or identical (SM) to that whereby the analyte yields its signal (see, again, the example for chemical interferences above) or **different** (DM) from it (see the example for physical interferences with a chemical determination). The effect on the calibration curve depends on the relationship of the interference concerned to the analyte. Thus, **additive** interferences are caused by species or effects that are independent of the analyte concentration: the calibration curve has a positive or negative – though blank-independent – intercept and is shifted up or down as a result. **Proportional** interferences arise when the perturbation depends on the analyte concentration level; the sensitivity is increased or decreased as a result of the change in the slope of the linear portion of the calibration curve. Finally, **multiple** interferences produce both additive and proportional errors.

The dramatic growth of multideterminations in the same sample has introduced new connotations in the selectivity concept. Thus, interferences with the

determination of a single analyte (or the total analyte concentration in a mixture) may arise from the sample matrix. The **discriminate determination of several analytes** can additionally be subject to mutual interferences between analytes.

_____ Box 2.12

In order to illustrate some of the more common types of interferences, let us examine the photometric determination of iron in wine.

The method used involves supplying an aliquot of the wine with a reductant (e.g. ascorbic acid or hydroxylamine) to reduce Fe^{3+} to Fe^{2+}, adjusting the pH by addition of a buffer and, finally, adding a ligand (1,10-phenanthroline, L) to form a soluble, deeply coloured chelate FeL_3^{2+}. An aliquot of the sample thus prepared is used to measure the absorbance at 510 nm.

The table below shows various types of potential interferences with the determination, classified according to source.

Source	Type of interference			
	Origin	Mechanism	Sign	Effect
Base colour of wine	Chemical	Different (DM)	Positive	Additive
Presence of Cu^{2+} ions, which give a coloured chelate (CuL_2^+)	Chemical	Same (SM)	Positive	Additive
Presence of F^-, which forms colourless complexes with the analyte (FeF^{2+}) in a competing process	Chemical	Same (SM)	Negative	Proportional

Quantifying the selectivity of a method is a technically complex task as it entails a systematic, time-consuming study of each individual source of interferences with the analytical process. The procedure is usually simplified by using approximations that (*a*) assume theoretically non-perturbing species not to interfere; (*b*) minimize the number of species potentially present in the sample (classified as white, grey or black depending on available knowledge of its composition); and (*c*) enforce strict control of those physical and instrumental factors that might introduce some perturbation.

The selectivity of a method can be expressed in quantitative terms via two different parameters. The **maximum tolerated ratio** of an interference is the concentration ratio of interferent (C_{int}) to analyte (C_{anal}) that introduces a perturbation in the result (an error) of such magnitude that the result falls at the upper or lower limit of the range spanned by the datum obtained from the determination of the analyte in isolation and its specific uncertainty

$(R_{C_{\mathrm{anal}}} \pm U_{C_{\mathrm{anal}}})$:

$$TR = \frac{C_{\mathrm{int}}}{C_{\mathrm{anal}}}$$

Thus, an interferent concentration slightly greater than C_{int} will cause a positive or negative perturbation –if the analyte concentration, C_{anal}, is kept constant. When the interferent exerts its perturbation via a mechanism similar to that which produces the analytical signal – and hence gives its own calibration curve –, the selectivity of the method concerned can be expressed as a **sensitivity ratio**:

$$SR = \frac{S_{\mathrm{anal}}}{S_{\mathrm{int}}}$$

In both cases, each interferent possesses a specific relation. Consequently, the selectivity of the method will be combination of TR and SR values.

The selectivity of two methods 1 and 2 for each potential interferent can be expressed via the so-called **selectivity factor**, which is the quotient of the tolerated ratio for each method:

$$SF = \frac{(TR)_1}{(TR)_2}$$

The selectivity of analyte multideterminations is expressed in global terms via **Kaiser's selectivity parameter**, K, which considers n signals $(x_1, x_2, \ldots x_n)$ at each of n instrumental parameters $(P_1, P_2, \ldots P_n)$. Each parameter possesses a specific sensitivity for each analyte $S_{m,n} = x/C$. If $m = n$, then the following sensitivity matrix is obtained:

$$K = \begin{bmatrix} S_{1,1} & S_{1,2} & S_{1,3} & \cdots & S_{1,n} \\ S_{2,1} & S_{2,2} & S_{2,3} & \cdots & S_{2,n} \\ \cdot & \cdot & \cdot & \cdot & \cdot \\ \cdot & \cdot & \cdot & \cdot & \cdot \\ \cdot & \cdot & \cdot & \cdot & \cdot \\ S_{m,1} & S_{m,2} & S_{m,3} & \cdots & S_{m,n} \end{bmatrix}$$

2.6 Accessory Analytical Properties

Accessory analytical properties (expeditiousness, cost-effectiveness and personnel-related factors) are at the lowest step in the hierarchy of Fig. 2.1. They are all attributes of the analytical process and determine laboratory productivity. Despite their seemingly lesser significance, these properties are frequent priority targets. All three are related to one another and to the other two groups of analytical properties.

2.6.1 Expeditiousness

Expeditiousness is associated to analysis time (the time needed to obtain the result), which is usually expressed as "throughput", *i. e.* as the number of samples that can be fully processed per unit time (typically hour, h, or day, d):

$$\delta = n \cdot h^{-1} \qquad \delta = n \cdot d^{-1}$$

This analytical property is of a high practical significance to analytical problems. In fact, any analytical information that is delivered beyond the scheduled deadline will be useless and hence of poor quality.

In quantifying throughput, one should consider all the steps of the analytical process (calibration operations included). The literature abounds with unrealistic throughput values based on incomplete computations.

The preliminary operations that separate the untreated, unmeasured sample from signal measurement constitute the single greatest contribution to throughput – in most cases, they fill 70–90% of the overall analysis time. As a result, automation, miniaturization and simplification are three key technical approaches to increasing expeditiousness in performing these operations. Analytical systems capable of providing on-site analytical information (*e.g.* pollutant monitors, industrial process analysers, *in vivo* sensors) are effective responses to many current analytical problems.

Expeditiousness is usually increased at the expense of other analytical properties (both capital, basic and even accessory properties). These contradictory dependences are discussed in a subsequent section.

Expeditiousness, which is time-related, should never be confused with the number of samples to be analysed. Although both are almost invariably related, a given analytical problem may not demand delivery of results within a particularly tight deadline.

2.6.2 Cost-effectiveness

The applied facet of Analytical Chemistry cannot evade the economic cost of each analysis performed since this is an essential element of the productivity of a laboratory – and of the analytical process it implements –, and often a priority. Analytical costs are expressed as a sum of money per result (or sample) and, less often, in man-hours. The lower analytical costs are, the higher will be cost-effectiveness, which can thus be considered an analytical property.

Lowering costs usually entails sacrificing on some other analytical property (whether capital, basic or accessory). The contradictory relationships that emerge are exposed in the next section.

Overhead costs arise from laboratory set-up and maintenance, the purchase and maintenance of equipment and the salaries of the staff (managers, secretaries, cleaning crew). As a rule, they are considered fixed costs even though they can certainly vary to some extent.

Specific costs are those to be paid by the laboratory's clients, *i. e.* by those who demand the information delivered. The difference between what the client actually pays and specific costs can be used as a criterion to establish the quality of an analytical process. Specific costs can also be fixed or variable. In any case, they should be balanced in relation to the number of analyses required, the throughput and the desired level of accuracy (uncertainty) in the results. This entails individual optimization in each case. Automation raises fixed costs (those of equipment) but decreases variable costs (it reduces man-hours); this is only acceptable when a large number of samples is to be processed or high expeditiousness is demanded.

_____ Box 2.13

Choosing an analytical technique on the basis of two complementary analytical properties: cost-effectiveness and throughput

The following example is intended to illustrate the influence of two accessory analytical properties (expeditiousness and cost-effectiveness) on the choice of the analytical technique to be used in an analytical process, which will obviously affect its performance.

The determination of inorganic nitrogen in a fertilizer can be carried out by using a large variety of analytical techniques that use a wide range of equipment from a modest burette to

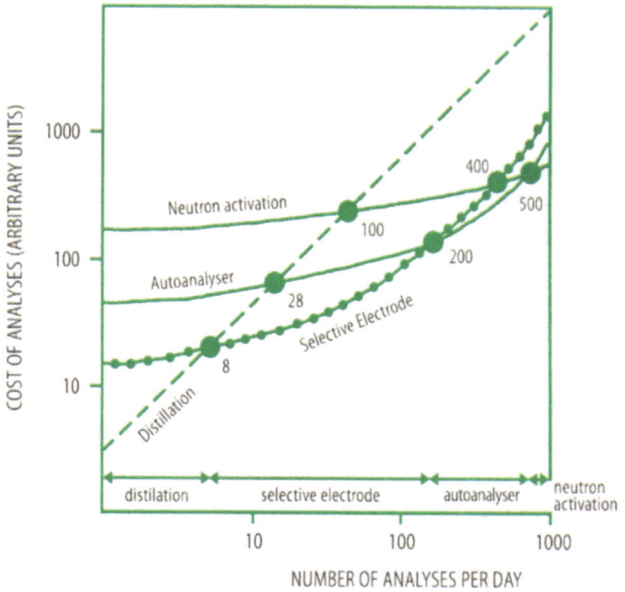

a highly sophisticated, expensive, operationally complex instrument such as a neutron activation detector. The technique of choice will be that resulting in the minimum possible cost; this will depend strongly on the number of samples per day that the laboratory is expected to process.

The figure above uses a log–log scale to represent costs (in an arbitrary currency) as a function of the number of analyses per day to be performed with each analytical technique.

As can be seen:

(a) The initial cost (1 sample/day) differs markedly among choices and is related to technical complexity.

(b) The way cost increases with increase in the number of samples per day also varies strongly among techniques. Thus, only for distillation is the variation linear. Also, the cost per sample remains constant over some ranges (*e. g.* it costs the same to analyse 10 or 100 samples by neutron activation). As a rule, the use of a sophisticated technique is only justified when a large number of samples per day are to be processed.

In practice, the goal is to minimize costs in the throughput zone considered. Cost points allow one to bound such zones. Thus, between 1 and 8 samples/day, distillation is the technique of choice for our problem; from 8 to 200, the selective electrode is to be preferred; between 200 and 500, the autoanalyser is the most cost-effective choice; finally, above 500 samples/day, neutron activation becomes the most inexpensive choice in the long run. The other cost points come into play when the laboratory concerned has no access to a more affordable analytical technique.

2.6.3 Personnel-related Factors

The human factor is no doubt crucial to any activity involving personnel, as is the case with laboratory operations. While not properties or attributes in a strict sense, personnel-related factors are practical aspects of a high significance to the analytical process. There are two broad types of factors, namely:

(*a*) Those related to the **safety** of the operator, laboratory staff and environment (waste production), which have promoted major strategic changes in modern laboratories and have a direct a impact on costs.

(*b*) Those related to the **comfort** of the laboratory staff, which should be aimed at avoiding time-consuming tasks that might lead to stressing situations, and at fostering creativity and competitivity via effective technical and economically rewarding incentives.

2.7 Relationships among Analytical Properties

As noted in Sect. 2.1, it is a glaring error to deal with analytical properties in isolation. Such an approach is rather unrealistic. In fact, the properties are related to one another in ways that can be as important as the properties themselves. Some comment on such relationships is thus warranted at this point, where each individual property has been described in some detail. Figure 2.14 shows a

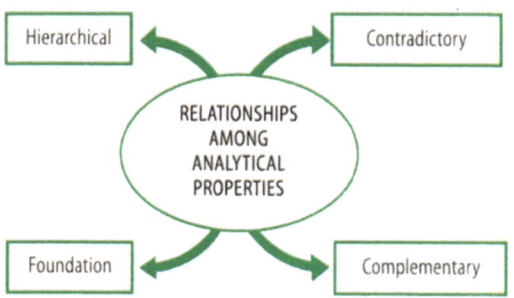

Fig. 2.14. Principal types of relationships among analytical properties

simple scheme of the main types of relationships between the analytical properties depicted in Fig. 2.1; such relationships can be of the hierarchical, foundation, complementary or contradictory type.

Figure 2.15 provides a global schematic depiction of the potential relationships among analytical properties. The tetrahedron on the left represents capital properties; at its vertices are three basic analytical properties. Unfortunately, representativeness (a capital property) cannot be included as such in this geometric body; however, it is represented as a combination with another capital property to signify that both determine the quality of the results. The tetrahedron on the right represents the accessory analytical properties, which fall at its three free vertices and characterize productivity.

2.7.1 Hierarchical Relationships

Because quality of the results is the ultimate theoretical expression of analytical quality, the properties assigned to results (accuracy and representativeness) can be placed at the top of the analytical hierarchy and referred to as **capital**

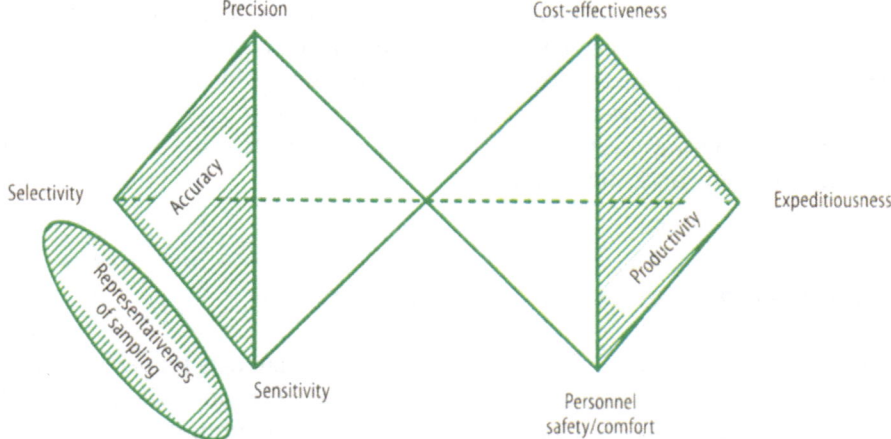

Fig. 2.15. Analytical properties can be represented by means of two tetrahedra sharing one vertex. For details, see text

properties (see Fig. 2.1). At the next level down the ranking are **basic** properties (precision, sensitivity, selectivity and proper sampling), which sustain capital properties and characterize the analytical process in conjunction with **accessory** properties (expeditiousness, cost-effectiveness and personnel-related factors, which are at the lowest level in the ranking). However, this situation is only consistent with a theoretical, basic approach since, as shown below, the analytical problem often requires quality compromises to be made that may even reverse this theoretical hierarchy.

2.7.2 Foundation Relationships

As noted above, basic properties provide support for capital properties. A method cannot be accurate if it is not precise (*i.e.* subject to little uncertainty), sensitive and selective enough. With poor precision, sensitivity and selectivity, the results depart markedly from the value held as true. Similarly, no representativeness can be expected unless sampling is conducted properly and consistently with the analytical problem addressed.

Productivity is an indirect analytical feature common to other applied disciplines that is sustained by the three accessory analytical properties (expeditiousness, cost-effectiveness and personnel-related factors). Maximizing productivity entails ensuring that the analytical process is expeditious, cost-effective, safe and convenient to perform.

2.7.3 Contradictory Relationships

The two tetrahedra in Fig. 2.15 are essentially intended to facilitate the visualization of contradictory relationships among analytical properties. Their edges represent "stress" between vertices. In each case, the tetrahedra are distorted in a way dependent on the property or properties that are favoured over the rest in accordance with the quality compromises adopted in addressing the analytical problem. Some relevant examples are commented on below.

Relationship between accuracy and productivity
Overall, basic and accessory analytical properties bear a contradictory relationship via the two properties that sustain them. As can be seen from Fig. 2.16, the two tetrahedra in Fig. 2.15 can exist in a balanced situation (where the maximum possible accuracy and productivity are obtained through a mutual sacrifice) or in distorted form (one tetrahedron grows at the expense of the other when, for example, the analytical problem demands a high accuracy – and the theoretical quality of the results prevails – or productivity). A high accuracy – and low uncertainty – cannot be achieved without some sacrifice in productivity – the analytical process is slower or more expensive, or involves more staff members. If, on the other hand, productivity is to be favoured, high accuracy and low uncertainty will be two elusive goals.

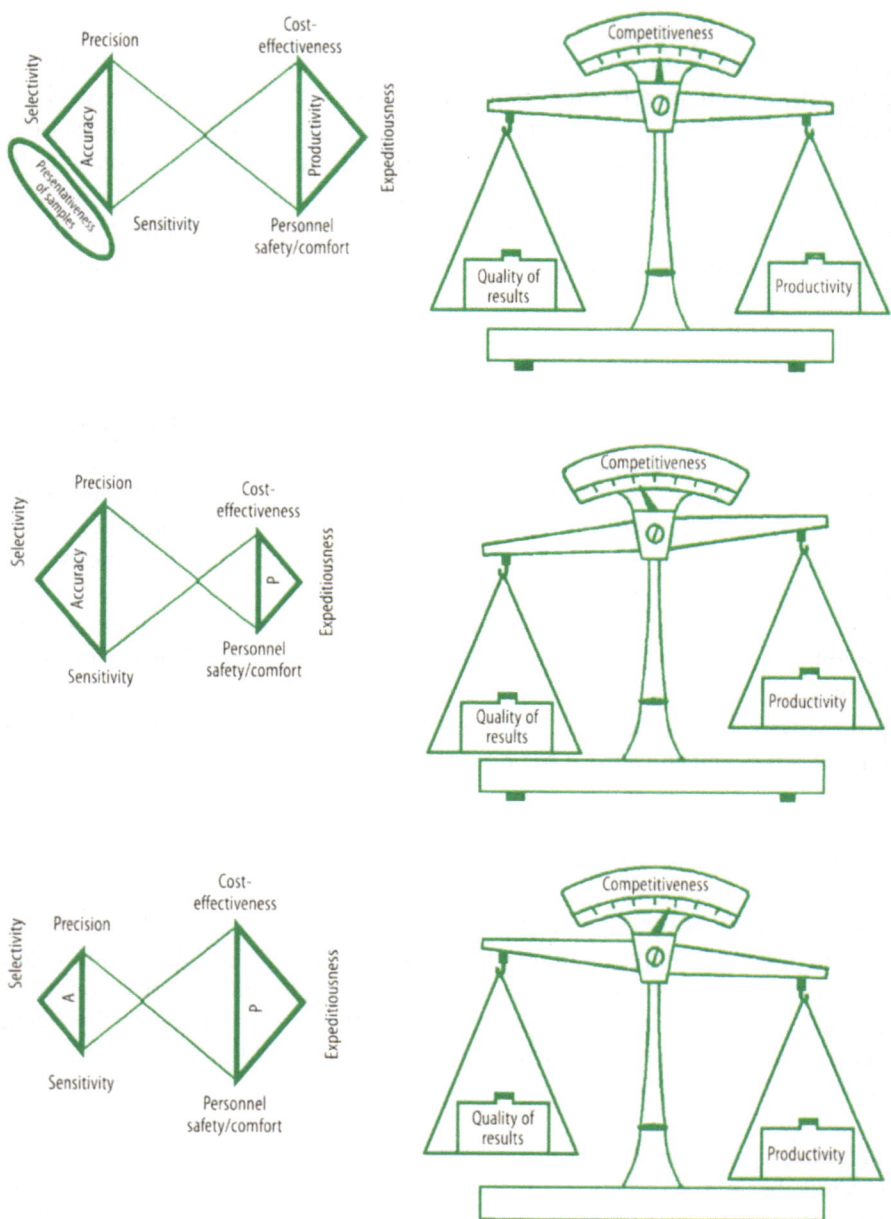

Fig. 2.16. Contradictory relationships among basic and accessory properties established on the basis of the accuracy–productivity pair, which defines laboratory competitiveness

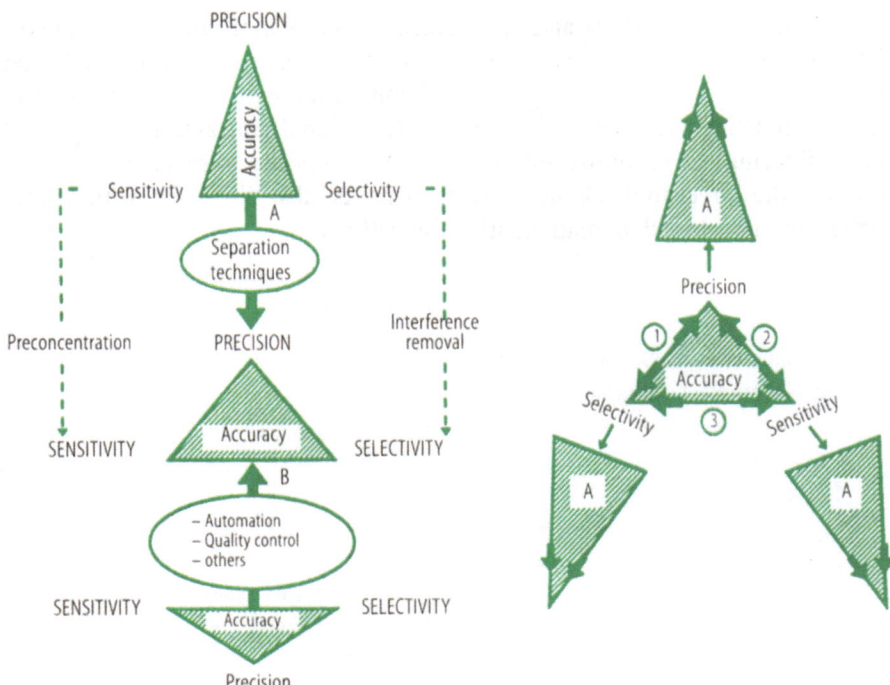

Fig. 2.17. Contradictory relationships among basic analytical properties

Relationships among basic analytical properties

These relationships can be envisaged by distorting the triangle that bounds accuracy in Figs. 2.15 and 2.16. The equilateral triangle represents a balance among the three properties; if one prevails over the other two, however, an isosceles triangle of the same area (accuracy) results where the edge joining the other two properties is smaller than the rest (see Fig. 2.17). If, under conditions of high precision, one wishes to increase sensitivity and selectivity by using a separation technique (*e.g.* one involving preconcentration and interference removal), it will obviously have to be at the expense of precision. Similarly, if precision is to be raised by automating the analytical process, then sensitivity and selectivity will suffer. Below are discussed some binary contradictory relationships between these analytical properties.

There are at least three different types of contradictory relationships between **precision and sensitivity**. Thus, the lower the analyte concentration and the more sensitive the method required for its determination are, the lower will be the precision achieved. Also, using a separation technique (*e.g.* preconcentration) to boost sensitivity will complicate the analytical process and increase the number of variability sources, thus decreasing precision. Finally, the less sensitive a method is, the larger will be the amount of sample it will have to use and the lower the relative random error made in the initial weight or volume measurement of the sample (*i.e.* the higher will be the precision).

Precision and selectivity also bear contradictory relationships. Increasing the latter by use of a separation technique (*e.g.* interference removal) complicates the analytical process and decreases precision, similarly to the precision–sensitivity pair. Using a chemometric technique (*e.g.* signal subtraction, background noise filtering) to minimize interferences increases the uncertainty (and decreases the precision) of the results. This is also the case with multi-determinations based on mathematical algorithms.

_____ Box 2.14

Here are some examples of contradictory relationships among analytical properties based on the scheme and nomenclature used in Fig. 2.18.

enhanced property or properties	Example
Cost-effectiveness	The first step usually taken to cut laboratory expenses is to reduce personnel costs. This results in a heavier workload for the personnel remaining and in slower delivery of results (Fig. 2.18, A)
Expeditiousness	Increasing throughput entails raising costs (by investing in new equipment or infrastructure) and/or compelling personnel to work harder (Fig. 2.18, C)
Personnel-related factors	If political or union trade pressure increases personnel benefits, then costs rise and throughput falls (Fig. 2.18, E)
Cost-effectiveness and expeditiousness	If costs are to be cut and throughput boosted, the analytical process must be partly or fully automated. Automation results in personnel dismissals and in the assignment of greater responsibilities to those remaining, who thus necessitate retraining (Fig. 2.18, B)
Expeditiousness and personnel-related factors	For personnel not to work under stressing conditions and the required throughput to be achieved, the analytical budget must be increased (Fig. 2.18, D).

Relationships among accessory properties

Productivity is bounded by an equilateral triangle whose edges represent a balance among expeditiousness, cost-effectiveness and personnel-related factors. However, the triangle can be distorted by pulling one or two of its vertices (see Fig. 2.18), *i.e.* by expanding one edge, thus sacrificing on two or one of the accessory analytical properties. If expeditiousness prevails, then the process will

Fig. 2.18. Contradictory relationships among the accessory analytical properties that define productivity

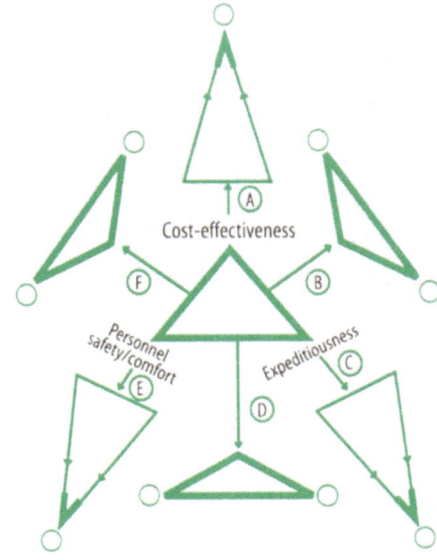

be more expensive to implement and demand more intensive human participation. If the last two properties are favoured simultaneously, then the results will inevitably have to be less expeditiously delivered.

2.7.4 Complementary Relationships

The adjective "complementary" is used here to emphasize favourable mutual influences of analytical properties. Some complementary relationships can be readily inferred from the above-described contradictory relationships.

Basic properties such as **precision** and **sensitivity**, for example, can bear two different types of complementary relationships, namely: (*a*) statistically, the greater the slope of the linear portion of the calibration curve is (and the higher the uncertainty is as a result), the less uncertain (*i.e.* the more precise) will be the result at a given signal level and standard deviation of the estimate for the calibration curve; and (*b*) the limits of detection (C_{LOD}) and quantitation (C_{LOQ}) are defined via a precision-dependent parameter (*viz.* the standard deviation of the blank, σ_B).

The **sensitivity–selectivity** pair is complementary in many respects. Thus, selectivity can be defined as a (multiple) relation between sensitivities (S_{anal}/S_{int}). Also, the difference between additive and multiplicative interferences is based on the way the slope (sensitivity) of the calibration curve is affected by each. The approaches through which both properties are boosted are usually similar and used simultaneously (*e.g.* separation techniques). Finally, the more sensitive a method is, the higher will be the dilution required and, as a result, the lower will be the concentration of interfering species (*e.g.* macromolecules in the determination of analytes of low molecular weight in biological fluids).

_____ Box 2.15

The purpose of this example is to establish a specific relationship between a basic (precision) and two accessory analytical properties (expeditiousness and cost-effectiveness), in addition to one between the latter two.

The figure below is a log–log plot of standard deviation S as a function of the analysis time, t. Both precision and throughput increase as the origin is approached; the number (0.0) is the unreachable ideal. Each cost level, in arbitrary units (1000, 500, 200 and 100), gives a parabola and the resulting curves are all parallel.

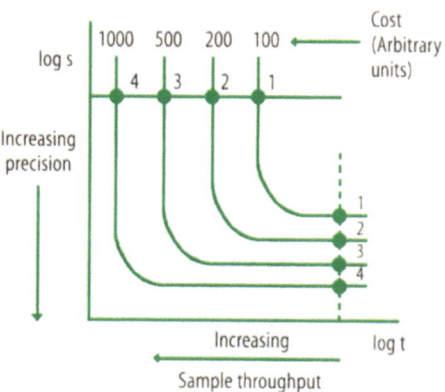

At a given precision level, raising throughput increases costs. Also, when higher precision is required at a given throughput level, costs also increase with decreasing s. Cost differences vary depending on cost-effectiveness and expeditiousness.

From the plot it is clearly apparent that high precision and throughput at a low cost are quite impossible.

_____ Box 2.16

Relationships between precision and sensitivity

This box discusses a contradictory and a complementary relationship between these two analytical properties.

The Horwitz graph below is a paradigm of **contradictory relationships** between precision (as the coefficient of variation) and the increasing sensitivity required as the analyte concentration decreases. In fact, at analyte concentrations above 0.1%, CV = ±5%; however, CV grows exponentially as the analyte concentration is decreased below that level. This arises from the fact that variability sources become much more significant in trace analysis.

The **complementary relationship** between precision and sensitivity materializes in the variation of the uncertainty in a result with the slope (sensitivity) of the linear portion of the

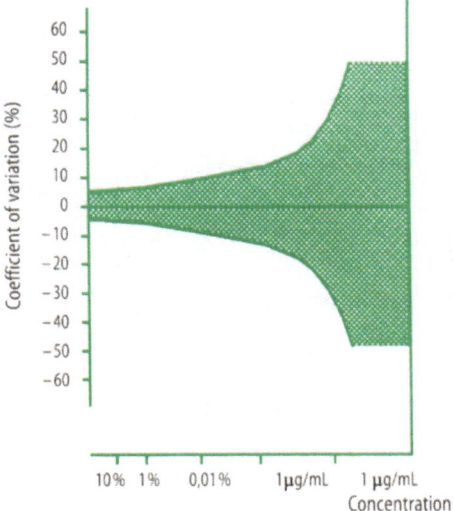

calibration curve. The curve in turn is bounded by two limits imposed by the uncertainty of the calibration process that are wider at the ends. At a given signal level x, the greater is the slope (A), the smaller is the uncertainty of the extrapolated result; hence, the precision is greater than when the sensitivity is smaller (B).

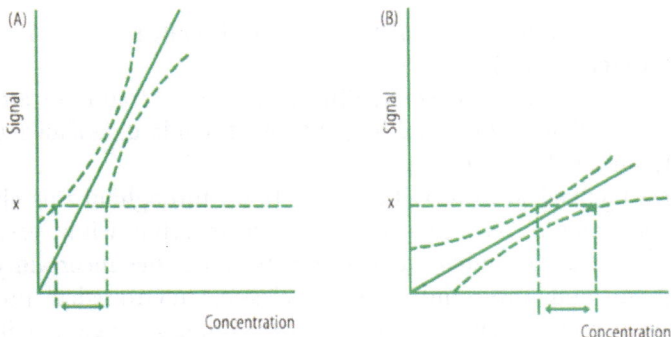

2.8 Other Analytical Properties

In addition to the conventional or classical analytical properties of Fig. 2.1, there exist other analytical features and attributes related to the previous ones that complete and enrich their meaning. Figure 2.19 establishes relationships between these "other" features and classical analytical properties.

Trueness represents the ideal of accuracy, which is reached when a result coincides with the intrinsic (absolute) information latent in the object; therefore, it can only be ascribed to the true value, $\hat{X}$. While trueness may also be reached by accident, an experimental result is always subject to some uncertainty, whereas the true value ($\hat{X}$) is not. The more realistic concept "relative trueness" can be

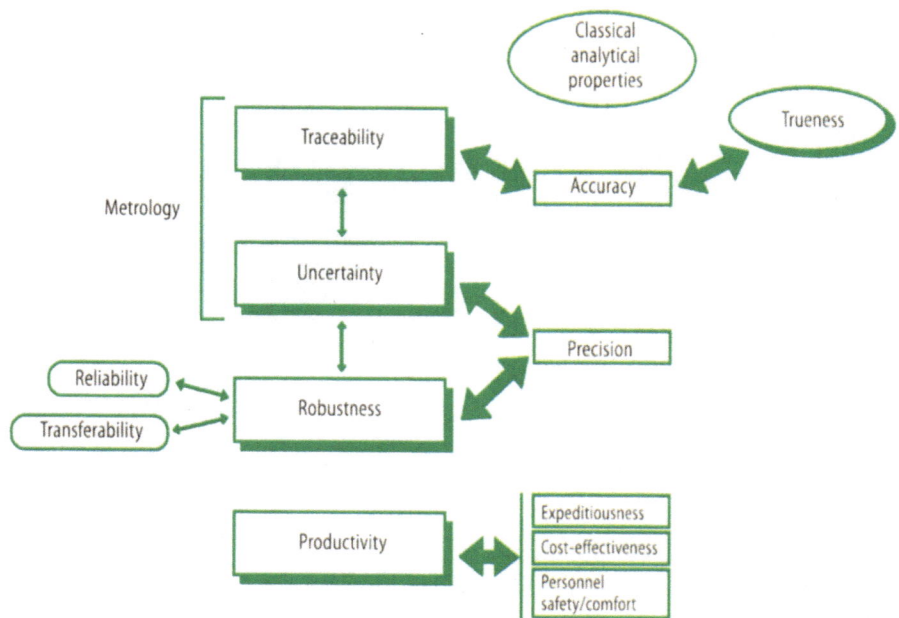

Fig. 2.19. Miscellaneous analytical features and their relationships to conventional analytical properties

expressed as the difference $\hat{X}' - \mu'$, which is the ultimate reference for the degree of accuracy (see Fig. 2.4).

Accuracy is related to the **traceability** of a result, a feature or attribute that relates the result to the values for different types of standards as shown in Chap. 3 – devoted to this analytical feature.

We have referred to **uncertainty** many times throughout this chapter (see Sect. 2.2, for example). As shown above, specific uncertainty is inversely proportional to precision: the higher the precision, the lower the uncertainty.

Traceability and uncertainty are two classical metrological properties of widespread use in the physical measurement domain. Their use in chemical metrology enriches the statistical concepts of accuracy and precision.

Robustness is an analytical feature of growing interest that has been aroused by the late systematic introduction of quality systems in the analytical laboratory. The robustness of an analytical method represents its resistance to changes in the response (result) when applied to individual aliquots of the same sample under slightly different experimental conditions. The experimental alterations introduced in determining the robustness of a method are used to identify potential sources of variability in routine practice. The ultimate purpose is to detect and quantify the experimental "weaknesses" of the method so that any critical factors can be anticipated and controlled in order to ensure that the operating conditions will fall within an undisturbed range. For example, a method that tolerates temperatures between 15 and 20 °C will be more robust than other that provides acceptable results at (20 ± 0.01) °C only.

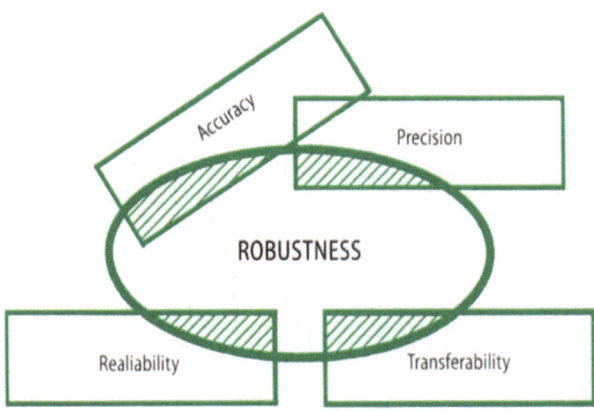

Fig. 2.20. Relationships between robustness and other features of analytical methods

Figure 2.20 relates robustness to other analytical features. The most immediate relation is to precision (and, indirectly, to accuracy). In fact, robustness is measured by the difference produced by random changes in the experimental factors that introduce variability in the results; consequently, robustness is also related to uncertainty.

The **reliability** of a method is defined as its ability to preserve its accuracy and precision over time. This feature, which is directly related to robustness, was formerly referred to as "safety". At present, this word is used in the context of accessory analytical properties to describe the effects of a method on human health or the environment.

The **transferability** of a method is a practical property that represents its capacity to provide consistent results when used by different laboratories. It relies heavily on robustness and on a detailed description of the method (the "procedure" included in the hierarchy of Fig. 1.16). Transferability ensures **comparability** among laboratories, an ancillary property that leads to mutual acknowledgement of their results.

_____ Box 2.17

A study of robustness should begin with a careful identification of those experimental (physical, physico–chemical and chemical) factors that are crucial to the analytical method.

A univariate study, where each individual factor is altered while all others are kept constant, is labour-intensive, time-consuming and thus inadvisable. For example, with $n = 7$ factors, an overall $2^7 = 128$ experiments are required.

It is preferable to conduct a multivariate study by using an experimental design approach or even an expert system.

The YOUNDEN–STEINER procedure is widely used in this context on account of its simplicity. It involves the following steps:

1. The variables or factors to be processed ($n = 7$) are strategically selected.
2. A high value (A–G) and a low value (a–g) for each factor are chosen.

3. An overall $n + 1$ $(7 + 1 = 8)$ experiments are designed in which the factors are simultaneously changed between their high and low values.
4. The eight results thus obtained (y_1 to y_8) are used to construct a Younden table as follows:

Experiment	Factor values							Results
1	A	B	C	D	E	F	G	y_1
2	A	B	c	D	e	f	g	y_2
3	A	b	C	d	E	F	g	y_3
4	A	b	c	d	e	F	G	y_4
5	a	B	C	d	e	F	g	y_5
6	a	B	c	d	E	f	G	y_6
7	a	b	C	D	e	f	G	y_7
8	a	b	c	D	E	f	g	y_8

5. The effect of each factor is calculated by using two means, namely:
   - That of the four analyses which contain the high value of the factor

   For the first factor: $\bar{Y}_A = \dfrac{y_1 + y_2 + y_3 + y_4}{4}$

6. Each of the following differences,

   $$(\bar{Y}_B - \bar{Y}_b), (\bar{Y}_C - \bar{Y}_c), (\bar{Y}_D - \bar{Y}_d), (\bar{Y}_E - \bar{Y}_e), (\bar{Y}_F - \bar{Y}_f)$$

   is compared with

   (a) the rest, in order to identify the most influential factor in qualitative terms; and
   (b) the experimental (random) error, calculated from replicate measurements at any level.
7. The robustness test is crucial with a view to describing the protocol of the method. It includes the ranges over which the critical experimental factors may vary in order to ensure adequate accuracy and precision. Such factors are of the physical (pressure, temperature, flow-rate), physico–chemical (solvent polarity, specific surface) or chemical type (pH, reagent concentration).

Questions

(1) Relate the chemical metrological hierarchy to the three types of analytical information.

(2) Which analytical facets are characterized by analytical properties?

(3) Which type of error characterizes accuracy? Which precision?

(4) Why can accuracy not be correctly established in the absence of precision?

(5) Distinguish precision of a result and precision of a method.

(6) Distinguish accuracy of a result and accuracy of a method.

(7) Comment on the different connotations of analytical uncertainty and its relationships to other analytical aspects.

(8) Same, with regard to analytical trueness.

(9) Relate uncertainty, precision, accuracy and trueness.

(10) Why are both capital analytical properties indispensable with a view to characterizing analytical results? Give several practical examples.

(11) Discuss the different connotations of accuracy of a method.

(12) Describe the term "uncertainty of an individual result".

(13) Can accuracy be achieved if only systematic errors are considered? Why?

(14) Comment on the different connotations of the definition of precision.

(15) Distinguish repeatability and reproducibility.

(16) Discuss the different types of errors encountered in Analytical Chemistry according to source. Use all the adjectives that describe them.

(17) How many types of systematic errors are there?

(18) Why are accuracy and representativeness deemed capital properties?

(19) Comment on the word "independent" in the definitions of repeatability and reproducibility.

(20) Why if the analytical results are consistent with a given industrial information demand will they also be representative of the analytical problem, the object and the samples received by the laboratory?

(21) Comment on the direct and inverse relationships among precision, dispersion and uncertainty.

(22) Comment on the rigour of the different uses of the word "trueness" in Analytical Chemistry.

(23) Relate accuracy and trueness.

(24) Use a graphical example to relate the definitions of sensitivity ($\Delta s/\Delta C$) and the ability to discriminate between similar analyte concentrations.

(25) In what way are the limits of detection, quantitation and decision related to sensitivity ($\Delta s/\Delta C$) ?

(26) Why is statistics required to define sensitivity?

(27) What are the upper and lower limits of the dynamic and linear ranges of the calibration curve?

(28) In what way is uncertainty related to the maximum tolerated ratio used to define selectivity?

(29) Can sensitivity be used to define selectivity? When? How?

(30) Discuss selectivity in the multidetermination of analytes in the same sample.

(31) Comment on the different types of interferences and their mutual relationships.

(32) Explain different degrees of rigour in describing expeditiousness.

(33) Justify the choice of an analytical method in terms of costs and throughput requirements.

(34) Distinguish overhead and specific costs, and fixed and variables costs. Relate them.

(35) Define personnel "safety" and "comfort".

(36) Are hierarchical relationships among analytical properties unchangeable? Why?

(37) Use several examples to illustrate binary and ternary relationships among precision, cost-effectiveness and expeditiousness.

(38) How can quality compromises be materialized?

(39) Explain the contradictory relationship between accuracy and productivity.

(40) Are representativeness and productivity related? In what way? Give some examples.

(41) Explain the relationships between sensitivity and selectivity.

(42) Comment on foundation relationships among analytical properties.

(43) What are classical metrological properties?

(44) Define competitivity of an analytical laboratory.

(45) To which other properties is robustness of a method related?

(46) What is productivity? To which analytical properties is it related?

(47) Why can precision be highly rigorously defined via robustness?

(48) Relate classical analytical properties to other analytical features mentioned in the text.

(49) Why are analytical properties parts of the intrinsic fundamentals of Analytical Chemistry?

Seminars: Numerical Problems

Group 1

■ **Learning objective**
   ● To introduce the basic statistical concepts required to characterize accuracy, precision and uncertainty.

(1) Samples of a certified reference material (CRM) were used to assess the accuracy and precision of a method for the determination of a pesticide. Six individual applications of the method provided the following results: 3.21, 3.30, 3.35, 3.28, 3.40 and 3.25 µg/kg. The CRM was certified to contain an amount of pesticide of 3.45 ± 0.02 µg/kg. Calculate and discuss the parameters that define accuracy, precision and uncertainty.

Precision

● The data set is statistically processed, whether by hand or using a pocket calculator. First, the mean of the six results is calculated to be $\bar{x} = \sum x_i/n$. Then, $|d| = |x_i - \bar{X}|$, and d^2 are calculated as follows:

x_i	$|d|$	$|d|^2$
3.21	0.09	0.0081
3.30	0	0
3.35	0.05	0.0025
3.28	0.02	0.0004
3.28	0.02	0.0004
3.40	0.10	0.010
3.25	0.05	0.0025

$$s = \sqrt{\frac{\sum |d|^2}{n-1_2}} = \sqrt{\frac{0.0271}{5}} = 0.0685$$

$$\sum |d|^2 = 0.0235$$

Using the calculator to perform the statistical calculations provides the following parameter values:

$$\bar{X} = 3.29833 \approx 3.30 \qquad s = 0.06853 \approx 0.07 \qquad n = 6$$

● The limits of confidence at the 95% confidence level ($P = 0.05$) are calculated from $\bar{x} \pm (ts/\sqrt{n})$, where t is Student's parameter. With $U = 5 - 1$ degrees of freedom and $P = 0.05$, $t = 2.57$

$$3.30 \pm 2.57 \frac{0.07}{\sqrt{6}} = 3.30 \pm 0.070 \begin{cases} 3.23 \\ 3.37 \end{cases} \text{µg/kg}$$

● The **mean** characterizes the method.
● The absolute **uncertainty** is $\pm U = 0.072$ µg/kg. It can also be calculated from the expression $U_R = R \cdot S_R$; since $R = 2$ ($P = 0.05$) $\Rightarrow \pm U_R = 2 \times 0.07 = 0.14$ µg/kg. The divergence arises from the fact that n is small (distant from the proposed Gaussian distribution).
● The relative uncertainty will be $\pm \dfrac{U}{3.30} \times 100 = 2.18\%$.

Accuracy

- $\hat{X}'$, the value held as true and the result of operating under near-excellence conditions, will be given by

  $$\hat{X}' = 3.45 \pm 0.02 \text{ µg/kg}$$

- One can refer to accuracy since an appropriate reference, $\hat{X}'$, is available.
- It is accuracy of a method since one intends to obtain the mean $\bar{X}$ of 6 results and, specifically, to quantify the bias ($n < 30$).
- The absolute systematic error will be $|\bar{X} - \hat{X}'| = 3.30 - 3.45 = -0.15 \text{ µg/kg}$.
- The relative systematic error will be $\pm \dfrac{|\bar{X} \cdot \hat{X}'|}{\hat{X}'} = -4.35\%$.
- The bias of the method is thus negative (results are underestimated).

Other considerations

- The precision of the method used is lower than that of the CRM since the uncertainty in the former is greater:

  $$|U_{\bar{X}}| > |U_{\hat{X}'}| \quad 0.07 > 0.02$$

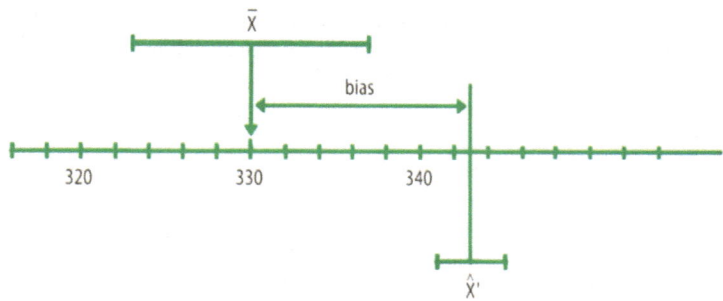

- If $U_{\hat{X}}$ is assumed to have been obtained at the same probability level, then the values of $\hat{X}'$ and $\bar{X}$ can be compared with their ranges.
- The method used is scarcely accurate because the ranges $\pm U_{\bar{X}}$ and $\pm U_{\hat{X}'}$ do not coincide. Therefore, the relative error in the negative bias (-4.35%) is misleadingly low.

(2) An overall sixteen analytical processes were applied separately to aliquots of the same sample of whole cow milk in order to determine its total fat content. The results obtained, as percentages (% mass/mass), were as follows: 3.95, 4.60, 3.75, 3.73, 3.90, 3.97, 3.74, 3.85, 3.79, 4.02, 3.86, 3.81, 3.41, 3.79, 3.96 and 3.78. Assess the precision of the method.

The first step is to arrange data in decreasing order: 4.60, 4.02, 3.97, 3.96, 3.95, 3.90, 3.85, 3.85, 3.81, 3.79, 3.79, 3.78, 3.75, 3.74, 3.73, 3.41.

As can be seen, the first and last value are rather different from the rest. Therefore, they are in principle dubious (*i.e.* subject to larger errors than the others).

The whole data set ($n = 16$) is subjected to a statistical study of precision with the aid of a calculator that provides the following results:

$$\bar{X}_1 = 3.867 \simeq 3.87\% \quad s = 0.241\% \quad n = 16$$

At the 95% confidence level ($P = 0.05$), $t = 2.12$. This allows the limits of confidence to be calculated as follows:

$$\bar{X}_1 \pm t\frac{s}{\sqrt{n}} \quad 3.87 \pm 2.12\,\frac{0.241}{\sqrt{16}} = 3.87 \pm 0.13\%$$

Then, Dixon's test is used to reject potential outliers (which both extreme data in the set are).

The implicit range for the first dubious datum excludes the second:

$$Q_{cal} = \frac{4.60 - 4.02}{4.60 - 3.73} = \frac{0.58}{0.87} = 0.66 \quad \text{at} = 15, \quad Q_t = 0.32 \; (P = 0.01)$$

Since $Q_t < Q_{cal}$, then the result should be rejected.

The implicit range for the second dubious datum excludes the first:

$$Q_{cal} = \frac{3.73 - 3.41}{4.02 - 3.41} = \frac{0.32}{0.61} = 0.52 \quad \text{at} = 15, \quad Q_t = 0.32 \; (P = 0.01)$$

Since $Q_t < Q_{cal}$, this result should also be rejected.

In addition, both potential outliers fall beyond the limits of confidence (4.00 and 3.74), which confirms that the decision to reject them was correct. More rigorous application of Dixon's test would lead to identifying 4.02 and 3.76 as two further outliers.

At this point, the remaining 14 results are evaluated (including both extremes in the whole set). The new results are

$$\bar{X}_2 = 3.85\% \qquad s = 0.095\% \qquad n = 14$$

At the 95% confidence level ($P = 0.05$), $t = 2.14$ and the confidence limits are

$$\bar{X}_2 \pm t\frac{s}{\sqrt{n}} \quad 3.85 \pm 2.14\,\frac{0.095}{\sqrt{14}} = 3.85 \pm 0.05\%$$

As can be seen, the precision is much better (s and the uncertainty are much lower) after the outliers are discarded.

As can also be seen, the means $\bar{X}_1$ and $\bar{X}_2$ are quite similar as a result of an overestimated and an underestimated result being discarded, and the effects on mean $\bar{X}_1$ being mutually countered. Should one or more results differing from the mean with the same sign have been rejected, the effect on the mean would have been rather different.

(3) The precision (uncertainty) of a fast immunoassay method (A) for the determination of phenol in water relative to that of the slower, classical chromatographic method (B) must be determined. For this purpose, two experiment series are conducted in order to apply both methods to aliquots of the same sample. The results obtained (in µg/L) are as follows:

Method A: 1.7, 1.8, 2.1, 2.2, 1.7, 1.9, 1.5, 1.3, 1.9, 1.7, 1.8, 1.4
Method B: 1.60, 1.82, 1.73, 1.81, 1.70, 1.73

The problem involves evaluating the dispersion of the two data sets, which consist of a different number of results owing to the slowness of method B. Inputing the previous data into a calculator provides the following statistical parameters:

Method A   ($n = 12$) $\bar{X}_A = 1.75$ µg/L $s = 0.26$ µg/L
Method B   ($n = 6$) $\bar{X}_B = 1.73$ µg/L $s = 0.08$ µg/L

The uncertainties can be calculated from the simplified expression $\pm U = R \cdot S$, with $R = 2$ at $P = 0.05$ (i.e. at the 95% confidence level).

$$\pm U_A = \pm 2 \times 0.26 = \pm 0.52 \ \mu g/l$$
$$\pm U_B = \pm 2 \times 0.08 = \pm 0.16 \ \mu g/l$$

The results for the two experiment series can thus be expressed in full form (including their respective uncertainties) as

A: $1.75 \pm 0.52 \ \mu g/L$
B: $1.73 \pm 0.16 \ \mu g/L$

The higher expeditiousness of method A results in decreased precision (a contradictory relationship between a basic and an accessory analytical property) and hence in increased uncertainty.

The accuracy of the method cannot be assessed because no appropriate reference is available. However, one can indeed state that the two methods have a similar bias – the means are quite similar – and that method A will be less accurate than B because it is subject to greater uncertainty.

(4) In order to evaluate instrumental precision and uncertainty in the gravimetric determination of nickel in a mineral sample, an amount of 0.8320 g of mineral is weighed and dissolved, and interferents removed. The remaining solution is supplied with NH_4Cl/NH_3 and dimethylglyoxime to precipitate nickel selectively as $Ni(DMG)_2$. The precipitate is filtered, dried at an appropriate temperature and, finally, weighed. The weight thus obtained is 0.391 g. (Random error in the weighing ± 0.2 mg; molecular weight of DMG 288.71 g/mol; atomic weight of nickel 58.71).

The amount of nickel in the unknown sample is calculated from the following expression:

$$W_{Ni} = F \cdot W_g \quad \left| \begin{array}{l} W_{Ni} = \text{nickel weight in the sample} \\ F \quad = \text{gravimetric factor} = \dfrac{W_{at} \cdot Ni}{W_M Ni(DMG)_2} = \dfrac{58.71}{288.71} = 0.2033 \\ W_g \ = \text{final gravimetric weighing} = 0.3391 \ g \end{array} \right.$$

The final gravimetric weighing is subject to the random errors in two weighings, viz. that used to tare the crucible and the gravimetric weighing proper. These errors materialize in the combined standard deviation from a subtraction:

$$s_{W_g} = \pm \sqrt{(0.2)^2 + (0.2)^2} = \pm 0.28 \ mg \approx \pm 0.28 \ mg \approx \pm 0.3 \ mg = 3 \cdot 10^{-4} \ g$$

which is the error in W_{Ni} since F is assumed to be subject to no random error.

$$s_{W_{Ni}} = 3 \cdot 10^{-4} \times 0.203 = 0.6 \cdot 10^{-4} \ g$$

The initial weighing, W_i, is subject to a standard deviation

$$s_{W_i} = \pm 0.20 = 2 \cdot 10^{-4} \ g$$

The proportion of nickel in the sample will be

$$\% Ni = \frac{W_{Ni}}{W_N} \cdot 100 = 8.31\%$$

Because the result is obtained from a quotient, its relative standard deviation will be

$$s_{\%Ni} = 100\sqrt{\left(\frac{S_{W_{Ni}}}{W_{Ni}}\right)^2 + \left(\frac{S_{II}}{S_{II}}\right)^2} = 100\sqrt{\left(\frac{0.6 \cdot 10^{-4}}{0.0671}\right)^2 + \left(\frac{2 \cdot 10^{-4}}{0.8320}\right)^2}$$

$$\frac{s_{\%Ni}}{\%Ni} = \sqrt{8.1 \cdot 10^{-7}} = \sqrt{81 \cdot 10^{-8}} = 9 \cdot 19^{-4}$$

$$s_{\%Ni} = 9 \cdot 10^{-4} \times 8.31 = 7500 \cdot 10^{-4} = 0.075$$

At $P = 0.05$ (the 95% confidence level),

$$R = 2 \quad \text{or} \quad \pm U_{\%Ni} = 2 \cdot 0.0075 = \pm 0.015 \approx \pm 0.02\%$$

so the result can be expressed as

$$\%Ni = 9.31 \pm 0.02\%$$

(5) The confidence level with which an individual result (12.378) obtained using the Karl–Fischer procedure for the determination of the percent water content in a rice batch can be accepted must be determined. Two preliminary precision studies under repeatability (A) and reproducibility conditions (B) were conducted that provided the following results:

A: 12.40%, 12.39%, 12.37%, 12.42%, 12.39%, 12.40%
B: 12.45%, 12.35%, 12.39%, 12.47%, 12.32%, 12.39%

First, the two data sets are subjected to a statistical study.

	n	$\bar{X}$	s
A (Repeatability)	6	12.395	0.016
B (Reproducibility)	6	12.395	0.037

Coincidentally, both means are identical; however, the standard deviation obtained in the reproducibility study (B) is about 3.5 times greater than that for the repeatability study (A).

Let us examine four different confidence levels, namely: 90% ($P = 0.1$), 95% ($P = 0.05$), 98% ($P = 0.02$) and 99% ($P = 0.01$). The corresponding t values for $6 - 1 = 5$ degrees of freedom are given in the following table:

% confidence	90%	95%	98%	99%
P	0.1	0.05	0.02	0.01
t	2.02	2.57	3.36	4.03

These values allow one to calculate $\pm U = t \cdot S$ for the two sets at different probability (confidence) levels.

% confidence	90%	95%	98%	99%
$\pm U_A$	0.013	0.016	0.022	0.026
$\pm U_B$	0.047	0.060	0.078	0.094

A plot of the means and their respective ranges allows one to assess the latter more easily on a concentration scale.

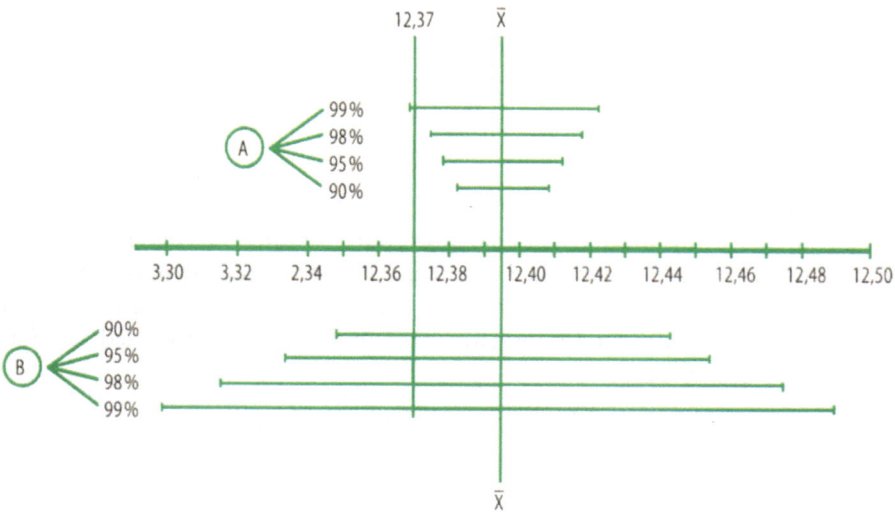

As can be seen, dispersion was greater, precision smaller and uncertainty greater in the reproducibility study (B). In each case, increasing the confidence level (and decreasing the probability) increased t and uncertainty (limits of confidence).

The datum 12.37, which is the key to the problem, occurs in all the ranges used in the reproducibility study (B), but only in the 99% confidence interval ($P = 0.01$) in the reproducibility study (A). In the more favourable – less rigorous – study (A), such a value should be rejected at $P > 0.01$.

Since reproducibility entails greater rigour (*i.e.* less precision and more uncertainty) and the result is acceptable in all cases, it can be deemed valid.

Group 2

■ **Learning objective**
- To define selectivity and sensitivity via the parameters discussed in the text.

(6) Determine the maximum tolerated ratio of nickel (interferent) to copper (analyte) in the photometric determination of the latter. These species form two complexes with a common reagent–ligand L that differ in their spectral features: λ_{max} is 540 nm for CuL_2^{2+} and 500 nm for NiL_2^{2+}.

The uncertainty is determined by assessing the precision for samples containing 1.0 µg/L copper; absorbances over the range 0.453 ± 0.020 AU are obtained at the 95 % probability level. Then, a series of samples containing 1.0 µg/L copper and increasing concentrations of nickel from 0 to 3 µg/L is prepared that provides the following results:

Sample no.	1	2	3	4	5	6
$[Cu^{2+}]$ (µg/L)	1.0	1.0	1.0	1.0	1.0	1.0
$[Ni^{2+}]$ (µg/L)	0.0	0.5	1.0	1.5	2.0	3.0
Absorbance (AU)	0.452	0.457	0.462	0.473	0.489	0.504

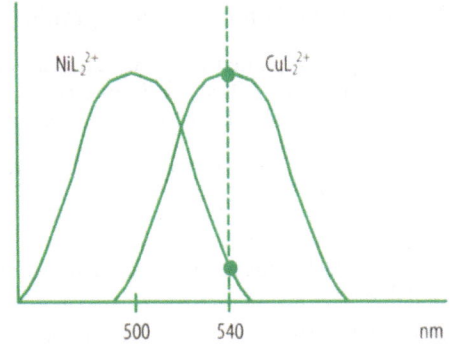

- Types of interferences involved:
  - positive
  - same mechanism
  - chemical origin
  - additive
- The limits of confidence (in absorbance units) are:
  0.433
  0.473
- If a response above 0.473 is obtained, then an interference will occur.
- A detailed study reveals that

Experiment	1	2	3	4	5	6
$[Cu^{2+}]$	1.0	1.0	1.0	1.0	1.0	1.0
$[Ni^{2+}]/[Cu^{2+}]$	0	0.5	1.0	2.5	2.0	3.0
AU	0.453	0.457	0.462	0.473	0.489	0.504

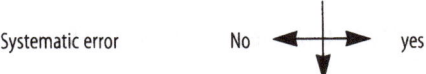

Systematic error   No ⟵┼⟶ yes

- Ni^{2+} is thus tolerated at concentrations 1.5 times higher than is Cu^{2+}.

(7) The sensitivity of a method for determining quinine in tonic water is assessed in two experiment series. In the first, a blank (tonic water containing no quinine) is measured 11 times and the following fluorescence intensities are obtained: 0.70, 0.74, 0.72, 0.71, 0.73, 0.71, 0.73, 0.74, 0.70, 0.69 and 0.75. In the other series, increasing amounts of quinine are used to prepare 7 synthetic samples (tonic water containing no quinine) that provide the following results:

[quinine], µg/L	1.0	2.0	3.0	4.0	6.0	8.0	10.0
Fluorescence intensity	2.51	4.60	6.52	8.60	12.71	16.68	20.81

Express the sensitivity of the method in the different possible forms and establish their mutual relationships via a graph.

- The function $x = \varphi(C)$ is plotted and an intercept is obtained. Although the blank value is unknown, it clearly falls above the intercept.
- The sensitivity is calculated either graphically (slope = 2.04) or as $S = x/\Delta C$:

$$s = \frac{4.20 - 2.51}{2.0 - 1.0} = 2.09 \; ; \quad s = \frac{12.71 - 4.60}{6.02 - 2.0} = 2.93 \; ; \quad s = \frac{16.68 - 6.52}{8.0 - 3.0} = 2.03$$

$$s = \frac{20.81 - 16.68}{10.0 - 8.0} = 2.09 \; ; \quad \text{the mean of these four determinations} = 2.05$$

- The units are: 2.05 $I_fU/\mu g/mL$ = 2.05 $I_fU \cdot mL \cdot \mu g^{-1}$, where I_fU denotes fluorescence intensity units.
- As can be seen, the slope varies slightly depending on the portion of the curve; this suggests that the study is subject to random errors.
- The intercept can be determined graphically, by extrapolation (0.45), or calculated via the sensitivity:

$$s = 2.05 = \frac{2.51 - y}{1.0 - 0.0} \; ; \quad y = 0.46 \, I_fU$$

- The statistical study of the fluorescence intensity values for $n = 11$ blanks provides the following results:

$$\bar{X}_B = 0.75 \, \mu g/mL \quad \sigma_B = 0.20 \, \mu g/mL$$

- The limit of detection is calculated from the following expression:

$$X_{LOD} = \bar{X}_B + 3\sum_B = 0.72 + 3 \cdot 0.02 = 0.78 \, UI_f$$

- The expression for the sensitivity can be used to derive the concentration C_{LOD}:

$$2.05 = \frac{2.51 - 0.78}{1.0 - C_{LOD}} \; ; \quad C_{LOD} = 0.16 \, \mu g/mL$$

- The intercepts (0.78, 0.16) are the lower limits of the linear portion.

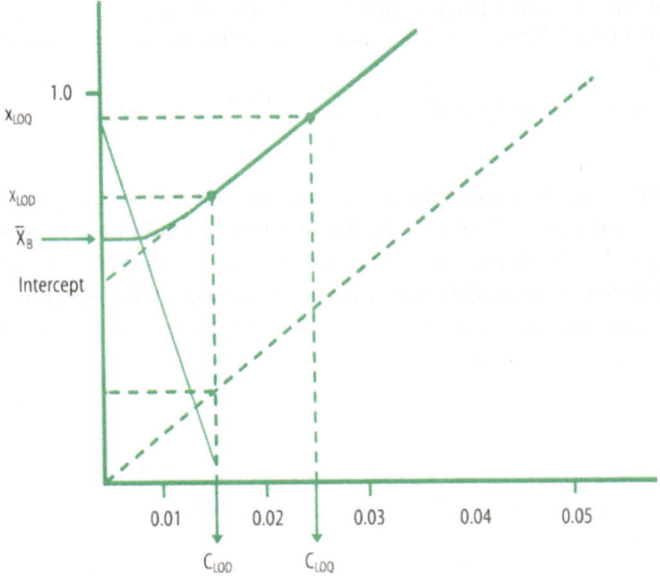

- The limit of quantitation is calculated from

$$x_{LOQ} = \bar{X}_B + 10\,\sigma_B = 0.72 + 10 \cdot 0.02 = 0.092 \; I_f U$$

and the concentration C_{LOQ} is derived from

$$2.05 = \frac{2 - 0.51 - 0.92}{1.0 - C_{LOQ}}; \quad C_{LOQ} = 0.22 \; \mu g/mL$$

- If the plot is extended near the origin, then the data pairs (x_{LOQ}, C_{LOQ}) and (x_{LOQ}, C_{LOQ}), the intercept and/or the blank value $(\bar{X}_B, 0)$, can be located.

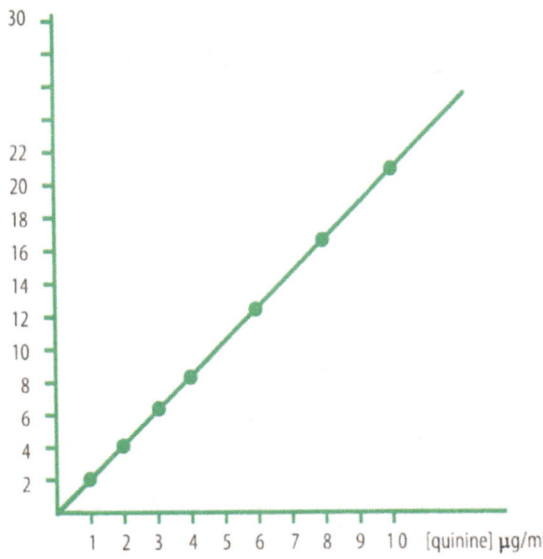

- If the analytical blank coincided with the intercept ($y = \bar{X}_B = 0.46$), and the standard deviation were the same ($\sigma = 0.02$), then the relations to sensitivity would be

$$C_{LOD} = \frac{3\,\sigma_B}{S} = 0.03 \text{ mg/ml} \quad \text{and} \quad C_{LOQ} = \frac{10\,\sigma_B}{S} = 0.10 \text{ mg/mL}$$

(8) In order to determine total calcium in canned orange juice by atomic absorption spectrometry, a synthetic sample (containing no calcium) was supplied with increasing concentrations of the analyte. The following absorbance data (together with their uncertainties) were obtained separately by analysing several samples containing the same analyte concentrations:

$[Ca^{2+}]$, mg/L	0	0.5	1.0	1.5	2.0	2.5	3.0	4.0
Absorbance (AU)	0.040	0.056 (0.014)	0.074 (0.012)	0.090 (0.010)	0.108 (0.010)	0.125 (0.011)	0.152 (0.11)	0.177 (0.015)

What is the standard deviation of the blank if the limit of detection for the method is 0.05 mg/mL?

A plot of Absorbance (AU) = $\varphi(C)$ is constructed on graph paper (uncertainty included). From the plot it follows that:

(a) All of the above concentrations fall within the linear portion of the curve.
(b) The blank ($C = 0$) coincides with the intercept of the linear portion.
(c) Uncertainty is greater at the ends. In fact, the calibration curve is an area rather than a line.

The sensitivity of the method can be derived from the slope of the calibration curve or from the following expression:

$$S = \frac{\Delta x}{\Delta C} = \frac{x_2 - x_1}{C_2 - C_1} = \frac{0.152 - 0.074}{3.0 - 1.0} = \frac{0.078}{3} = 0.026 \text{ AU} \cdot \text{mg}^{-1} \cdot \text{ml}$$

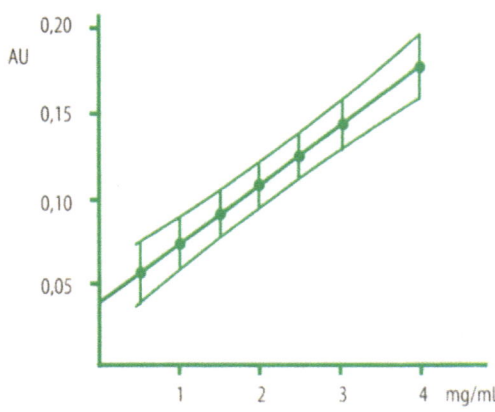

Since the blank, $\bar{X}_B$, coincides with the intercept, then the following expression can be applied:

$$S = \frac{2\sigma_B}{C_{LOQ}}; \quad \text{where} \quad \sigma_B = \frac{S \cdot C_{LOQ}}{3}$$

$$\sigma_B = \frac{0.026 \times 0.03}{3} \text{ AU} \cdot \text{ng}^{-1} \cdot \text{mL} \cdot \text{ng/mL} = 2.6 \cdot 10^{-4} \text{ AU}$$

Suggested Readings

(1) "The Hierarchy and Relationships of Analytical Properties", M. Valcárcel and A. Ríos, *Anal. Chem.*, 1993, 65, 781A.
    The first global approach to analytical properties, which are ranked according to significance. It establishes foundation, contradictory and complementary relationships among them. Based on suggested reading no. 3 in Chap. 1.

(2) "Teaching Analytical Properties", M. Valcárcel and A. Ríos, *Fresenius J. Anal. Chem.*, 1997, 357, 202.
    Explains how to introduce the teaching of analytical properties in the Analytical Chemistry curriculum. The present chapter materialized from this paper.

(3) "A Hierarchical Approach to Metrology in Chemistry", M. Valcárcel and A. Ríos, *Trends Anal. Chem.*, 1999, 18, 68.
    Discusses in detail the foundations of the metrological hierarchy of Fig. 2.2 and the establishment of the different conceptions of uncertainty and trueness in the chemical metrological domain.

(4) "Quality Control in Analytical Chemistry", G. Kateman and L. Buydens, *Wiley*, New York, 1993.
    This excellent monograph, which contains a well-balanced combination of Chemometrics and Quality, provides deep coverage of the properties discussed in this chapter.

(5) "Statistics for Analytical Chemistry", 3rd edn, J.C. Miller and J.N. Miller, Ellis Horwood, Chichester, 1994.
    A straightforward book that describes and applies the basic statistical tools available to the analytical chemist. Provides a comprehensive discussion of the analytical data processing approaches briefly explained in this chapter.

Traceability: Reference Materials

Objectives

- To introduce the integral concept of traceability, one of the cornerstones of Analytical Chemistry.
- To underline the role of standards in the metrological science Analytical Chemistry.
- To describe the different types of standards available, particularly analytical chemical standards, and their use in major analytical operations such as calibration, validation and quality control.
- To apply the traceability concept not only to results, but also to other analytical chemical entities that are thus conceptually enriched.

Table of Contents

3.1 Introduction

Analytical chemical information arises from the act of measuring and measuring involves comparing with established references so that the data obtained in the process can be delivered as universally understandable, equivalent, comparable items of information. Standards are the analytical references that constitute the fundamental technical cornerstones of Analytical Chemistry.

Traceability is a key word of Analytical Chemistry, to which it is related through two tentative definitions, *viz.* "the science of (bio)chemical measurements" and "the (bio)chemical metrological science"; both are ultimately equivalent as the word "metrology" comes from the Greek "metron" (measure).

Traceability has been an abstract concept inherent in Analytical Chemistry ever since the beginning of this science a few centuries ago. It is comparable to other generic concepts such as honesty in social relations and profitability in the economic realm. In the metrological field, traceability is related to quality of measurement processes and of the results they produce.

───── Box 3.1

The significance of standards to measurements of physical, chemical, biochemical and biological quantities is undeniable. What sort of suits would a tailor make if he used an elastic tape to take his clients' measurements? What sort of business relations could two firms have if they used two different mass standards of unknown equivalence? How could 200-metre runners worldwide possibly be ranked if a different time standard were used in each competition?

Evidence of the metrological impact of references is provided by the practical problems that arise from the use of different references, however closely related. Such is the case with temperature measurements, which are expressed in kelvins, degrees centigrade or degrees Fahrenheit, depending on the particular technical field or geographic location. Similarly, mass measurements are referred to different standards (kilogram, pound), as are length measurements (metre, foot), in different locations.

The availability of universally accepted standards is an essential ingredient of social, scientific, technological, industrial and trading relations; one that materializes in the traceability concept.

In recent years, the traceability concept has been revitalized by the growing interest in instituting quality systems in analytical laboratories and the expansion of international trade, both of which rely on mutual acceptance of the results produced by measurement processes carried out by different laboratories.

3.2 The Integral Concept of Traceability

The traceability concept cannot be approached in a unique manner. In fact, only an integral, combined view of its different aspects can provide a comprehensive definition for this term.

The most immediate meaning of traceability is related to tracing or tracking, and is included in its definition in the ISO Vocabulary for Quality Management and Quality Assurance (see Box 3.2). This meaning is also consistent with the definitions given in renown linguistic works such as the Oxford English Dictionary and Webster's Third New International Dictionary, which contain entries not only for the noun *traceability* ("the quality of being traceable"), but also for the adjective *traceable* ("capable of being traced" or "suitable or of a kind to be attributed") and for the verb *to trace* ("to follow or study out in detail step by step" or "to follow the course, development or history of").

_____ Box 3.2

The Quality Management and Quality Assurance Vocabulary (ISO 8402), issued by the International Standards Organization (ISO) in 1994, defines traceability in generic terms as "the ability to trace (know) the history, application or location of an entity (a product, process, service, organization) by means of recorded identifications." In the Vocabulary, this definition is completed by the following two technical explanatory notes:

(1) The word "traceability" has three major meanings, namely:
  - Traceability of a *product* is related or can be referred to the origin of the starting materials and pieces, the history of its production, and its distribution and location after it has been delivered or distributed.
  - Traceability of *calibration* applies to the relationship of measuring equipment to national or international standards, to primary standards, to fundamental constants or physical properties, or to reference materials.
  - Traceability of *data acquisition* is related to the data produced and computations made around the "quality loop", frequently in compliance with the "quality requirements" of a specific institution.
(2) Every aspect of the applicable traceability requirements (*e.g.* time intervals, source or identification point) should be clearly specified.

Traceability should thus be approached in a global manner in the technical field; using it for the sole purpose of characterizing individual aspects (*e.g.* a result in Analytical Chemistry) is definitely a poor strategy.

In the analytical chemical field, traceability is realized as the history of the production of a result or of the performance of a system (a laboratory, instrument, apparatus, reference material, *etc.*), which should be clearly, comprehensively documented by qualified personnel under internal supervision or the external control of national or international organizations.

In the metrological field, references are no doubt the landmarks of the "history" inherent in the etymological definition of traceability. Such a definition is thus closely related to the standards used in the comparisons involved in chemical, physical, biochemical and biological measurements. Figure 3.1 illustrates the traceability concept as a combination of the etymological standard-based approaches.

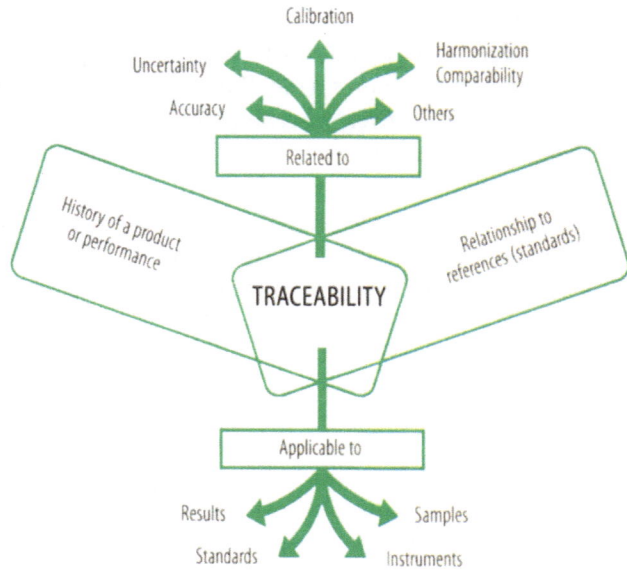

Fig. 3.1. The traceability concept in the metrological domain. Relationship to analytical notions and applicability to analytical entities

Traceability is also related to various analytical concepts (Fig. 3.1). One is the capital analytical property "accuracy of an analytical result" (see Sect. 2.4.1), which traceability enriches with additional facets such as the "comprehensive history" of the measurement process and standards employed, their quality and their use in calibration operations, among others. As noted in Chap. 2, traceability is also related to specific uncertainty; in fact, accuracy cannot be properly defined in its absence. Regarding applications, traceability is the basis for comparison of results produced by different laboratories; this entails their harmonization and the mutual acceptance of their results – a key to fluent social and trade relations, both national and international.

In the metrological context, traceability was formerly considered a property virtually exclusive of results (see Sect. 3.6.1); there is no plausible reason, however, to avoid applying it to other analytical concepts, especially if the generic definition of ISO is considered (see Box 3.2). Thus, as shown in Sect. 3.6, traceability can also be ascribed to a standard, an instrument or a sample (or aliquot), among others.

It is interesting to examine the different facets of the standard–traceability relationship schematically depicted in Fig. 3.2. Thus, traceability among the different types of standards, ranked in terms of nearness to the true value of the measured quantity, is crucial to analytical references (see Sect. 3.4). The ISO definition of traceability of a result includes standards (see Sect. 3.5.1), just like traceability of a measuring instrument is related to standards used in calibration operations (see Sects. 3.5.4 and 3.6.3).

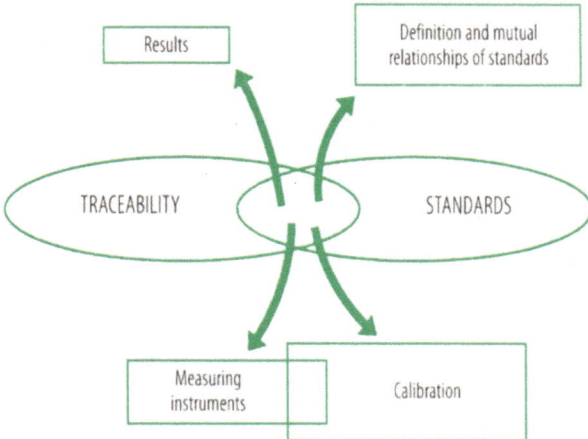

Fig. 3.2. Connotations of the relationship between traceability and standards. For details, see text

_____   Box 3.3

It is interesting to examine the translation of the English word "traceability" into some Romance languages. Thus, Spanish and French use an almost literal adaptation, *viz. trazabilidad* and *traçabilité*, respectively. In other languages, however, the translation reflects the special emphasis placed on some facets of the term. Such is the case with the Portuguese *rastreabilidade*, which focuses on the tracking facet. On the other hand, the Italian word, *riferibilità*, emphasizes the relationship to references (standards).

This linguistic variety in the translations of traceability reveals that the concept encompasses two complementary aspects without which it cannot be defined properly.

3.3 Physical and Chemical Traceability

Out of tradition or unawareness, Metrology has traditionally been used exclusively in connection with measurements of physical quantities. There is no reason, however, to exclude measurements of chemical and biological quantities. There are indeed strong technical differences in the way physical (PMPs), chemical (CMPs) and biological measurement processes (BMPs) are conducted.

The most salient differences between physical and chemical measurements as regards traceability can be summarized as follows:

(*a*) Physical measurement processes (PMPs) are virtually independent of the sample and object being tested, which, however, can to some extent dictate the type of instrument to be employed (*e.g.* to measure the width of a piece of furniture, the height of a tree or the distance between two places). On the other hand, chemical measurement processes (CMPs) are strongly dependent on the type of sample concerned. Thus, the determinations of a pesticide in a commercially available formulation, water and soil involve rather different analytical processes, even if the same measuring instrument (*e.g.* a gas chromatograph) is used in all cases.

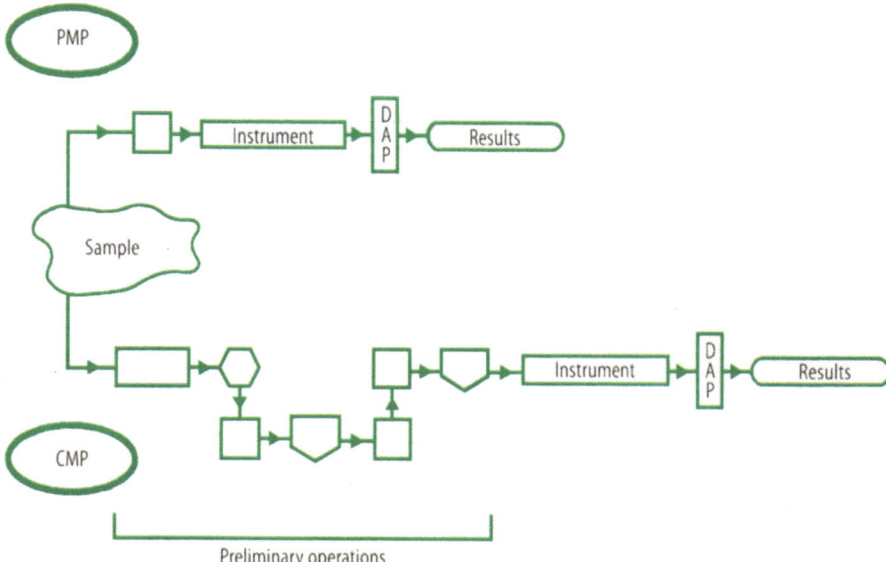

Fig. 3.3. Differences between physical measurement processes (PMPs) and chemical measurement processes (CMPs). PMP Physical Measurement Process; CMP Chemical Measurement Process; DAP Data Acquisition and Processing

(*b*) Chemical measurement processes are usually more labour-intensive and time-consuming than are physical measurement processes. As can be seen in Fig. 3.3, a chemical measurement process (CMP) involves several preliminary operations (*e. g.* dissolution, filtering, separation) intended to condition the sample for measurement by the instrument, operations that make CMPs complex to implement relative to PMPs. As a result, CMPs are more liable to variability (*i. e.* their results are subject to higher uncertainty) and also slower – occasionally, even tedious. The advent of novel CMP technologies based on automation and the use of sensors, among others, have significantly closed this technical gap between CMPs and PMPs. In any case, accomplishing chemical traceability remains a longer, less unequivocal process than realizing physical traceability.

(*c*) The standards on which traceability of PMPs rests are quite affordable. In fact, there are reliable reference standards for virtually every physical quantity. By contrast, available standards can hardly encompass the wide variety of CMPs possible. This is one weakness of chemical metrology and hence of its traceability (based on a recent estimate, only 8–10% of the present needs for CMP standards appear to have been fulfilled). The problem of finding new, suitable standards for CMPs lies mainly in the wide variety of samples to which they may be applied; even in those cases where pure analytes are available for use as standards in the measurement step proper, the specific nature of the sample dictates the type of pretreatment to be used – a wrong choice may significantly alter the quality of the final results. Obviously, assuring chemical traceability entails developing dedicated procedures such as those involving the use of samples spiked with the analytes, special calibration methods or reference matrices (standards), all of which are redundant in PMPs.

(*d*) Although both PMPs and CMPs require calibrating the measuring equipment, some chemical methodologies additionally entail relating the concentration (amount) of analyte to the signal it gives. The differences between equipment and method (analytical) calibration, both of which are closely related to traceability, are discussed in Sect. 3.5.4.

_____ Box 3.4

The characterization of a given sea water sample in physical (*e. g.* temperature), chemical (*e. g.* pH, concentration of mercury species) and biological terms (*e. g.* presence or absence of microbes) differs strongly in complexity and difficulty.

Measuring the water temperature is fairly simple: it suffices to use a pre-calibrated thermocouple. Measuring its pH is also quite simple: a pH-meter is also a sensor equipped with a combined potentiometric electrode which, similarly to the thermocouple, must be calibrated with standard solutions of known pH prior to immersion in the sample. On the other hand, the speciation of mercury is a longer process that involves using various types of containers to store the sample, withdrawing an accurately measured aliquot from it (*e. g.* by means of a calibrated pipette) and conditioning it (*e. g.* by preconcentration, clean-up), inserting the treated aliquot into a complex instrumental system (*e. g.* a gas chromatograph coupled on-line to an atomic spectroscopic detector the response of which should previously have been calibrated with standards of usually limited commercial availability) and processing the transient signals obtained. Obviously, the speciation of mercury will be subject to many more sources of variability and to larger systematic and random errors than will the previous measurements. Finally, the scarcity of certified reference materials for speciation will normally pose an additional difficulty.

The microbiological characterization of sea water entails the use of a scarcely robust measuring system for which available references (measurement standards) are also scant.

3.4 Types of Standards and their Traceability

Standards are closely related to traceability in all metrological fields. In fact, they are the basis for measuring physical, chemical, biochemical and biological quantities. It is thus important to know not only what types of standards are available, but also how they relate to one another.

Standards for chemical measurements can be ranked in terms of two different criteria, namely (Fig. 3.4): (*a*) nearness of their value in the chemical parameter of interest to the value held as true; and (*b*) availability or tangibility. Uncertainty in the reference value increases as one climbs down the ranking of Fig. 3.4. All the steps in this hierarchy are mutually related via traceability; thus, analytical standards are related to SI base units through chemical standards. Therefore, chemical and analytical chemical standards can also be referred to as "transfer standards", similarly as in physical metrology.

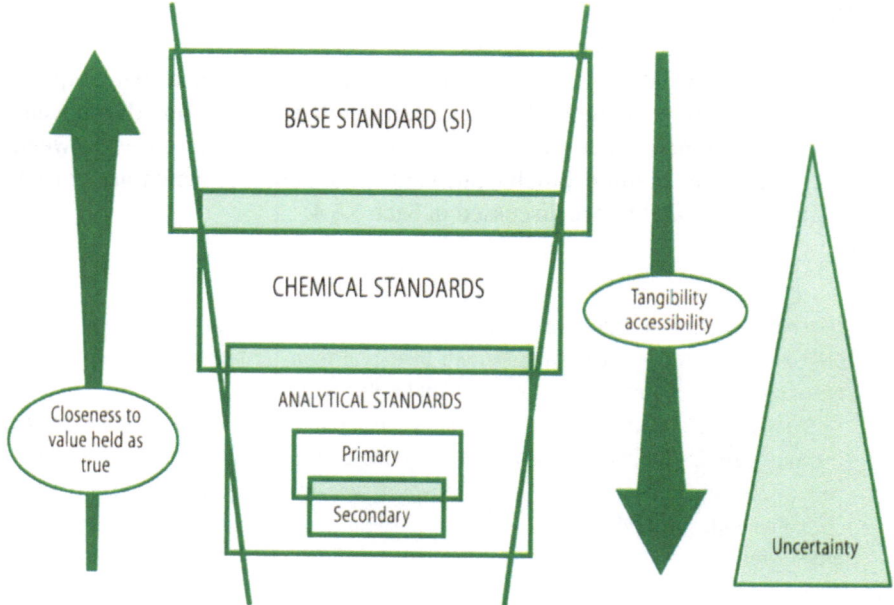

Fig. 3.4. General types of standards used in chemical metrology and traceability among them. Traceability between each standard and the next is realized through the intervening zone (interface)

____ Box 3.5

Physical measurements are also amenable to ranking in a manner similar – though not identical – to chemical measurements (see Fig. 3.5). Thus, the last link in the traceability chain for mass measurements is the kilogram prototype, which is kept by the International Bureau of Weights and Measures (BIPM) near Paris. The next standard type, the kilogram primary

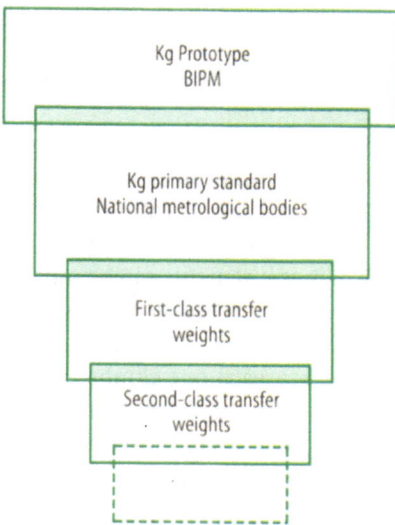

standard, is kept by a national body in each country and is periodically checked against the Paris kilogram. The mass of the Spanish prototype is 0.999999890 kg, with a combined uncertainty of 2.3 µg. First-class transfer weights are periodically checked against the national standard and used by competent public or private bodies engaged on assuring the quality of mass measurements. On another step in the ranking below are second-class transfer weights, which are periodically checked against first-class weights and used to calibrate balances (of wide use in the physical and chemical fields).

The traceability of the amount of mass in a weight used for calibration is established via the serial comparisons inherent in the above ranking. In addition to establishing the amount of mass, one should determine the combined uncertainty of the checking operations as a whole. Calibrating a balance without consideration of the uncertainty involved is rather a frequent mistake in practice.

Base standards coincide with the seven SI base units for measuring time (second), length (meter), electric current (ampere), thermodynamic temperature (kelvin), luminous intensity (candela), mass (kilogram) and amount of substance (mole) (see Fig. 3.5). From these base units, new, derived units such as the cubic metre (volume), hertz (frequency), pascal (pressure) or mole per kilogram (concentration) have been obtained via simple algebraic operations. Derived units are the references for other types of standards of use in both physical and chemical measurements. Note that a standard is not the same as a unit. In fact, units are all – except the kilogram prototype, which is tangible – abstract concepts with a name; on the other hand, standards are objects, devices or procedures that realize a unit in a tangible, not abstract, manner.

The kilogram and the mole are the two most significant base standards used in chemical metrology; this, however, uses additional standards (*e.g.* in distance and time measurements in chromatography or capillary electrophoresis, current

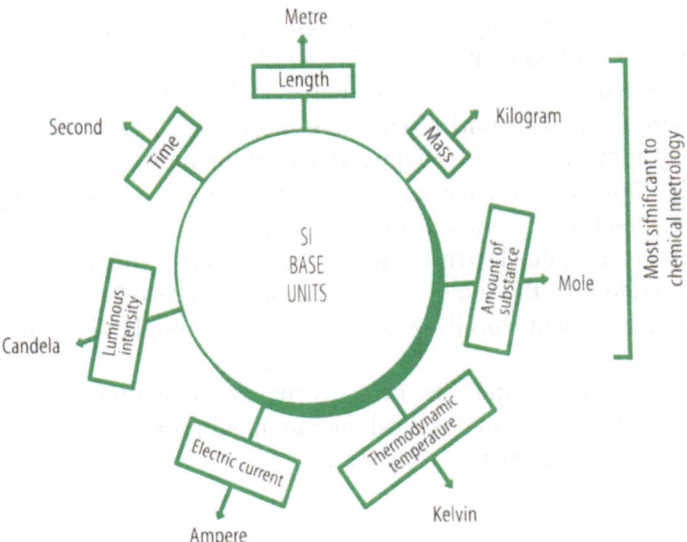

Fig. 3.5. Base units of the International System (SI)

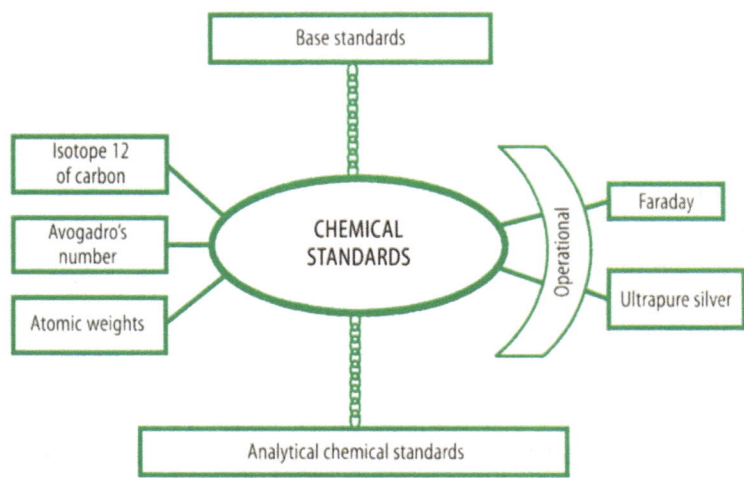

Fig. 3.6. Types of chemical standards and their relationship to the other types of standards

intensity measurements in polarography, frequency measurements in nuclear magnetic resonance spectroscopy). The kilogram is the sole material, tangible SI base unit; it is defined as the mass of an internationally accepted prototype of platinum-iridium alloy that is kept near Paris by the International Bureau of Weights and Measures (BIPM). The mole is the amount of substance that contains as many elemental units (atoms, molecules, ions, electrons or other individual particles or particle groups) as are in 0.012 kilograms of the isotope carbon-12.

Chemical standards can be considered the traceability links between base standards and analytical chemical standards (see Fig. 3.6). These standards can be classified into two groups in terms of tangibility. Thus, the isotope carbon-12 is a chemical standard related to the mole (a base standard) and to Avogadro's number, $N = 6.023 \times 10^{23}$, which is the number of atoms or molecules contained in one mole (or gram-atom) of any substance. Carbon-12 is the basis for the universally admitted atomic weights, which are the most widely used chemical standards. Operational chemical standards include the Faraday (96 487.3 coulomb), which is the amount of electricity needed for one equivalent of a redox substance to be electrochemically transformed – the equivalent is thus obviously related to the mole –, and ultrapure (> 99.999 %) silver, a tangible standard almost unequivocally related to the atomic weight of this metal.

Analytical chemical standards, of the primary or secondary type, are the practical standards used in chemical measurement processes. As such, they are dealt with at length in the following section.

_____ Box 3.6

The word "standard" bears a number of meanings that are frequently misleading. Thus, it can be used to refer to written documents; to codes, signs and languages used for communication between systems; to behavioral or operational patterns; *etc.* In the measurement domain, a "standard" is used (*a*) to define the minimum acceptable specifications (attributes) of an object or service (*e.g.* technical specifications, instructions of use or operation, safety levels, quality criteria) or (*b*) to establish references such as currencies for international trade.

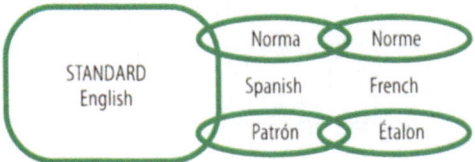

Therefore, the reference standards used in the metrological field should be called simply "standards". Such standards can be material (*e.g.* objects such as the kilogram prototype, ultrapure silver or potassium iodate, or devices such as voltammeter calibration circuitry) and operational or procedural (*e.g.* reference or standard methods).

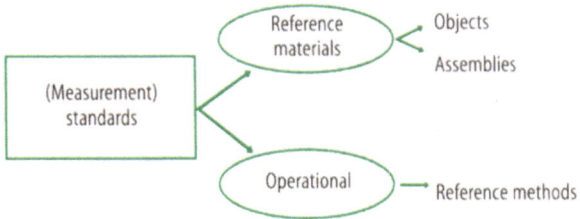

In 1996, Pierce suggested using "measurement standard" or "étalon" instead of simply "standard" in English.

Figure 3.7 shows the **traceability network** formed by the above-described types of standards. The definition of the mole includes the kilogram. By definition, the mole is related to isotope 12 of carbon and to Avogadro's number, which are also mutually related. Atomic weights are based on the mass of carbon-12. The operational chemical standard Faraday is related to two SI base units, *viz.* the ampere and the second – the number of coulombs produced is equal to the product of the intensity of the electric current by the time during which it is supplied. In addition, the Faraday is related to Avogadro's number and the mole. Because it is a tangible, material standard, ultrapure silver is on the boundary between chemical and analytical standards; it is related to the Faraday, atomic weights and primary standards through testing. Primary analytical standards are related to atomic weights and, through experimentation, to ultrapure silver. Finally, secondary standards are also related to primary standards through testing.

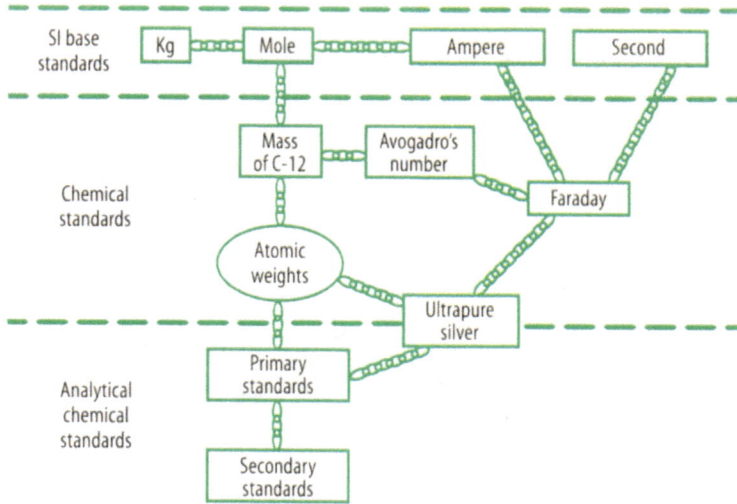

Fig. 3.7. Traceability network formed by the different types of standards. For details, see text

___ Box 3.7

The increasing use of the metrological approach to CMPs has raised a debate over what the last link in the traceability chain of the standards shown in Figs 3.4 and 3.6 is.

Experts at various international institutions such as EURACHEM and CITAC, which have played crucial roles in the revival of chemical metrology, claim that the last link in the chain is the mole (*i.e.* the most "chemical" of all base standards) and use the expression "traceability to the amount of substance".

Other experts believe that the last link in the chemical traceability chain continues to be the kilogram prototype, basically because defining the mole entails using the kilogram just as defining the metre requires using the time unit (a metre is the length travelled by light in vacuum during a time interval of 1/299 792 458 of a second).

In fact, the debate is absolutely sterile. The important thing is to clearly understand how standards are related (*i.e.* the network of Fig. 3.7). The manner in which one expresses traceability is insubstantial and contributes little to a better understanding of chemical metrology.

3.5 Analytical Chemical Standards

Analytical chemical standards are the standards ordinarily or extraordinarily used in chemical measurement processes. As such, they are tangible and almost always commercially available. This type of standard can be a chemical substance (a pure compound, a natural or artificial mixture) or an operating device (*e.g.* voltammeter calibration circuitry); the latter shares some features with physical standards.

3.5.1 Types

Analytical chemical standards can be classified into two, mutually non-exclusive, categories according to (*a*) intrinsic properties or (*b*) nature and purpose (see Fig. 3.8).

_____   Box 3.8

Determining the acidity of commercially available vinegar entails using a "standard" solution of sodium hydroxide (*i.e.* one of accurately known concentration, which will be one basis for traceability). Solid sodium hydroxide does not meet the purity and stability requirements for use as a primary standard. Therefore, an experimental traceability link must be established between the secondary standard (NaOH) and a primary standard of acid character (*e.g.* potassium hydrogen phthalate, which meets the previous requirements).

A solution about 0.1 M in NaOH is thus prepared. With due care, the solution is used to fill a conventional or automatic burette. In another operation, an amount of 0.8642 g of potassium hydrogen phthalate is weighed on a balance and dissolved in water. This latter solution is supplied with phenolphthalein as indicator and titrated; in the titration, which is equivalent to standardizing the titrant, a volume of 40.15 mL (practical volume) of the NaOH solution is used. A simple stoichiometric calculation reveals that the amount of primary standard initially weighed should have consumed 42.15 mL of an exactly 0.0001 M solution of NaOH (theoretical volume). The titrimetric factor, defined as the ratio

$$f = \frac{\text{theoretical volume}}{\text{practical volume}} = \frac{42.15}{40.15} = 1.0550$$

is the number by which the approximate NaOH concentration (0.1 M) will have to be multiplied in order to determine the actual titrand concentration. Once the NaOH solution is standardized, it can be used to titrate the vinegar samples – always exercising good care in handling the solutions.

Based on intrinsic properties, analytical standards can be of two types, namely:

(*a*) **Primary analytical standards,** which are stable, homogeneous substances of well-known properties (*e.g.* purity, species concentrations) that can be directly related to chemical standards (*e.g.* atomic weights) via a traceability chain provided the specific uncertainty involved in deriving the characteristic quantity is considered [*e.g.* a purity of $(99.5 \pm 0.2)\%$].

(*b*) **Secondary analytical standards,** also known as "working standards", are substances that lack the properties required for direct use as analytical standards (*e.g.* stability, purity, homogeneity) but have to used because an appropriate primary standard for the intended purpose either does not exist or is unavailable or expensive. Secondary standards must be traced to a primary standard through testing (*e.g.* by standardizing a titrant solution as described in Box 3.8).

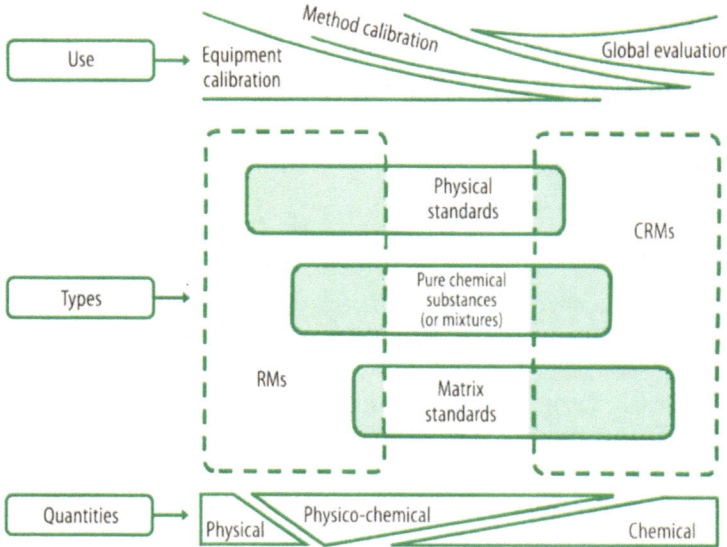

Fig. 3.8. Types of primary reference materials according to a combined criterion based on nature, purpose, reference quantities and endorsing body

Figure 3.8 classifies primary analytical standards according to a combination of criteria such as nature and associated quantity, quality and issuing body. Initially, one can distinguish between two types of primary standards:

(a) **Reference materials** (RMs), which are substances one or more properties of which are sufficiently uniform and well known for use in calibrating an instrument or apparatus, assigning values to materials and systems, or assessing analytical methods.

(b) **Certified reference materials** (CRMs), which are reference materials with certified values (specific uncertainties included) for one or more of their properties. These materials are obtained by special procedures (*e.g.* interlaboratory exercises) under the supervision of a competent, independent organization. The material's accompanying certificate includes detailed documentation of its traceability. The CRMs issued by the US National Institute of Standards and Technology (NIST) are called "standard reference materials" (SRMs).

The previous distinction, endorsed by ISO, does not encompass every possible type of analytical standard. As can be seen in Fig. 3.8, a number of pure and mixed substances can be used as RMs or CRMs. Based on this criterion, their use and the nature of the associated quantities, primary analytical standards can be further divided as follows:

(a) **Physical standards,** which are standards with accurately known physical or physico-chemical properties that are used almost exclusively to calibrate instruments. Transfer weights might also be included here on the grounds of their frequent use in analytical laboratories; however, they rather belong in a physical standard ranking (see Box 3.5). As a rule, physical standards are either RMs issued by manufacturers or, less often, CRMs.

(b) **Pure (and mixed) substances used as standards,** which are chemical materials consisting of one or more substances the identity and (high) purity of which is assured by a manufacturer or, less frequently, a certifying body. The certified properties can be physical (*e.g.* the boiling point of phenol), physico-chemical (*e.g.* the UV absorption spectrum for a holmium filter or the IR absorption spectrum for polystyrene film) or chemical (*e.g.* the redox equivalent of KIO_3). This type of standard can be used to calibrate an instrument, an apparatus or a relative analytical method, to determine an analyte (by titrimetry) or to experimentally standardize secondary standards. The purity of some standards [*e.g.* polycyclic aromatic hydrocarbons (PAHs) and their mixtures, dioxins] is certified by a national or international organization engaged on chemical metrology.

_____ Box 3.9

Calibration of a UV-visible molecular absorption spectrophotometer

A photometer can be calibrated via two parameters, namely: the signal (transmittance or absorbance) it provides and the wavelength it uses, which represent the quantitative and qualitative concepts, respectively, discussed in the text. Also, a photometer can be calibrated for a specific CMP by calibrating its response to a standard of the measurand (*viz.* establishing the absorbance-analyte concentration relation).

A wide variety of available measurement standards exists for the qualitative and quantitative calibration of photometers; such standards can be of two types depending on their state of aggregation (solids and solutions) and purpose (absorbance and wavelength standards).

Absorbance standards are used for the quantitative calibration of spectrophotometers by comparing the absorbance A or molar absorptivity ε [derived from Beer's law $(A = \varepsilon l c)$] experimentally obtained with the standard and its certified (CRM) or stated value (RM).

Specific liquid absorbance standards can be prepared by weighing appropriate amounts of pure or mixed substances and dissolving them in a given volume of a solution of controlled pH. Alternatively, they can be purchased as RMs in sealed vials.

One typical example of solid standard of this type is potassium dichromate, which is available as both an RM and a CRM (NIST); the latter has a certified ε value and comes with comprehensive documentation regarding preparation of its solutions and usage to make absorbance measurements. A solution of cobalt and nickel in 0.1 M $HClO_4$ that is sold in sealed ampoules as a CRM with certified absorbance values at four different wavelengths (302, 395, 512 and 678 nm) is a typical dissolved standard.

Wavelength standards are used to calibrate monochromaticity in order to ensure that the wavelength set and that actually employed by a spectrophotometer will coincide. There are various types of standards for this purpose. Glow discharge lamps emit at highly precise wavelengths (*e.g.* the low-pressure mercury lamp at 253.65 and 184.96 nm); as a result, spectra recorded under their light facilitate comparison of reference and actual wavelengths. Holmium oxide and neodymium-praseodymium glasses, commonly known as holmium and didymium filters, are also used for this purpose.

Reference wavelengths

Holmium filter	Didymium filter
241.5	573.0
279.4	586.0
287.5	685.0
333.7	
360.9	
418.4	
453.2	
536.2	

The holmium filter spans the whole UV-visible spectral region, whereas the didymium filter is used to calibrate wavelengths in the vicinity of the near-infrared region. The problem with these standards is the variability in reference wavelengths among manufactured glass batches. This requires each standard to be supplied with stated reference wavelengths.

(c) **Matrix standards**, also called "sample standards", are artificial, naturally occurring or modified natural materials intended to simulate as closely as possible the actual sample to be subjected to a CMP (e.g. sediment with a certified dioxin content, freeze-dried serum with a certified cholesterol content). The certified quantities are assured by the issuers in "internal reference materials" (IRMs) and also by national (e.g. NIST in the USA) or international organizations (e.g. SMT in Europe). The preparation, storage and use of matrix standards involve special procedures that are described later on. Table 3.1 lists some of the more representative materials of this type, endorsed by the European Union's SMT programme (former BCR). These standards are usually employed for the global assessment of CMPs but can also occasionally be used to calibrate an instrumental response. Their existence and significance to CMPs make two of the most salient differences between chemical and physical metrology.

3.5.2 General Properties

Analytical chemical standards should meet several requirements, some indispensable, some desirable. Of the following nine criteria, primary standards should meet the second to eighth, secondary standards the ninth, and both types of standard the first to fourth.

(1) The standard used in each application should be fit for the purpose, i.e. it should help one fulfill the intended metrological objective. Thus, its purity should be very high and its uncertainty very low if high accuracy and traceability are to be assured. Obviously, an acid-base standard is of no use for standardizing an oxidizing secondary standard with no acid-base properties. An artificial sample spiked with the target analyte can never be

Table 3.1. Selected matrix reference materials supplied by BCR

Field of use	BCR Code	Sample type	Certified analyte contents (mg/kg)	Remarks
Environmental analysis	CRM 141R	Calcareous soil	Cd (14.6 ± 0.5), Co (10.5 ± 0.4), Cr (195 ± 7), Cu (46.4 ± 1.8), Hg (0.25 ± 0.02), Pb (103 ± 3)...	
	CRM 100	Beech leaves	Al (0.435 ± 0.004), Ca (5.30 ± 0.05), Cl (1.49 ± 0.06), K (9.94 ± 0.20), Mg (0.878 ± 0.017), N (26.29 ± 0.25), P (1.55 ± 0.04)...	Used to assess the effects of acid rain
	CRM 463	Fish (tuna)	Total Hg (2.85 ± 0.16), methyl-Hg$^+$ (3.04 ± 0.16)	Sample standard for mercury speciation
Microbiological analysis	CRM 257	Powdered milk	No. aliquots = 34; uncertainty range 29–40 *Enterobacter cloacae*	Used with ISO 9308.1
Food analysis	CRM 466	Rice	Amylose (23.1 ± 1.3 g/100 g)	
	CRM 283	Powdered milk	Aflatoxin M$_1$ (0.09 ± 0.04 µg/kg)	
	CRM 391	Cattle urine	Hexoestrol (13.3 ± 3.1 µg/l)	Freeze-dried urine
	CRM 430	Pork fat	Pesticide contents (µg/kg): HCB (392 ± 26), α-HCH (140 ± 12), β-HCH (259 ± 21), dieldrin (82 ± 0.60), o,o'-DDT (3400 ± 180)...	
Biomedical analysis	CRM 347	Human serum	Progesterone (10.13 ± 0.21 nmol/l)	Freeze-dried serum
	CRM 573	Human serum	Creatinine (68.66 ± 1.10 nmol/l)	

as suitable as a sample standard prepared from real material; in fact, the former may even be quite unfit for the intended purpose (see Box 3.10).

―――― Box 3.10

The "history" behind the preparation of a CRM is as important as are its certified value and associated uncertainty. The form in which the analytes are present in a sample standard is decisive. There are two commercially available sample standards of soil for the determination of traces of polychlorinated biphenyls (PCBs) that are endorsed and supplied by two renown organizations; their certified values are at the same concentration level (in the microgram-per-kilogram range). When aliquots of the two sample standards are subjected to supercritical fluid extraction under different experimental conditions (pressure, temperature, presence or absence of methanol as modifier, *etc.*), the results are rather disparate. Thus, one of the sample standards provides extraction yields of 99% under all experimental conditions tested, whereas the other barely reaches 25–30%. What is the origin of the differences between these two analytical references, deemed excellent by their issuers? The answer is quite simple: the two materials were prepared in a different way. In one, a PCB mixture was spiked to clean soil; matrix-analyte interactions were thus minimal and extraction of the analytes was very easy as a result. In the other, samples were collected from soil under the polluting effects of an industrial area, so the PCBs were strongly bound to the matrix and thus very difficult to remove.

Using one or the other standard as reference will make a significant difference. In fact, only the sample prepared from polluted soil is fit for the purpose. How many analyses and assessments will clearly have been spurious if the sample obtained by spiking clean soil is used as the sample standard?

(2) Analytical measurement standards should be varied enough to meet the wide range of needs posed by CMPs, which are discussed in the next section.

(3) The way a standard is to be stored and used should be carefully described in its accompanying documentation.

(4) Analytical standards should be affordable. Currently available materials of this type are largely fine chemicals and thus of a high added value that ultimately results in an also high price to the end user. In fact, even though the starting material (*e.g.* cement, sea water, plant material) is usually inexpensive, its processing and providing certified values (and their uncertainties) for one or more quantities entails extensive, painstaking work that results in the typically high added value of fine chemicals.

(5) The reference properties of primary standards should be well established and, preferably, certified – particularly the accuracy and uncertainty in the numerical features (quantities) that are to be used as references.

(6) Primary analytical standards should be stable, *i.e.* their properties should not vary with handling or time (*e.g.* through interaction with atmospheric agents). There are two types of stability in this context that relate to the time elapsed from preparation (stated as an expiration date) and from use (*viz.* from the moment the container is opened). Some standards can only be used once.

(7) Primary analytical standards should obviously be homogeneous and their homogeneity assured.

(8) Primary analytical standards should be easy to prepare (condition) for use and also to store.
(9) Secondary analytical standards should provide a clear, unequivocal traceability pathway to primary standards. Once prepared and checked, they should exhibit adequate, accurately known stability for a fixed period of time at the least.

3.5.3 Preparation and Storage

These two operations are crucial with a view to ensuring proper handling and usage of primary analytical standards. Obviously, the more complicated these operations are the more expensive to conduct they will be and the wider will be their sources of variability – and, also, obviously, the higher will be the uncertainty in the reference quantities. The specific preparation and storage procedures to be used vary with the nature of the particular standard.

The **physical standards** used to calibrate instruments are prepared by commercial manufacturers or competent bodies. They require little conditioning prior to use (*e.g.* transfer weights need only be cleaned with ethyl ether to remove any dust or grease accidentally left by the operator). Each standard should be stored as per the exact instructions usually supplied by the manufacturer.

Because **primary standards** are pure or mixed substances, their preparation for use involves procedures of variable complexity dependent on their state of aggregation and the properties that are to be used as references. Gaseous standards are especially complex. At present, they are supplied in permeation tubes impregnated with them; on insertion into specially designed assemblies, the tubes release a given amount of SO_2, NO_x, *etc.*, into a carrier gas that is circulated at a flow-rate carefully adjusted through temperature. Volatile substances used as standards (*e.g.* pure and mixed aliphatic hydrocarbons for calibrating gas chromatographs) are usually available in sealed vials. Solid standards (*e.g.* titrimetric standards such as $CaCO_3$, potassium hydrogen phthalate, Na_2CO_3, KIO_3) usually require drying at a given temperature over a preset time to remove moisture; they are usually stored in weighing bottles that are placed in containers the atmosphere in which is kept dry by means of a desiccating substance (*e.g.* P_2O_5). This type of standard is usually prepared by a manufacturer; its price is proportional to the complexity of the operations involved – the purer the standard, the higher the price. The homogeneity and stability of CRMs are assured by competent bodies and only after thorough testing.

_____ Box 3.11

A result of a CMP conducted by a routine laboratory can be made traceable in various ways.

First, one must decide which will the last link in the comparison chain be. Whenever possible, it should be an SI unit (the kilogram or mole); however, this is usually impossible. Alternatively, one can use a substitute reference such as a CRM or a primary method. The important thing is that this new reference or standard be accurately specified so that results can be reliably compared.

The most thorough approach possible involves realizing traceability via equipment and method calibration, using primary standards allegedly linked to SI units. However, what is usually available is one or more substances of purity or composition certified exclusively by the supplier that must be linked to CRMs or SI units via **transfer standards** in much the same way as in physical metrology.

The use of **primary methods** (gravimetry, titrimetry, coulometry, isotope dilution mass spectrometry), which are at the top of the metrological ranking, to check the results of routine CMPs is one alternative to realizing traceability.

The results obtained by subjecting **certified reference materials** (CRMs) to CMPs provide a means for realizing global traceability in the latter.

Sample (matrix) standards can be prepared by individual institutions (*e.g.* a scientific society, a laboratory group) or by international bodies. Those standards that meet the following stringent requirements are endorsed as CRMs:

(*a*) The matrix should be as similar as possible to that of the real sample. Natural samples are better than partially natural ones (*e.g.* samples spiked with the target analytes), which in turn are preferable to artificial samples.

(*b*) The material should be homogeneous and stable.

(*c*) The reference quantities should all have been determined with a high accuracy and a low uncertainty by different laboratories using various CMPs; also, they should be traceable to base standards.

(*d*) A complete "history" of the procedures used to prepare the standard and measure its certified uncertainties should be included in the accompanying documentation.

Obtaining a standard stock (*e.g.* 1000–3000 samples containing 100 mL or 100 g of a CRM) usually involves the following operations:

(1) Selecting the natural starting material.

(2) Preparing the selected material by freeze-drying, grinding, sieving, blending, *etc.*, depending, among others, on the state of aggregation of the raw material; thus, some types of liquid samples (*e.g.* milk) are freeze-dried while others (*e.g.* sea water) are not.

(3) Preparing an artificial material. Artificial samples are simply spiked with the analyte as uniformly as possible – both on the surface and inside the sample if it is a solid (*e.g.* soil soaked in a pesticide solution in a volatile solvent that is allowed to freely evaporate with time). Liquid artificial samples are easier to prepare.

(4) Checking for sample stability in various containers under different conditions (*e.g.* a variable temperature, presence and absence of light) over a period of a least one year, using the same CMP throughout.

(5) Distributing the material among small containers of the selected type in order to check for homogeneity (both in the bulk material prior to splitting and in its portions) by using statistical criteria to process the data provided by CMPs applied to different aliquots.

(6) Conducting a preliminary campaign to check the suitability of the laboratories that are to take part in the certification of the reference parameters.

Tests can be based on standard solutions of the pure analyte or on matrix solutions (extracts). While not indispensable, this step provides highly valuable preliminary information about the identity of the participating laboratories and the different methodologies (CMPs) they use, and helps establish mutual relations that will no doubt favour collaborative work in subsequent steps of the process. If the results of this campaign differ markedly among laboratories, the exercise is repeated until closer agreement is reached.

(7) Conducing the certification campaign proper. Sample standards stored in appropriate containers are sent to the participating laboratories for subjection to their preferred CMPs. The more different the CMPs used and the more similar the interlaboratory results are, the higher will be the quality of the ensuing CRM. The results of each laboratory are submitted to the supervising body for statistical processing. Subsequently, officials from the participating laboratories meet to reach a consensus on the certified quantities and their uncertainties, which occasionally entails discarding disparate results from one or more of them. Finally, a certified value to be held as the true value, and its uncertainty, $\hat{X}' \pm U_{\hat{X}'}$, are issued.

The need for a competent, independent body to scrutinize, supervise and sanction all these operations is self-obvious – otherwise, the CRM in question will be subject to generic uncertainty. Consequently, not all CRMs are equally reliable.

_____ Box 3.12

Preparation of two simulated natural water sample standards (CRMs) with certified nitrate contents (CRM 479 and CRM 480 from BCR)

Objective: To obtain CRMs for the determination of nitrate at concentrations below and slightly above the maximum tolerated level in the European Union directives for underground and drinking water.

Preliminary studies: Three solutions with an $NaNO_3$ concentration of 1, 10 and 50 µg/L in ultrapure water were supplied with variable amounts of Na_2CO_3 and HCl to obtain a final pH of 6.8, $CaCl_2$ and $MgSO_4$ to simulate the hardness of natural water, solid lauryl sulphate to simulate the presence of humic acids and phenylmercury acetate as sterilizer. The stability of the three solutions was checked by storing them at -20, $+20$ and $+40\,°C$ in containers of different materials (glass, plastic). The stability of the three solutions over a period of 3 months was also examined. A preliminary interlaboratory study showed that the solutions were suitable for certification; two of them were selected to prepare as many CRMs.

Preparation of the CRMs: Two solutions of 150 L each containing different nitrate concentrations and the same ingredients used in the preliminary experiments at their stated levels were prepared in PVC containers. Under continuous stirring, each solution was used to fill 1200 glass vials, which were heat-sealed after loading. All these operations were carried out under strictly clean conditions. By using a highly reproducible FIA method, the samples were shown to be homogeneous and to remain stable for at least one year.

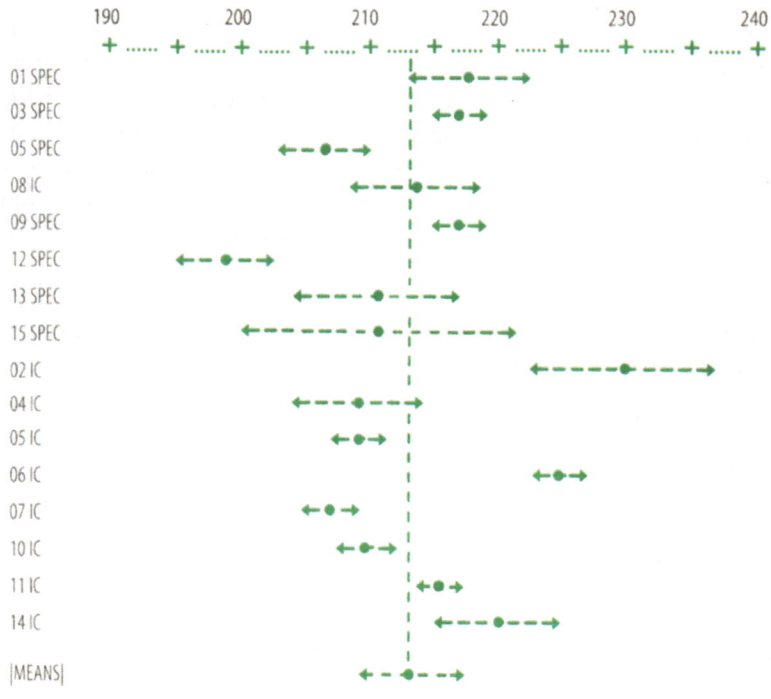

Certifying campaign: Fourteen laboratories from various European Union countries took part in the interlaboratory exercise by using ion chromatography with different types of detectors, as well as continuous automatic methods (FIA, SFA) with spectrophotometric detection. Each laboratory performed 5 – 10 individual analyses of each material, the mean results and their standard deviations being submitted to BCR for processing. The results for CRM 479, in µmol/kg, are shown in the figure above (the uncertainty was calculated at the 95 % confidence level). Each datum and its uncertainty contributed to the mean and its associated uncertainty, which were the ultimately certified data:

CRM 479	214 ± 4 µmol/kg	CRM 480	885 ± 13 µg/kg
	13.3 ± 0.3 mg/kg		54.9 ± 0.8 mg/kg

Availability: The European Union's Institute for Reference Materials and Measurements (IRMM) supplies both CRMs in glass vials containing about 100 mL each. Both are accompanied by a document of 100 pages describing the material's integral traceability (preparation, stability, homogeneity, certification, instructions of use, *etc.*).

3.5.4 Uses

Analytical chemical (measurement) standards are intended primarily to serve as references for the measurements (comparisons) involved in CMPs. Different types of standards have also different applications in CMPs (see Fig. 3.8).

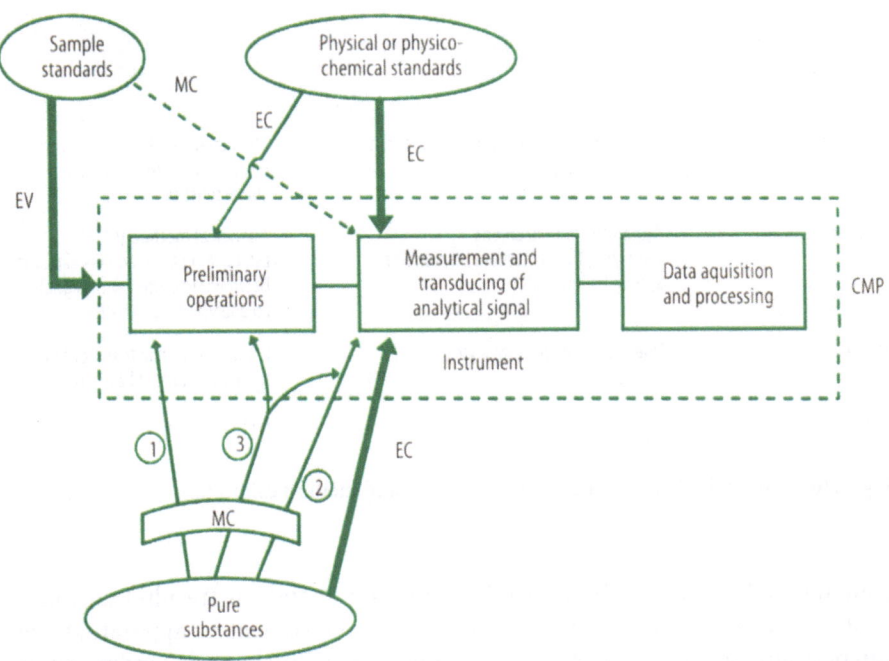

Fig. 3.9. Generic, non-exclusive approach to the use of standards in relation to a chemical measurement process (CMP). EV global evaluation. EC Equipment calibration. MC method calibration (MC1 stoichiometric methods, MC2 relative methods, MC3 internal standardization). For details, see text

Analytical standards are used in both classical and instrumental analysis, and are essential for both identification and quantitation purposes. Their more relevant uses are schematically illustrated in Fig. 3.9 and commented on below.

3.5.4.1 Calibration of Equipment and Methods

Equipment and method calibration are two different, complementary aspects of crucial significance to chemical traceability and yet the two terms are often used indifferently. As can be seen from Fig. 3.10, the two types of calibration differ not only conceptually, but also in their goals, the types of standards they use and the information they produce.

Equipment calibration is the process by which an instrument or apparatus is checked to operate correctly (*i.e.* as expected); as such, it is a verification operation. The purpose is to correct the response of an instrument or the readings of an apparatus until the value held as true for the standard being used (usually an RM or even a CRM with well-known physical or physico-chemical properties) is reached. Calibration is specially relevant to the second step of the analytical process, but also, occasionally, to instruments (balances, pipettes) used in its preliminary operations (see Fig. 3.9). As a rule, a standard used for this purpose

	EQUIPMENT CALIBRATION	METHOD CALIBRATION
Definition	Checking that an instrument or apparatus operates as expected	Characterizing the instrument response in terms of analyte characteristics
Purpose	Apparatus or instrument Correcting the instrument response or apparatus indication	Analytical method (CMP) Establishing an unequivocal relation between measurements (signals) an analyte characteristics
Standard	Does not contain the analyte	Contains the analyte (except in titrimetric standardization)
Information	Analytical other	Analytical

Fig. 3.10. Essential differences between equipment and method calibration

contains no analyte. Calibration relates to both analytical (results) and non-analytical information (the performance of an instrument or apparatus), and can be a support for both qualitative and quantitative analysis, depending on the way the instrumental response is used.

Analytical method calibration involves characterizing the response of an instrument – never an apparatus – in terms of the properties of one or more analytes. It relies on the establishment of an unequivocal relation between instrument signals and analyte concentrations. Consequently, it is essentially quantitative in nature, even though it can occasionally have qualitative connotations (*e.g.* when using Oster relations or Kovats retention indices in gas chromatography). This type of calibration applies to both preliminary operations (*e.g.* standardizing solutions by titration with a secondary standard such as MC-1 in Fig. 3.9) and the second step of the CMP (*e.g.* MC-2 in Fig. 3.9). Depending on the particular analytical technique used in the second step of the analytical process, calibration is approached in different ways. Thus, method calibration makes no sense in gravimetry (an absolute method) as only the balance used need be calibrated. On the other hand, titrimetry (a stoichiometric method) requires calibration of the burette employed and the use of a primary standard (or a standardized secondary one) as titrant. UV-visible molecular photometry requires calibrating the photometer (see Box 3.9) and finding the absorbance-concentration relation for each individual CMP. Finally, gas chromatography calls for calibration of the chromatograph and the establishment of peak area-analyte concentration and retention time-analyte relations for quantitative and qualitative purposes, respectively. Box 3.12 shows selected examples of equipment and method calibration for both purposes.

_____ Box 3.13

Examples of equipment calibration

1. FOR QUALITATIVE PURPOSES

Instrument	Parameter	Standard
UV spectrophotometer	λ_{max}	Holmium and didymium filters
Spectrofluorimeter	$\lambda_{ex}, \lambda_{em}$	Plastic-dispersed fluorophore

2. FOR QUANTITATIVE PURPOSES

Instrument	Parameter	Standard
Balance	Mass	Transfer weights
Burette	Volume	Transfer weights
Spectrofluorimeter	I_F	Quinine solution
Potentiometer	pH	Buffer solutions
Polarograph	$I(V)$	Dummy electronic cell

Examples of method calibration

1. QUALITATIVE ANALYSIS

1.1 Relationship of the signal to analyte properties

Technique	Signal	Analyte
Polarography	$E_{1/2}$	• Metal ions • Organic substances • Chelates
IR spectroscopy	$v_1, v_2, \ldots v_n$	Organic substances
Mass spectrometry	n/z	Organic substances

1.2 In a dynamic instrumental system

Measured signal	Technique
Retention (migration) time	• Column chromatography • Capillary electrophoresis
Travelled distance	• Classical electrophoresis • Planar chromatography

2. QUANTITATIVE ANALYSIS

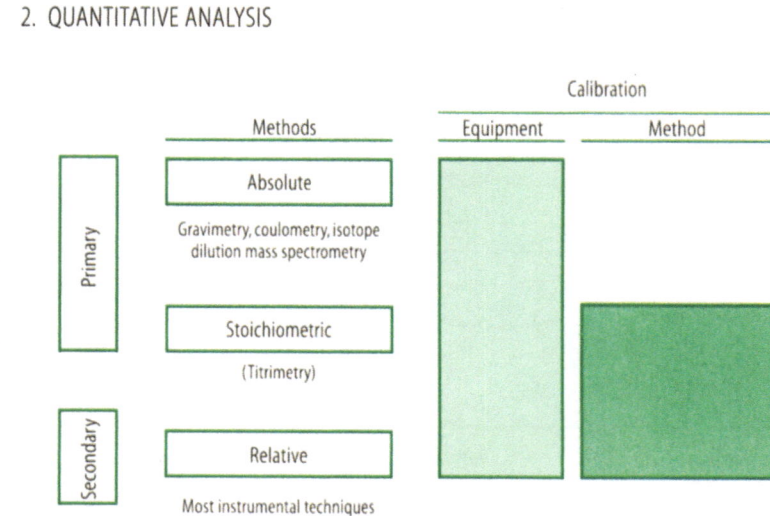

As shown in Fig. 3.11, method calibration applies to methods and differs among CMPs, whereas equipment calibration pertains to instruments (or apparatuses) and can be shared by many CMPs. In practice, the word "calibration" is used indifferently to refer to both chemical metrological goals. This improper usage has been promoted by assimilation from physical metrological processes, where no such distinction exists. Note the equivalence in this respect between PMPs and absolute methods (*e.g.* gravimetry).

One special metrological operation in this context is so-called "internal standardization", intended to avoid fluctuations in instrument readings. Box 3.13 summarizes its features and provides a representative example corresponding to the notation MC-3 in Fig. 3.9.

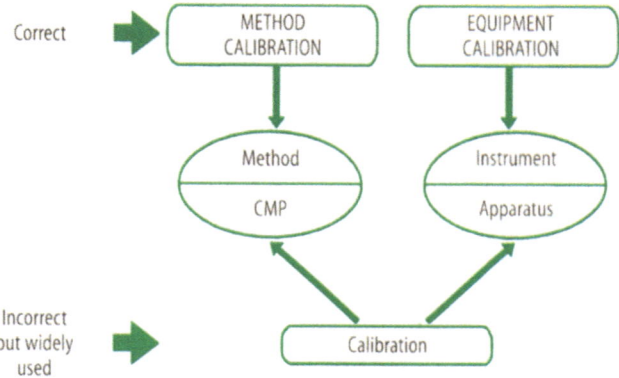

Fig. 3.11. Usage of the word "calibration" in chemical metrology

_____ Box 3.14

Use of an internal standard

Internal standards are related to both equipment and method calibration as they are primarily intended to minimize fluctuations in instrument readings and to assist in the development of qualitative and quantitative references. The internal standard (IS) is always different from that used for conventional calibration and is added to both samples and standards. Measurements (of analytes and standards) are **standardized** with respect to the value provided by the IS.

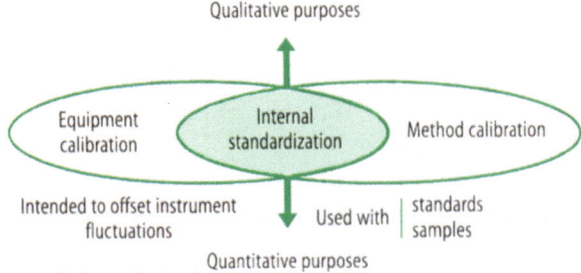

Internal standards should be compatible with the sample (*e.g.* miscible with it); also, they should be chemically and physico-chemically similar to the analytes but depart clearly from them as regards the signals produced.

Example: Use of internal standards in chromatography: (*a*) Three analytes (1, 2 and 3) in the same sample; (*b*) standard solutions containing variable concentrations of each analyte are prepared.

CONVENTIONAL CHROMATOGRAM. A sample aliquot is inserted into the chromatograph after standards of each analyte at different concentrations have been injected.

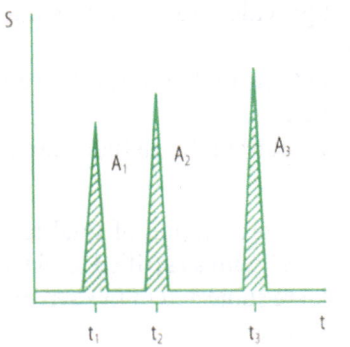

Qualitative: retention times t_1, t_2 and t_3 (identification by coincidence of times)

Quantitative: areas A_1, A_2 and A_3 following individual calibration

CHROMATOGRAM OBTAINED IN THE PRESENCE OF THE INTERNAL STANDARD. An aliquot of a sample containing the internal standard (IS) is inserted into the chromatograph following injection of various standards (also spiked with IS).

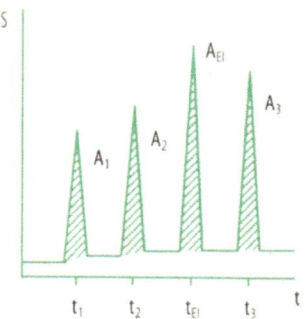

Analyte	Standardized signals	
	Retention times	Areas
1	$t_{N_1} = \dfrac{t_1}{t_{EI}}$	$A_{N_1} = \dfrac{A_1}{A_{EI}}$
2	$t_{N_2} = \dfrac{t_2}{t_{EI}}$	$A_{N_2} = \dfrac{A_2}{A_{EI}}$
3	$t_{N_3} = \dfrac{t_3}{t_{EI}}$	$A_{N_3} = \dfrac{A_3}{A_{EI}}$

3.5.4.2 Evaluation of Analytical Processes

In addition to assuring metrological quality in CMPs via equipment and method calibration, measurement (analytical) standards can be used to assess the CMPs themselves.

The **global evaluation of a CMP** involves using a sample standard – a CRM if available, which is fairly rarely the case (see EV in Fig. 3.9). The purpose is to verify the global analytical quality of a specific CMP in the following situations: (*a*) when an analytical method is implemented for the first time or has remained unused by a laboratory for a long time; (*b*) when a new CMP is developed; or (*c*) when laboratory quality is to be assessed.

The evaluation process comprises the following four steps:

(1) Selecting an appropriate CRM – one similar to the real sample – with certified values (and uncertainties) in the quantities of interest ($\hat{X}' \pm U_{\hat{X}'}$).
(2) Subjecting several aliquots of the CRM to the CMP concerned. An appropriate chemometric treatment will provide a result and its uncertainty ($\bar{X} \pm U_{\bar{X}}$) for each quantity examined.
(3) Comparing the certified values and results (with their respective uncertainties) by using a suitable statistical method.
(4) Assessing the quality (accuracy and precision) of the method studied on the basis of statistical criteria.

Consider, for example, a new CMP for the determination of total fat in milk, which is to be evaluated using a CRM from NIST with a certified total fat content of $(12.31 \pm 0.08)\,\%$. Five aliquots of the milk are subjected to five separate CMPs to obtain five data statistical processing of which yields a content of $(12.30 \pm 0.10)\,\%$. As can be seen, the uncertainty of the laboratory is slightly higher than that of the CRM. Also, the results of both quantities are quite similar – their difference, $\hat{X} - \bar{X}$, is smaller than the uncertainties. Finally, the uncertainty ranges are virtually the same. Based on these quantitative inferences, one can conclude that the result assessed is acceptably accurate and precise. There are, however, more rigorous chemometric methods available for comparing results.

Using a pure substance (an RM or CRM) not subjected to the whole but only part of the CMP, and a single, treated aliquot of the standard sample to obtain several portions for insertion into the measuring instrument, are two frequently made mistakes in evaluating CMPs; the results will obviously be spurious as they will be based on incomplete information.

The quantitative **evaluation of a laboratory** involves examining the results it produces by using suitable RMs or CRMS to implement specific CMPs. Control graphs (internal evaluation) and interlaboratory exercises (*proficiency testing*) are two typical approaches to the problem (see Box 3.15).

_____ Box 3.15

The role of standards in assessing the quality of an analytical laboratory

Standards can be used in three different ways to evaluate the performance of a laboratory depending on the **type of personnel** engaged in the evaluation – in any case, the analytical process will always be conducted by laboratory staff.

Internal evaluation: This is carried out by laboratory personnel.

Example: Control graphs, which can be constructed by using a CRM (expensive) or an internal reference material (IRM) prepared by the laboratory itself. The CRM (or IRM) is subjected to the whole analytical sequence, under reproducible conditions, and the statistical parameters that allow calculation of

$$\bar{X} = t \frac{s}{\sqrt{n}}$$

are obtained. Next, the control graph is constructed between two limits:

1) WARNING $\quad \pm 2 \frac{s}{\sqrt{n}}$ 2) ACTION $\quad \pm 3 \frac{s}{\sqrt{n}}$

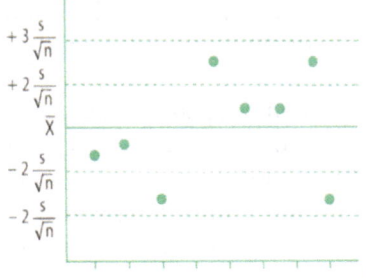

Batches (Days)

The same sample (CRM or IRM) in each batch is subjected to the CMP each day and the result obtained is plotted on the graph.

Mixed (internal and external) evaluation. This is carried out by the Quality Management Department of the parent body, which submits samples (blind or otherwise) to the laboratory for delivery of specific results.

Example: CRMs, IRMs, previously analysed samples.

External evaluation. This is entrusted to personnel belonging to neither the laboratory nor its parent body. There are two possible alternatives:

(a) Use of a CRM supplied by an external institution proficient in laboratory accreditation processes.
(b) Use of a CRM or IRM for multiple analyses by various laboratories, as in inter-laboratory studies aimed at testing the proficiency of the participants by comparison with one another (and with the certified value if the sample is a CRM).

3.6 Specific Meanings of Traceability

Although, metrologically, traceability is almost invariably used to refer to a property of a result, there is no plausible reason to exclude application to other analytical entities. In the following sections, the traceability concept is extended to various analytical concepts in accordance with ISO's recommendations in defining traceability in relation to quality management and quality assurance.

3.6.1 Traceability of an Analytical Result

In 1993, ISO defined traceability as a "property of the result of a measurement or the **value of a** *standard* whereby it can be related to stated references, usually national or international standards, through an unbroken chain of comparisons *all* **having stated uncertainties**". The boldfaced phrases in this definition are additions introduced after the original ISO definition of 1984 was issued, additions that extend application to standards – as done in this chapter – and underscore the significance of the uncertainty arising in the comparisons, which must unavoidably accompany every analytical result, whether partial or global.

The official definition of traceability of a result only considers one of its aspects (*viz.* its relationship to standards). Although the other two, "history" and "comparability" (Fig. 3.12), are implicitly included, they are highly relevant and warrant some emphasis.

The "stated standards" in the definition are primary analytical standards that should be related to chemical and base standards through the unbroken chain of comparisons, where equipment and method calibration play crucial roles. Reference materials (RMs) do not ensure traceability, so certified reference materials (CRMs) must be used instead. If, for example, the response of an atomic absorption spectrometer is calibrated with a commercially available $Pb(NO_3)_2$ solution, then an extrapolated result will have dubious traceability since the comparison chain may have broken at some link. In fact, one should ensure that the balance has been calibrated and that the instrument response has been calibrated with a CRM (or an RM previously checked against a CRM). The result of a CMP will have the stated traceability provided the process has been applied to a certified sample standard (a CRM).

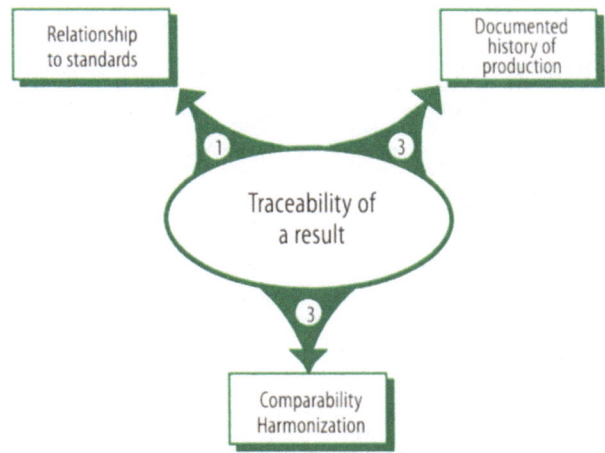

Fig. 3.12. Connotations of the traceability concept as applied to an analytical result

Based on the theory of error propagation, the uncertainty in each comparison (measurement) accumulates in the uncertainty of the result. Each standard should be supplied with a reference value and its uncertainty. Also, each measuring operation (*e.g.* weighing of the sample, measuring the volume of treated sample, recording the signal for a sample or standard) is subject to some uncertainty. Consequently, the shorter the traceability chain is, the lower will be the uncertainty of the result – with obvious exceptions. In principle, an absolute method will provide less uncertain results than will a relative method.

The "strength" of each link (comparison) in the chain depends on the quality of the measurement operations and that of the standards used. It should be assured by means of internal quality controls and external evaluations (audits) of the laboratory concerned.

One other implicit component of traceability of a result (Fig. 3.12) is the comprehensive, documented "history" of its generation (production). As shown in Fig. 3.13, the history should include every element involved in the CMP and answer the following questions: who carried out the CMP? What materials or instruments were used? How and under which conditions was the CMP conducted? When were the operations involved performed? The history should comprise not only the sample measurement process but also any calibration operations.

Realizing the traceability of a result to base standards via solid, reliable comparisons (path A in Fig. 3.14) involves (*a*) assuring traceability among the standards in the unbroken measurement chain used by each laboratory (path B in Fig. 3.14) and (*b*) assuring consistent, reliable relationships among the results produced by various laboratories (path C in Fig. 3.14). This latter condition has a major practical implication as it ensures comparability among the results supplied by different laboratories, which facilitates harmonization and mutual acceptance of their proficiency (a crucial ingredient of currently booming international social and trade relations).

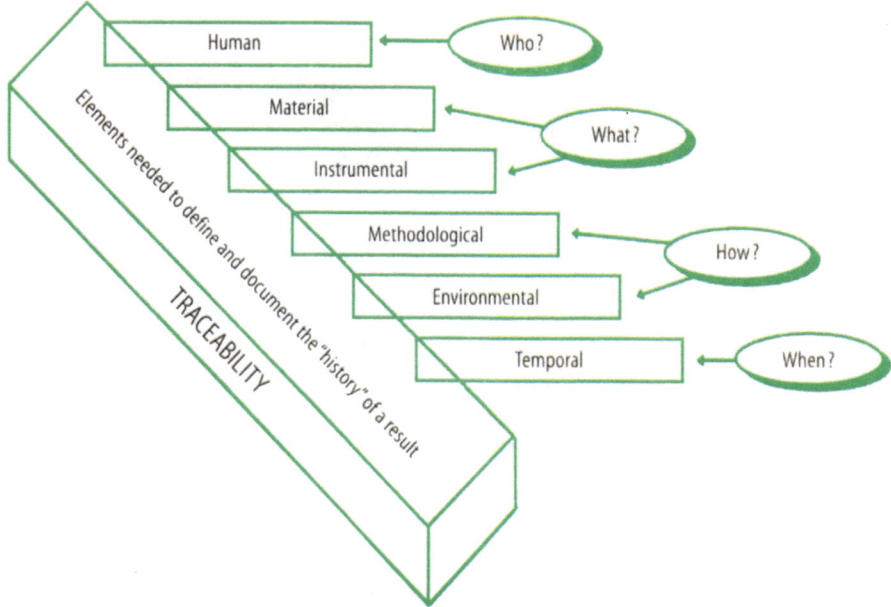

Fig. 3.13. Tracking facet of traceability of an analytical result

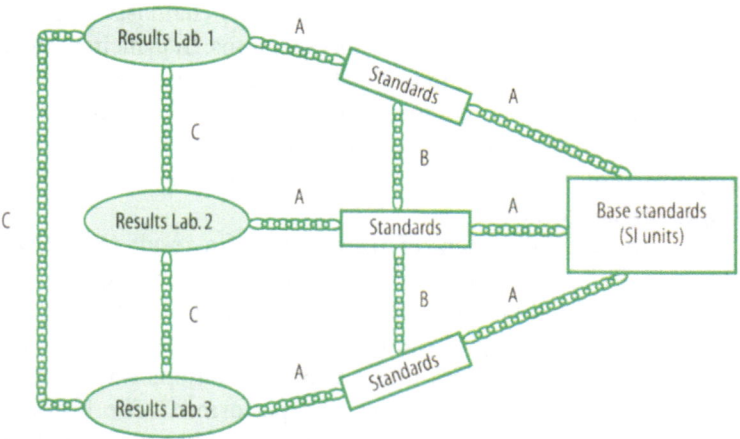

Fig. 3.14. Practical implications of traceability: comparability among three laboratories (C), which rests on traceability of their results to primary standards of mutually related traceability (B) arising from their individual relationships (A) to base standards

3.6.2 Traceability of a Sample (Aliquot)

The sample aliquots that are ultimately subjected to the analytical process exhibit two intertwined traceability aspects that materialize in an unbroken, cyclic chain of relations (see Fig. 3.15).

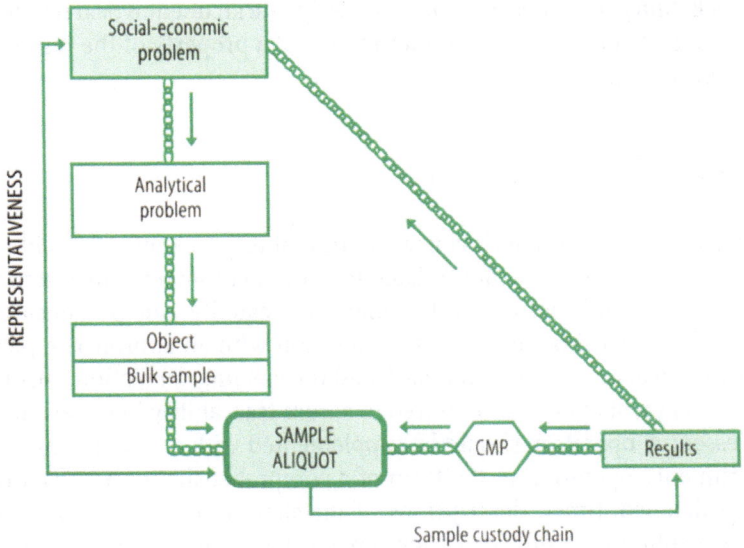

Fig. 3.15. Traceability of a sample aliquot subjected to an analytical process (a CMP): relationship to the information required and the results supplied. For details, see text

Thus, the samples should be representative of the economic or social problem addressed (*e.g.* pollution in a given place). This translates into an analytical problem, which, among others, requires defining the object (*viz.* the polluted location, in space and time). The object comprises a set of "bulk samples" collected at appropriate points and times from which aliquots (samples) are set aside for subjection to the analytical process. The difficulty of assuring representativeness decreases in this sequence. However accurate, a scarcely representative result will always be a poor result.

The samples subjected to the analytical process and the results they provide must be unequivocally linked, particularly in automated determinations of large numbers of samples. The so-called "sample custody chain" is an essential element of the Quality Manual of every analytical control laboratory. In this way, representativeness of the results, and hence their consistency with the economic or social problem addressed, are assured and the sample traceability circle is closed.

This cyclic approach to traceability, which is essential for Analytical Chemistry to fulfil one of its primary goals, has the severe constraint that, unlike traceability of a result, it cannot rest on the use of stated standards via a comparison chain. In fact, the sample traceability chain consists of weaker links (relations) than does the result traceability chain. On the one hand, representativeness can be assured by using chemometric methods and links to other social and technical areas. On the other, the strength of the sample custody chain can be assured by using modern identification (*e.g.* bar coding) and tracking procedures (*e.g.* Laboratory Information Management Systems, LIMS).

Although this approach to traceability relies on two properties of the sample aliquot, its greatest practical implication is representativeness of the result. The word "trackability" has lately been used to designate an unequivocal relationship between a result and a sample aliquot in terms of a property of the result rather than one of the sample.

3.6.3 Traceability of an Instrument

As shown in Box 3.2, the generic concept of traceability encompasses calibration, whereby measuring equipment is related to national or international standards. Consequently, in the metrological domain, traceability can be considered a property of a measuring instrument. Calibration with an appropriate physical, physico-chemical or chemical standard, and the ensuing correction, ensure that the instrument will operate as required to assure traceability in its results.

This essential operation should be supplemented with a detailed "history" of the instrument's operation in the form of a comprehensive, customized record including detailed, timed descriptions of installation, malfunctioning, repair, servicing, calibration and correction operations, hours of use, number of samples processed, *etc.*

These activities are considered and organized in the quality systems of analytical laboratories. They constitute an essential chapter in the Quality Manual mentioned in the standards of the EN-45000 family and are the subjects of Standard Operating Procedures (SOPs) in Good Laboratory Practices (GLPs). The traceability of each instrument and apparatus should be visually and documentarily inspected in pertinent audits.

Any apparatus (oven, extractor, reactor) should also be the subject of continual inspection in order to ensure proper functioning as scheduled. Thus, a refrigerator set to maintain a temperature of 4 °C should be periodically checked to ensure that it keeps the desired temperature. Malfunctioning in an apparatus used in a CMP can lead to alterations affecting the traceability of the final results.

3.6.4 Traceability of a Sample Standard

The more relevant properties of certified reference materials (CRMs) can also be traceable. Thus, the certified value and its uncertainty should be traced to base standards through the results produced by the laboratories participating in the certifying campaign and through the standards used for calibration in each case. It is essential for the user to know why given results, if any, were discarded, what the most significant sources of error (*e.g.* sample treatment) were, which methods provided the highest accuracy and precision, *etc.* Also, the reference material should be documented with technical details about its preparation, homogeneity and stability. For example, the user should be able to know whether the analyte was added (spiked) to the sample or initially present in the bulk material since some techniques (*e.g.* supercritical fluid extraction for PCBs in sediment) can lead to rather disparate results in these two situations. Assuring traceability in a CRM entails documenting it in all these respects and also providing basic information about its certified value. Only if such comprehensive documentation is made available can a standard be properly used. In fact, a standard with no traceability is like a building without foundations. This should never be ignored, particularly by RM and CRM issuing bodies, but also by their clients (analysts), who should demand such information as a matter of course.

3.6.5 Traceability of a Method

An analytical method is the description of a chemical measurement process, the product of which is a result. Obviously, if a result is traceable to base standards, the CMP that produces it can also be traceable and hence a quality process.

Until fairly recently, the use of the words "official", "standard" and "reference" to qualify recommended methods was a source of widespread confusion regarding the quality of their results. At present, analytical methods are evaluated in every single technical respect in interlaboratory exercises; if the final results are acceptable, then the method in question is technically validated. However, the mere use of a good method does not ensure excellence, which must be demonstrated.

As a result, the quality methods of today are not official or systematically revised methods reported by renown institutions, but methods that can be traced to base standards. The traceability of a method should be assured in experimental and documental terms, not only on issuance, but also when first implemented by a laboratory. Traceability is unequivocally realized when the results and associated uncertainties obtained by subjecting aliquots of a CRM to the CMP concerned are statistically consistent with the values of the certified quantities (and their uncertainties). A less unequivocal realization can be accomplished by comparing the results of the potentially traceable method with those provided by aliquots of the same sample subjected to a primary or absolute method (*e. g.* gravimetry, coulometry, isotope dilution mass spectrometry).

3.6.6 Relationships among Traceability Meanings

The primary metrological meaning of traceability is that which relates it to a result. As can be seen from Fig. 3.16, traceability of a result rests on other, specific traceability notions.

The definition of traceability of a result places special emphasis on its relationship to standards. The linear or branched chains leading to base standards (the kilogram, the mole) implicitly include the mutual traceability of the standards (Fig. 3.7), without which no result can be traceable.

For a result to be traceable, the equipment (instruments and apparatuses) used in the CMP whereby it is obtained must also unavoidably be made traceable through appropriate standards. One should bear in mind that equipment and method calibration require different types of standards. Similarly, the traceability of the analytical method that produces the result rests on a CRM.

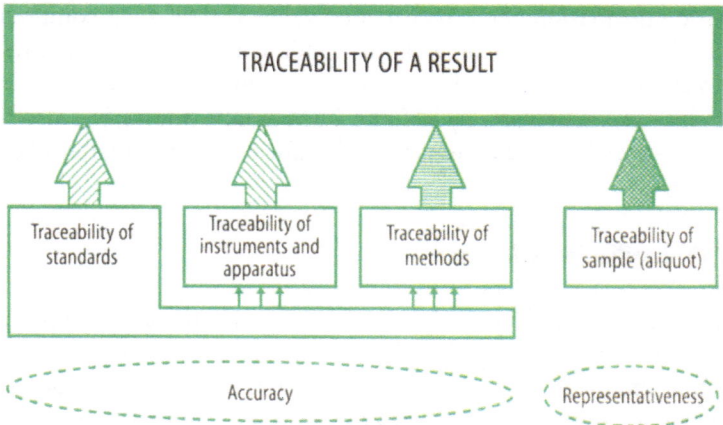

Fig. 3.16. Secondary meanings of traceability in chemical metrology, on which the primary concept "traceability of a result" rests, and relationships to capital analytical properties

The traceability of a sample aliquot bears no direct relation to the traditional concept of traceability in the results; however, it provides solid technical support for the traceability of a result inasmuch as it relates it to representativeness, a capital analytical property as crucial as accuracy to the characterization of the result. This is consistent with the comprehensive, integral approach to traceability presented in this chapter.

Questions

(1) Name the translations of the English word "traceability" in other languages and discuss their meaning.

(2) Describe the traceability links between the following pairs of concepts:
   (a) Atomic weights and the kilogram prototype.
   (b) The kilogram prototype and the mole.
   (c) Secondary standards and atomic weights.

(3) Give several examples to illustrate the fact that traceability is applicable in many situations of everyday life.

(4) State the basic meanings of traceability.

(5) What analytical chemical notions are related to traceability?

(6) Comment on the role of standards in Analytical Chemistry.

(7) Give some examples of tangible and intangible standards in the analytical chemical domain.

(8) Which SI units are the most relevant to Analytical Chemistry?

(9) In what types of operation are analytical chemical standards involved?

(10) Comment on the different connotations of the word "standard".

(11) What type of standard is ultrapure silver?

(12) What is the difference between an RM and a CRM?

(13) What general advantages do physical measurement processes have over chemical measurement processes?

(14) What is the traceability network?

(15) Give three examples to establish three traceability chains to as many working standards.

(16) Distinguish qualitative and quantitative calibration of an analytical instrument.

(17) Discern between equipment calibration and method calibration as applied to the same instrument.

(18) Give several examples of calibration of apparatus typically used in chemical measurement processes.

(19) Why are homogeneity and stability two important properties for standards?

(20) What specific properties should primary standards possess?

(21) What properties should a certified sample standard have?

(23) How is a burette related to transfer weights?

(24) What is internal standardization?

(25) Why does an absolute (primary) method require no method calibration?

(26) Why are metrological concepts applicable to Analytical Chemistry?

(27) Give some examples of equipment calibration in optical and electroanalytical techniques.

(28) Can the certified value for a CRM be traceable to SI units? Is the purity of a "pure" substance certified by its supplier? Why?

(29) Give some examples of method calibration in dynamic instrumental techniques.

(30) If method calibration in a CMP is performed with a pure substance (analyte), will the result be traceable?

(31) An instrument provides a response that depends critically on its operational variables. Through which operation can consistency between calibration results and sample results be assured?

(32) What type of CMP resembles PMPs most closely? Why?

(33) Are the chains of comparisons involved in realizing traceability always of the linear type? Can they be branched? Cyclic?

(34) Give some examples of method calibration.

(35) What is the sample custody chain?

(36) Comment on the different aspects of traceability of (a) a result, (b) a CRM, (c) an instrument and (d) a CMP.

(37) Explain the different connotations of the term "chain of comparisons" in the definition of traceability of a result.

(38) Which meaning of traceability is not based on standards?

(39) Relate traceability to capital analytical properties.

(40) How can the different aspects of traceability in the analytical chemical domain be ranked?

(41) Define the word "trackability".

(42) How can a CMP be evaluated in global terms?

(43) Why is uncertainty relevant in establishing the traceability of a result?

(44) Comment on the special connotations of traceability of a CMP.

(45) What role do standards play in control graphs?

(46) Explain how the results of different laboratories can be made comparable through traceability.

(47) Give several examples to demonstrate that traceability is an incomplete property if it is not accompanied by its "history".

(48) Is traceability a new approach to analytical quality? Why?

(49) Explain the paths a routine laboratory can follow to realize traceability in the results it produces.

(50) What is the most relevant aspect of traceability of an instrument or apparatus?

Seminar 3.1

Traceability in Calculable Analytical Methods

- **Learning objective**
  To find ways to realize traceability in CMPs based on calculable methods, with and without the use of analytical chemical standards.
- **Specific examples**
  - Gravimetric determination of the iron content in a mineral by precipitation as a basic salt and weighing as ferric oxide.
  - Determination of an oxidant in a food sample by addition of iodide ion and titration of the iodine formed with a thiosulphate solution (a secondary standard) previously standardized with KIO_3 (a primary standard).
- **Literature**
  Examples in suggested reading no. 1.
- **Traceability to what?**
  Based on the metrological features of the methods, the last links in the chains of comparisons should be SI units (the kilogram or mole).
- **Traceability paths**
  Explain the different available alternatives: equipment calibration, use of CRMs and comparison with other primary methods.
- **Equipment calibration**
  Check that the equipment used to make the measurements (mass, volume, *etc.*) operates as expected. Note that no method calibration is necessary.
- **Standards involved**
  What types of standards are required in each case? How are they linked to SI units?
- **Traceability chart**
  Draw a scheme showing the chains of comparisons as far as the SI units in both cases.
- **State traceability**
  Write a report describing how traceability to each CMP was realized.

Seminar 3.2

Traceability in Relative Analytical Methods

- **Learning objective**
  To find ways to realize traceability in a CMP by establishing the signal-concentration relation (calibration curve).
- **Specific example**
  Determination of cadmium traces (ng/g) in sea water. The CMP of choice preconcentrates the metal traces on a chelating resin, elutes the metal with HNO_3 and quantifies Cd^{2+} by atomic absorption spectroscopy. Find an

official or standard method for this purpose in the literature and draw its flow chart.

■ **Literature**
Example in suggested reading no. 3.

■ **Traceability to what?**
Choose the last link in the chain of comparisons, which will be used as reference. Preferably, it should be a base standard (an SI unit such as the kilogram or mole). When the traceability chain cannot be extended as far as such units, a standard with a value more distant from that held as true can be used instead.

■ **Traceability paths**
Explain the different available alternatives: use of primary standards, use of sample standards, comparison with primary methods.

■ **Equipment calibration**
Check that the equipment used to make the measurements (mass, volume, atomic absorption, *etc.*) operates as expected.

■ **Method calibration**
Find the relation between the cadmium concentration and the instrument's absorbance reading. Determine derived analytical parameters (linear range, limits of detection and quantitation).

■ **Use of a CRM**
Are there any sea water matrix standards with certified cadmium contents? How can they be used to realize traceability in the results of the CMP?

■ **Standards involved**
What type of standard is required for each type of calibration? How are the standards linked to SI units?

■ **Traceability chart**
Draw a scheme showing the chains of comparisons whereby traceability of the CMP can be realized. Include the different types of standards involved (secondary, primary, chemical, base).

■ **State traceability**
Write a report describing how traceability to the CMP was realized.

Seminar 3.3

Sample Standards: Preparation, Certification and Usage

■ **Learning objective**
To demonstrate, via several examples, the different ways in which a matrix CRM can be prepared, its homogeneity and stability assured, the certifying campaign conducted, the standard supplied and used, and its documentation completed.

■ **Sources**
  • CRM catalogues from different suppliers (NIST, BCR).

- Documents accompanying at least three different certified sample standards (*e.g.* three of the CRMs in Table 3.1).
- **Procedure**
Critically discuss the different steps involved in the process for each material: (*a*) identifying the actual needs; (*b*) choosing the material; (*c*) preparing it; (*d*) examining its stability (formulation and storage); (*e*) studying its homogeneity following splitting into bottles or vials; (*f*) conducting a certifying campaign for one or several analytes (if applicable, running a preliminary interlaboratory exercise using analyte solutions or extracts); (*g*) deciding whether the chosen material warrants certification; and (*h*) delivering the material and its accompanying documentation.
- **Concluding discussion**
  - What critical factors dictate the quality of a CRM?
  - Similarities and differences among the CRMs considered in this seminar.
  - General problems encountered in using CRMs in chemical metrology.

Suggested Readings

About Traceability in Analytical Chemistry

(1) "Traceability in Analytical Chemistry", M. Valcárcel and A. Ríos, *Analyst*, 1995, 120, 2291.
Provides an integral view of traceability consistent with the approach discussed in this chapter.

(2) "Sense and Traceability", M. Thompson, *Analyst*, 1996, 121, 285.
Presents an approach of highly practical interest.

(3) "Traceability of Chemical Analysis", B. King, *Analyst*, 1997, 122, 197.
Summarizes the basic, official approach to the traceability concept.

(4) "Mass Traceability for Analytical Measurements", M.W. Hinds and G. Chapman, *Anal. Chem.*, 1996, 68, 35A.
Discusses traceability in relation to mass measurements in chemical analysis.

(5) "An Extended Approach to Traceability in Chemical Analysis", M. Valcárcel and A. Ríos, *Fresenius' J. Anal. Chem.*, 1997, 359, 473.
Examines traceability of instruments, aliquots, methods and samples, and provides an integral approach to the concept.

About Reference Materials

(6) "Glassware and Primary Standards", A. Ríos, *Encyclopedia of Analytical Science*, Academic Press, New York, 1996, p. 439.
Summarizes the features of the main types of primary and secondary standards used in chemical metrology.

(7) "Impact of Reference Materials on the Quality of Chemical Measurements", Ph. Quevauviller, *Analusis*, 1993, 21, M47.
Discusses the significance of the quality of standards and their proper use for chemical measurements.

(8) "Quality System Requirements for the Production of Reference Materials", R. F. Walker, *Trends Anal. Chem.*, 1997, 16, 9.
Summarizes the primary considerations to be taken into account in producing reference materials for use in chemical metrology.

Standards and Guides

(9) "International Vocabulary of Basic and General Terms in Metrology", 2nd edn, International Standards Organization, Geneva, 1993.
(10) "Quality Management and Quality Assurance–Vocabulary", ISO 8402, International Standards Organization, Geneva, 1994.
(11) "Terms and Definitions Used in Connection with Reference Materials", ISO 30, International Standards Organization, Geneva, 1992.
(12) "Uses of Certified Reference Materials", ISO 33, International Standards Organization, Geneva, 1989.

4 The Measurement Process in Chemistry

Objectives

- To introduce the general procedural landmarks required for the reliable measurement of chemical parameters and delivery of results.
- To define the general features of analytical processes conducted inside and outside the laboratory.
- To provide a global description of the preliminary operations of the analytical process and underscore their significance.
- To provide a global description of measurement and transducing of the analytical signal, and of acquisition and processing of analytical data.
- To illustrate the process by which analytical results are obtained via suitable examples.

Table of Contents

4.1 Definition of Chemical Measurement Process

The chemical analysis of a sample invariably raises a series of questions such as those of Fig. 4.1. First, one must know "what" is to be analysed, which entails identifying the sample and its analytes. The human component of the analysis, *i.e.* "who" is to perform it, is also very important; in fact, the agent of the analysis is a crucial element in management-analyst-operator and operator-automated system relationships, among others. "How" the analysis in question is to be conducted materializes in a chemical measurement process, which is the subject matter of this chapter.

Each chemical analysis has a purpose that is normally stated as an analytical problem ("why" the analysis is needed). The answer to this question is dictated by that to "what" and to those related to the system under study ("where" and "when"); also, obviously, it dictates "how" the analysis is to be carried out. Such a straightforward scheme (Fig. 4.1) exposes the relationships among the answers to the previous questions and, specifically, the fact that all influence the chemical measurement process, as shown in greater detail in Chap. 7, which is devoted to the analytical problem.

A chemical measurement process (CMP) is defined as *a set of operations that separates the uncollected, unmeasured, untreated sample from the results it provides, expressed in a manner consistent with the analytical problem addressed* (Fig. 4.2). Based on this definition, a CMP is a sort of "analytical black box", a term coined by Chalmers and Bailescu. As discussed in Sect. 3.3, the essential difference between a CMP and a physical measurement process (PMP) is that the former involves not only measurement (and calibration) and data processing, but also other, equally important, operations.

As shown by the Analytical Chemistry ranking discussed in Sect. 1.6, a chemical measurement process is the materialization of an analytical technique

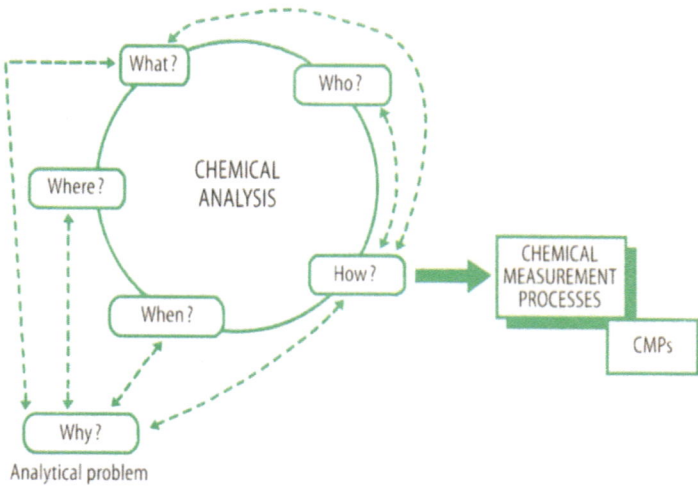

Fig. 4.1. Questions that arise in addressing a chemical analysis. For details, see text

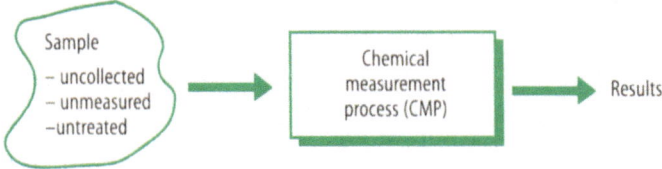

Fig. 4.2. Schematic definition of chemical measurement process (CMP)

(through an instrument) in one or more specific analyses of a given sample. A CMP in turn materializes in a *method*, which has a *procedure* as its most detailed expression.

_____ Box 4.1

The body of operations involved in obtaining a result in the analytical chemical field can be referred to, among others, as "analytical process", "analytical chemical process", "chemical process of measurement" and "chemical measurement process". Which term should be used? Based on the universal scope of the word "analysis", the second designation is obviously more suitable than the first. The other two terms differ in the way the qualifier "chemical" is applied (to the process or its measurement). In this book, we shall use the fourth, *i.e.* "chemical measurement process" (CMP) to underscore the chemical (or biochemical) nature of the measured parameters as the process frequently involves measurements made with physical tools as well.

Chemical measurement processes are implemented by means of the **analytical tools** described in Chap.1. These can be physical (instruments, apparatus), chemical (reagents, solvents, standards), mathematical (*e.g.* chemometric treatments of signals), biochemical (immunoassay reagents, immobilized enzymes) or biological (*e.g.* animal or plant material for constructing sensors) in nature. The operations through which traceability in measurements is realized (*e.g.* **calibration**, of equipment and methods) play crucial roles in CMPs.

Chemical measurement processes vary widely in scope and complexity depending on a number of factors that can be classified into the different categories of Fig. 4.3. Thus, a CMP depends strongly on the particular **analytical problem** addressed, the type of analysis (qualitative, quantitative, structural, surface, global) to be performed and the quality compromises to be adopted (*e.g.* to favour expeditiousness at the expense of accuracy or *vice versa*). The design of a CMP is also strongly dependent on the characteristics of the target **sample** (*viz.* its availability, state of aggregation, stability, *etc.*) and on those of the **analytes** it contains (*e.g.* number, nature, concentration). The choice of CMP is also dictated by the **tools** at hand; for example, the availability of a gas chromatograph decreases the number and complexity of preliminary operations needed relative to other technical alternatives. Finally, the use of **measurement methods** of the primary (absolute and stoichiometric) or relative type (based on a calibration curve) influences the complexity of the calibration operations involved. Box 4.2 gives some representative examples in this context.

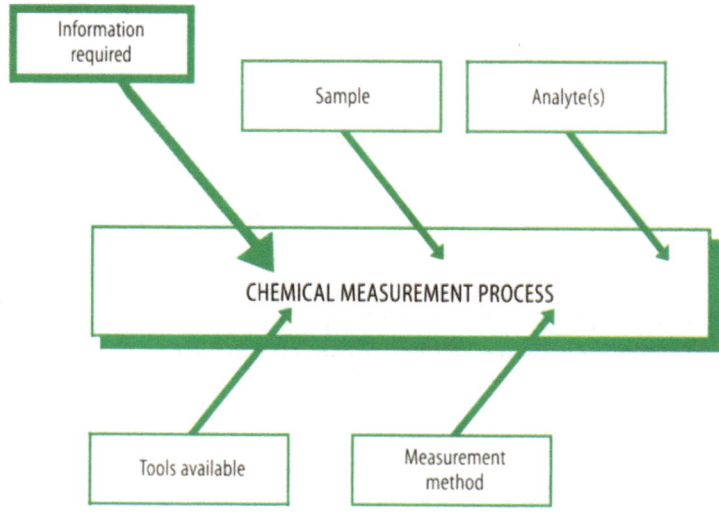

Fig. 4.3. Factors that determine the number of steps in a CMP and their difficulty

_____ Box 4.2

Chemical measurement processes (CMPs) depend critically on the factors discussed in the text and illustrated in Fig. 4.3, which are realized here with the following selected examples:

- The CMP of choice to determine arsenic traces in a given type of sample (water, soil, a biological fluid) will vary markedly depending on whether the total analyte content or discriminate information (*e.g.* the presence and/or concentration of inorganic or organometallic species of varying toxicity) is required.
- Determining polycyclic aromatic hydrocarbons (PAHs) will call for rather different CMPs depending on whether the sample that contains them is underground water, industrial waste water, atmospheric gas, river or marine sediment, oil crude, *etc.*
- A laboratory possessing laser ablation equipment incorporated into an atomic spectrometer for the determination of metals in solid samples will be able to choose a much more simple CMP (one that will afford direct insertion of aliquots of the solid sample) than those involving sample treatments such as dissolution, disaggregation, leaching, *etc.*.
- The moisture content (% H_2O) in animal feed or a solid fertilizer can be rapidly determined from direct near-infrared (NIR) spectroscopic measurements; these provide acceptable, expeditious results for untreated samples. In fact, the results may be subject to errors of 5–10%, which are acceptable for this type of problem. However, if high accuracy and precision are required, one will have to resort to a slower, more complex CMP such as the Karl-Fisher method, which involves the iodometric titration of the sample moisture.

Based on the concepts dealt with in Chap. 2, the analytical process can be related to two types of properties, namely:

(*a*) **Basic analytical properties** (precision, sensitivity, selectivity and proper sampling), on which capital analytical properties (accuracy and representativeness) rest.

(*b*) **Accessory analytical properties** (expeditiousness, cost-effectiveness and personnel-related factors), which determine laboratory productivity.

Because the analytical problem (Chap. 7) inevitably involves quality compromises, one should always bear in mind complementary and contradictory relationships among analytical properties in adopting a specific CMP.

4.2 General Steps of a Chemical Measurement Process

A chemical measurement process can be divided into the three generic steps of Fig. 4.4.

Preliminary operations, the first step in the CMP, are virtually absent from physical measurements. Their primary purpose is to make the sample ready for measurement proper. **Measurement and transducing of the analytical signal,** the second step, is directly related to the nature and concentrations of the analytes in the sample, and involves the use of an instrument (an hence of an analytical technique). **Data acquisition and processing,** the third, final step, provides the results of the process in the required format.

This generic CMP scheme is merely a simplistic approach and, as such, subject to frequent exceptions. Thus, the initial mass or volume measurement in a CMP involves the use of an instrument (a balance, pipette) in the preliminary operations of the process. These steps are frequently integrated in whole or in part, so their interfaces are rather diffuse (see Box 4.3).

_____ Box 4.3

The three steps of a CMP, schematized in Fig. 4.4, are frequently integrated and preclude distinction. Thus, **sensors** (*e. g.* pH electrodes) integrate the first two steps into a single one. In classical **qualitative analysis,** the last two steps are integrated by the human body (its senses

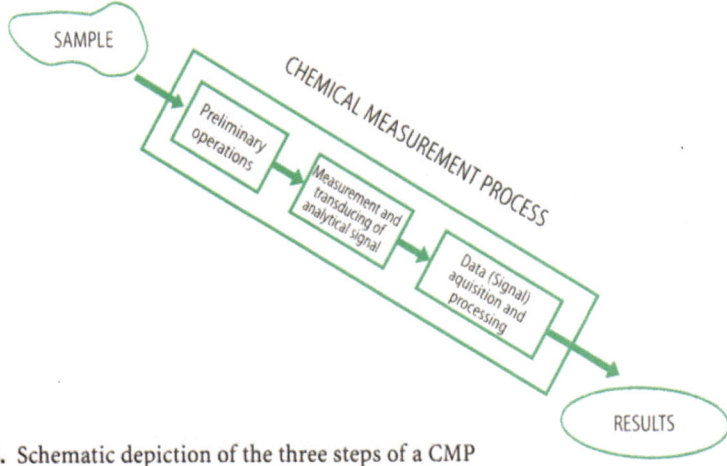

Fig. 4.4. Schematic depiction of the three steps of a CMP

and brain) in much the same way as in instruments equipped with a microprocessor and printer. A **gas (or liquid) chromatograph** performs a key preliminary operation. Thus, it separates the analytes from one another and from their interferences; also, the column output is coupled to an on-line detector. Therefore, the chromatograph carries out part of the first step in addition to the second. An **analyser** performs, theoretically in a automatic manner, the three steps of the analytical process; in fact, some sub-step of the preliminary operations is usually performed by hand (*e.g.* in clinical chemical autoanalysers, where samples are collected, stored and transferred to the analyser's autosampler in a manual manner).

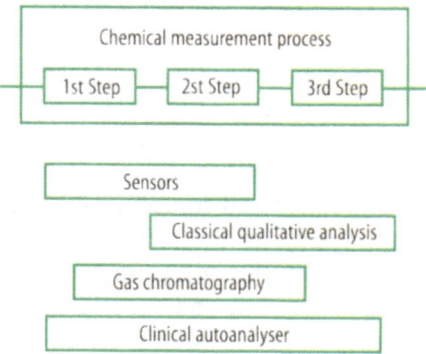

Metrological operations are also integral parts of a CMP; in fact, they constitute an essential landmark inasmuch as realizing traceability in the results entails checking both the performance of the instruments and apparatus used, and the signals they produce. Figure 4.5, which is complementary of Fig. 3.9, illustrates

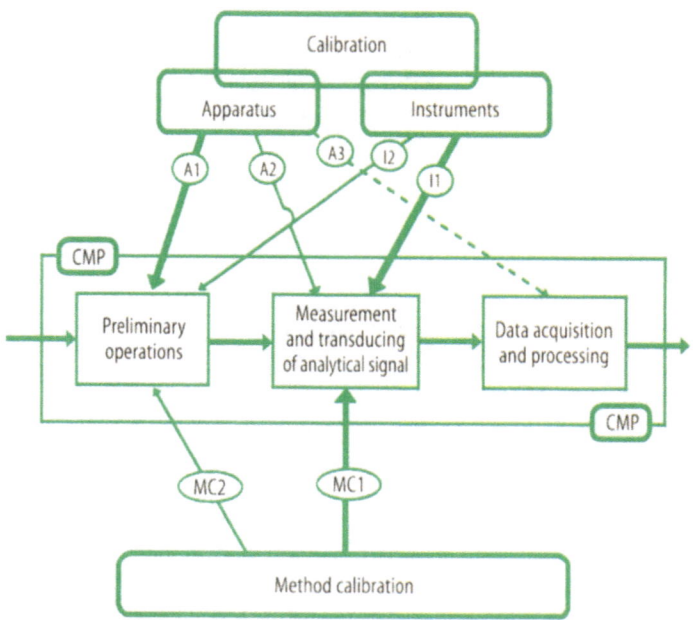

Fig. 4.5. Calibration operations involved in a CMP. For details, see text

the use of standards for equipment and method calibration in CMPs; thus, it describes the use of the different types of standards in CMPs based on primary (absolute and stoichiometric) and relative methods for qualitative and quantitative purposes.

Equipment (apparatus and instrument) calibration encompasses the three steps of a CMP. As shown in Fig. 4.5 (A1), apparatus (*e.g.* stoves, reactors, extractors) must be checked to operate as scheduled by calibrating parameters related to physical quantities (*e.g.* temperature, pressure). When an instrument (*e.g.* a chromatograph) integrates some preliminary operation, the apparatus or devices involved must also be calibrated (Fig. 4.5, A2). The dramatic advances in computer science in recent years have promoted the use of a variety of electronic interfaces to instruments and apparatus that must also be subjected to calibration (Fig. 4.5, A3). Equipment calibration proper affects the second step of the analytical process, both in absolute (*e.g.* the balance in gravimetries), stoichiometric (*e.g.* the burette in titrimetries) and relative methods (*e.g.* photometers, fluorimeters, polarographs, mass spectrometers); also, it assures proper performance of the instruments and apparatus used (Fig. 4.5, I1). As stated above, any instruments used in the first step of the analytical process (*e.g.* balances, pipettes) must be calibrated as well (Fig. 4.5, I2).

Method calibration (MC) pertains essentially to the second step of the CMP (see Fig. 4.5, MC1) in relative methods based on an unequivocal relationship between the concentration (or some other property) of the analyte and the analytical signal it produces. In titrimetries, one must additionally consider standardizing the titrant (see Box 3.7), which corresponds to link MC2 in Fig. 4.5.

4.3 Preliminary Operations

The generic term "preliminary operations" is used to designate a wide variety of actions that connect uncollected, unmeasured, untreated samples to the main measuring instrument in the second step of a CMP.

The first step of a CMP makes the strongest difference from physical measurements, where it rarely exists. Because of the special technical features of preliminary operations – particularly their difficulty –, the first step of a CMP is especially relevant as it has major implications on the quality (accuracy and representativeness) of the results and a strong impact on accessory properties (expeditiousness, cost-effectiveness and personnel-related factors).

Spectacular breakthroughs in equipment technology and computers have dramatically raised quality in the last two steps of the CMP. By contrast, on the edge of the XXI century, the first step continues to demand generous R&D endeavours despite the brilliant – though still insufficient – advances in this field in recent years.

The first step of a CMP can involve the use of materials, methods and approaches of widely varied nature (see Fig. 1.2) depending on the factors shown in Fig. 4.3.

This section provides a general picture of the first step of the CMP. Following a brief description of its general features, it discusses its major sub-steps (sampling and sample treatment).

4.3.1 General Features

Figure 4.6 provides a schematic depiction of the most salient features of preliminary operations. As can be inferred from its contents, preliminary operations are **widely variable in nature.** In fact, the factors potentially affecting a CMP (Fig. 4.3) are especially influential on these operations. The examples of Box 4.2 confirm that the type of information involved (sample, analyte, available equipment and measurement method) dictate the design of the preliminary operations to be performed. Figure 4.7 illustrates the wide variety of situations that can be encompassed by preliminary operations depending on the particular state of aggregation and nature of the sample matrix, and also of the nature and concentration of the analyte(s).

Box 4.4 discusses several typical examples. Such wide variability in the preliminary operations of a CMP obviously has some disadvantages including, but not limited to, the following:

(*a*) the need to individually design and optimize each analytical sample–analyte pair;

(*b*) the difficulty of applying available ordinary strategies;

(*c*) the risk of adopting seemingly obvious extrapolations (*e.g.* the preliminary operations to be used in a CMP to determine traces of a metal ion are different from those to be employed if the analytes are present in macro amounts); and

(*d*) the scantiness of commercially available technology – most apparatuses for these purposes have traditionally been aimed at niche markets, even though

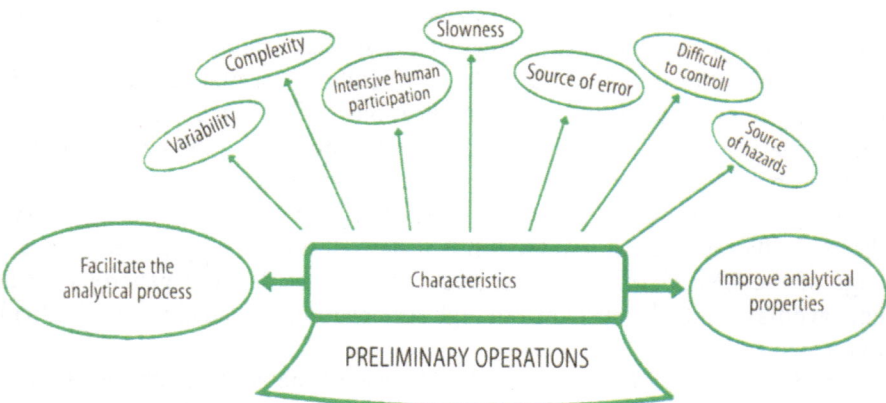

Fig. 4.6. General features of the first step of a CMP

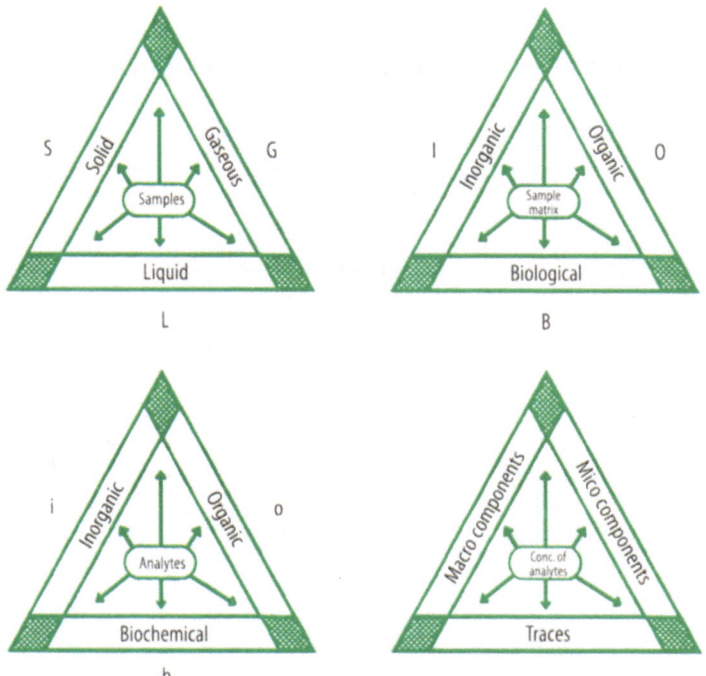

Fig. 4.7. Sample-related (state of aggregation and nature) and analyte-related factors (nature and concentration) that dictate the design of preliminary operations in a CMP

this trend is gradually being reversed as manufacturers realize that sales of instruments (the agents of the second step) are significantly boosted when these can be equipped with options such as on-line modules for the implementation of specific preliminary operations.

That preliminary operations are **complex** is obvious. As a rule, they are multi-step processes that use a variety of analytical tools (instruments, apparatus, devices).

In contrast to the other two steps of a CMP, preliminary operations are **labour-intensive**. Also, because of their high complexity, they are **difficult to automate** and only a few partially automated approaches to the off-line (digesters, extractors) and on-line coupling (chromatographs) to instruments (in the second step of the CMP) exist.

Between 70 and 90% of the time taken to perform a CMP is used in its preliminary operations, which are thus markedly **slow** relative to the second and third steps of the process.

Preliminary operations are the sources of gross **accidental and systematic errors** that have a decisive influence on the quality of the results (*viz.* their accuracy) and of the CMP itself (its uncertainty and precision). In fact, preliminary operations are Achilles' heel to a CMP. An inappropriate choice of sample, incomplete dissolution, poor storage conditions, incomplete removal of potential interferents, gains or losses of analytes present in trace amounts, *etc.*, can give

Box 4.4

Selected examples illustrating variety in preliminary operations

Sample type	Analytes	Combination			Usual sample treatment
		State of aggregation of sample	Nature		
			Of matrix	Of analyte	
Soil	Metals	S	I	I	• Leaching
					• Disaggregation
	Pesticides	S	I	O	• Solvent extraction
Serum	Urea	L	B	O	• Dialysis
	Enzymes	L	B	B	• Dilution
	Lead	L	B	I	• Destruction of organic matter
	Drugs	L	B	O	• L−L and SP extraction
Air (particulates)	Metal traces	G(S)	I	I	• Ion exchange
					• S−L extraction
	Organic pollutants	G(S)	I	O	• L−L extraction
Pharma-ceutical (vitamin complex)	Vitamins	S(L)	O/I	B	• Leaching: H_2O (water-solubles), o.s. (fat-solubles)
	Salts	S(L)	O/I	I	• Destruction of organic matter
	Excipients	S(L)	O/I	O	• L−L extraction
Animal tissue	Metals	S	B	I	• Freeze-drying
	Additives (human consumption)	S	B	O	• Leaching
	Proteins	S	B	B	• Leaching
Fresh orange juice	Ascorbic acid	L	B	B/O	• None needed
	Artificial sweeteners	L	B	O	• L−L extraction
	Metal traces	L	B	I	• Ion exchange/ elution

S solid, L liquid, G gas, SP solid-phase, I inorganic, O organic, B biological, o.s. organic solvent.

rise to gross errors – much greater than those typically made in the other steps of the CMP. Such errors arise both from the technical complexity of the operations and from the so-called "human factor". This potentially adverse implication of preliminary operations is the origin of their significance.

Systematic control of preliminary operations is not as affordable or convenient as is that of an instrument (in the second step of the CMP) through calibration. Checks during preliminary operations involve the painstaking testing of all apparatus and instruments used (*e.g.* with mass or volume measurements), and also confirming that they operate as scheduled. There are three ways of assuring quality in preliminary operations, namely:

(*a*) By systematic application of quality assurance systems in the analytical laboratory concerned, in the form of internal or external assessment (described in detail in Chap. 8).

(*b*) By global evaluation of the CMP, using samples of known composition (*e.g.* blind samples, certified reference materials) and matrices similar to those of the real samples. If the differences between the results and the known or certified values for the samples are not statistically significant, then the CMP will have been properly carried out and the preliminary operations correctly conducted.

(*c*) By using a combination of the previous two approaches, which is bound to provide greater assurance of quality in the operations.

The preliminary operations carried out in the first step of a CMP are a potential source of human and environmental hazards. Thus, the use of analytical tools such as acids, solvents, pressurized gases, digesters, distillers, *etc.*, can have adverse hygienic and health effects on laboratory personnel in the short to long run. For example, the specific preliminary operations involved in some CMPs produce variable amounts of toxic waste that requires careful handling and disposal.

For all these reasons, preliminary operations play a crucial role in CMPs. As such, they warrant special attention on the part of both laboratory personnel and quality control and assessment officials. As can be seen from Fig. 4.8, these operations have a strong impact on the three types of analytical properties (capital, basic and accessory) and thus call for dedicated research and development work. Thus, preliminary operations exert a direct influence on accuracy (through systematic errors), representativeness (through sampling errors), precision (through random errors) and accessory analytical properties (expeditiousness, cost-effectiveness and personnel-related factors). Also, they have indirect effects on selectivity (through interference removal) and sensitivity (through preconcentration), as well as on accuracy as a whole (through their individual effects on precision, sensitivity and selectivity).

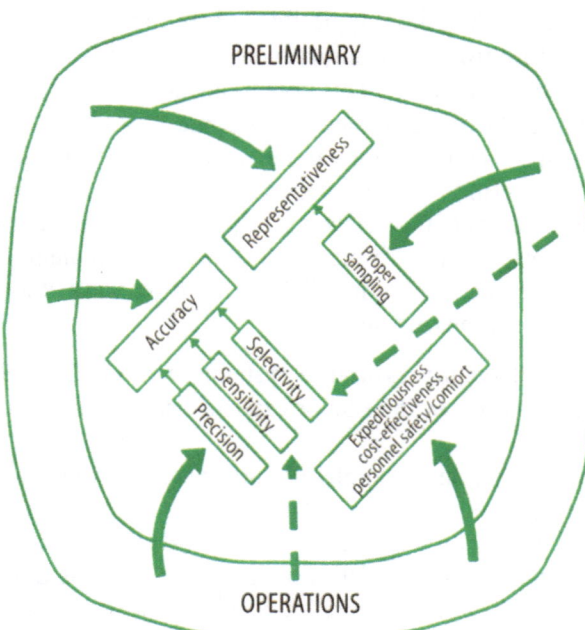

Fig. 4.8. Direct and indirect effects of preliminary operations of a CMP on the different types of analytical properties

---

Box 4.5

"Clean methods" are CMPs where the potential adverse effects of the analytical tools to be used in the preliminary operations on laboratory personnel and the environment have been considered. This approach, which is becoming increasingly popular and endorsed in Analytical Chemistry, can be realized in two different ways, namely:

(a) By minimizing or avoiding the use of toxic and environmentally hazardous reagents (*e.g.* mercury salts) and solvents (*e.g.* chlorofluorocarbons). This entails the use of new materials (*e.g.* novel chemicals of decreased toxicity relative to traditional reagents) and methodologies (*e.g.* supercritical fluid extraction instead of conventional solvent extraction).

(b) By developing new sub-steps for inclusion in the first or second step of the CMP in order to destroy or inactivate harmful materials.

Both strategies are hot topics in current analytical research, which must respond to the increasing need to manage laboratory waste in a cost-effective, environmentally friendly manner.

4.3.2 Sub-steps

Each CMP must be consistent with the analytical problem addressed. Preliminary operations are the greatest source of variability in number and nature of CMP sub-steps. Figure 4.9 shows some of the more common sub-steps involved in such operations. The sequence of this scheme is not always followed in

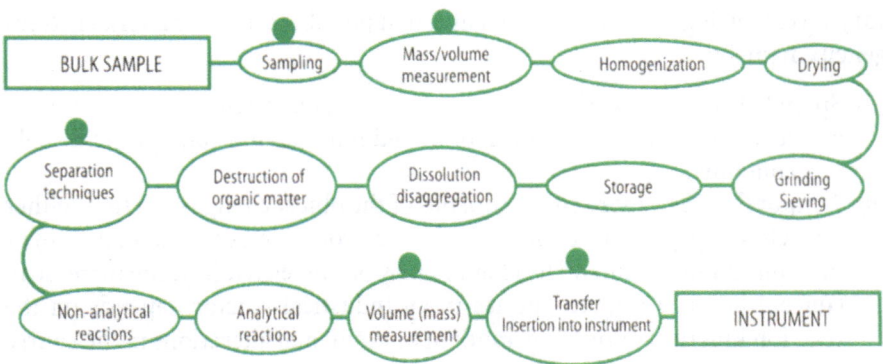

Fig. 4.9. Sub-steps of preliminary operations of the CMP. Those with a solid circle at the top are the most common

practice. Nor are all the operations shown always required. In fact, the specific sub-steps to be performed are dictated by the factors of Fig. 4.3 and the particular analytical problem. Thus, solid samples generally call for more operations – and of increased complexity. There is a current trend to simplifying CMPs by reducing the number of sub-steps, and also to automating and miniaturizing assemblies in order to minimize the adverse effects noted in the previous section. Typical examples in this respect include *in vivo* sensors (*e.g.* those for the continuous monitoring of pO_2 and pCO_2 in the blood stream of Intensive Care Unit patients) and environmental analysers (*e.g.* those for the continuous monitoring of atmospheric SO_2 and NO_x). Figure 4.9 shows the more frequently used and important types of preliminary operations, which are briefly commented on in the following sections.

4.3.3 Sampling

Sampling is the first sub-step of preliminary operations. It allows one to **select** a single or several portions (aliquots) from a given material (the object of the analysis) for subjection to the other operations of the CMP.

The sampling scheme used in each case should be consistent with the analytical problem addressed (Chap. 7) and, especially, with the nature of the analytical information required. This sub-step can be viewed as the link between the object and the analytical problem.

Proper sampling is a basic analytical property and also a cornerstone for the capital property representativeness (see Sect. 2.4.2). An accurate result that is not representative of the problem concerned will a poor result anyway; sampling is thus a crucial sub-step of the chemical measurement process.

The principal hindrance to proper sampling is spatial and temporal heterogeneity. Any portion withdrawn from an object that is 100% homogeneous will be representative of it, so the analytical problem need not be affected by the particular aliquot used. In practice, however, materials are heterogeneous to a

varying extent. Figure 4.10 shows the different possible types of material heterogeneity, namely:

(*a*) **Spatial.** The object varies in thickness, depth, *etc.* Such is the case, for example, with a steel sheet, a stack of mined mineral or a solid pesticide held
in a container.

(*b*) **Temporal.** One or more characteristics of the object change with time, either
discretely (*e.g.* pharmaceutical tablets on a production or packaging line) or
in a continuous manner. The changes may be correlated (*e.g.* the increase in
concentration of a given species in an industrial reactor depends on the
reaction kinetics) or not (*e.g.* random or accidental alterations in the course
of a process).

(*c*) **Spatial and temporal.** The object changes in space and time (*e.g.* a river,
which may vary widely between source and mouth, and also with time of
year).

Frequently, the sampling operation involves not only the mere selection of an
appropriate sample portion but also other sub-steps such as storage or preconcentration. Thus, water samples obtained at a variable depth from a lake are
usually supplied with a preservative (*e.g.* nitric acid or chloroform) to minimize
the loss of trace species. Also, atmospheric hydrocarbon concentrations are
determined by using a suction system including a sorbent cartridge where the
analytes contained in a given volume of air (*e.g.* $1\,m^3$) are concentrated. Pre-
serving the stability of the collected sample until it is eventually analysed is
especially important in some determinations (*e.g.* speciation).

The **sampling plan** is the scheme to be applied in order to ensure that the
results of a CMP will be consistent with the analytical problem confronted. The
plan should consider all the factors of Fig. 4.3, in addition to other, more specific
ones such as the size of the aliquot(s) to be subjected to the CMP or the capacity of the autosampler to be employed. The sampling scheme should rest on a

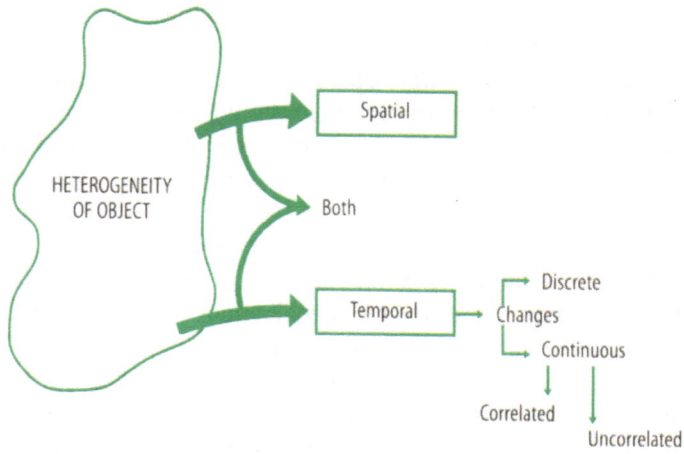

Fig. 4.10. Facets of object heterogeneity as related to sampling

compromise between the number of samples (size) to be processed and the costs/efforts incurred. Ideally, one should use infinite portions of the sample, which, however, is an unrealistic approach. In any case, the sampling plan should use the minimum number of samples that will ensure representativeness at the intended level. Statistics (chemometrics) provides invaluable technical support for this purpose. The ultimate goal is to maximize representativeness and precision with the smallest possible number of samples and sample size.

Depending on the global approach adopted in the plan, sampling can be of one of the following types:

(a) **Intuitive.** The analyst, by personal choice, selects the object portion to be processed (*e.g.* when a texture or colour change is observed in the studied material or some isolated alteration in a production process is suspected).

(b) **Statistical.** The selection is based on statistical rules that are applied after the analytical problem has been defined (*e.g.* when the minimum number of samples to be used is determined on the assumption of a Gaussian distribution in the object composition, characterized by a standard deviation, and in compliance with the accepted levels of error in the results and standard deviation in the measurements).

(c) **Directed.** This is the preferred choice when the analytical problem demands special information (*e.g.* when, in analysing natural water, the concentration of metal species in suspended particles of a specific size is the only parameter needed).

(d) **Protocol-based.** By legal mandate or by the client's demand, the sampling procedure to be used in some instances is one regulated by a standard or specified in an official publication.

Notwithstanding this classification, two or more of these sampling types are virtually indistinguishable in practice.

Broadly speaking, the sample is a portion of the object that contains the measurand or analyte; all are parts of the ranking discussed in Sect. 1.6. This definition of sample is impractical and, especially, oversimplified. One way of making it more explicit is by using different qualifiers to avoid confusion. The two approaches summarized in Fig. 4.11 illustrate the usage of the word "sample" in Analytical Chemistry.

Depending on the way it is collected, a sample can be qualified with different attributes. Thus, a **random sample** is one selected in such a way that any portion of the object will have had a given likelihood (*e.g.* 95%) of being withdrawn. A **representative sample** is one selected by applying a sampling plan consistent with the definition of the analytical problem addressed. A **selective sample** is a sample collected by following a selective sampling procedure. A **stratified sample** is one randomly chosen from a previously collected portion, stratum or well-defined zone of the object. Finally, a **convenience sample** is one selected in terms of accessibility, cost-effectiveness, efficiency, *etc.* (*i.e.* because it possesses favourable sampling characteristics).

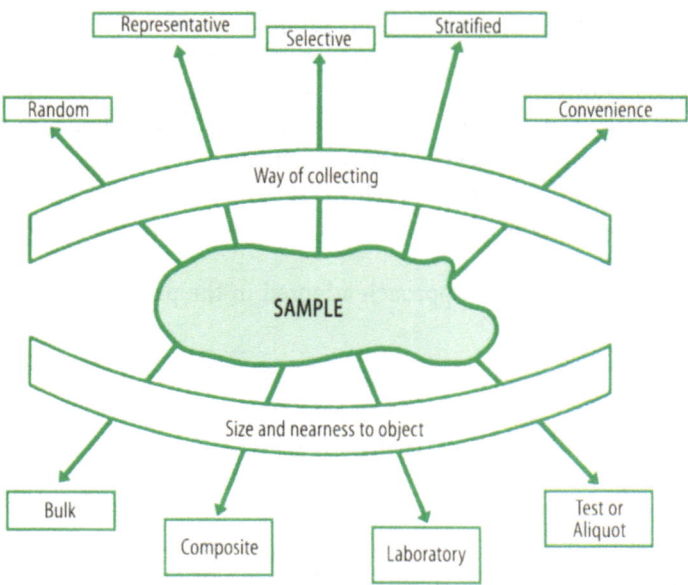

Fig. 4.11. Sample "qualifiers", arranged according to two non-mutually exclusive criteria. For details, see text

Samples can vary widely depending on size and nearness to the object. Thus, the **primary** or **bulk sample** is the usually large specimen obtained in the initial selection from the object. A **composite sample** results from the combination of several portions of the bulk sample. A **laboratory sample** is one resulting from the reduction of a composite sample – when global information is required – or the bulk sample – if discriminate information is needed – submitted, in an appropriate container, for analysis by a laboratory. Finally, a **test sample** or **aliquot** is a portion of a laboratory sample which, following reduction and conditioning (*e.g.* grinding), is eventually subjected to the other operations of a CMP. Figure 4.12 illustrates these concepts as applied to mineral stacked in the vicinity of a mine; a distinction is made according to whether the average composition (A) or spatially discriminate information (surface, inside and bottom of the mineral stack, B) is to be determined.

The sampling process can be the source of various types of error. **Accidental errors**, which materialize in individual samples, arise when the sampling tools are defective or improperly used. **Systematic errors** result from repeated non-compliance with the specific procedure specified in the sampling plan. **Random errors** or "sampling errors" proper, originate from incorrect extrapolation of an individual result or the means of those for several aliquots to the object as a whole. Accidental and systematic errors affect accuracy of the results as they lead to (positive or negative) bias; also, as shown below, they add to or cancel with errors of the same type.

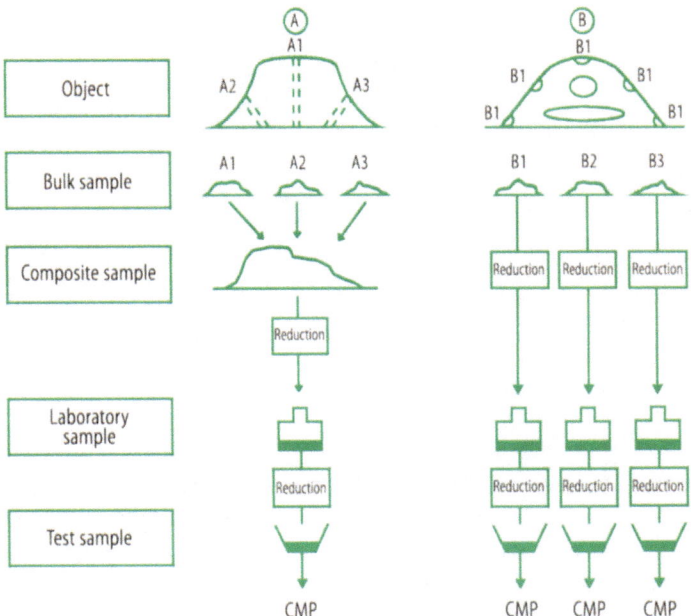

Fig. 4.12. Characterization of a sample in terms of size and nearness to its object: stacking of a mineral to obtain global (A) or discriminate information (B). For details, see text

_____ Box 4.6

Selected examples of environmental sampling

The design of a sampling procedure is dictated by both the characteristics of the sample, the types of analytes it contains and the analytical problem faced. Below are briefly described several typical examples of environmental sampling.

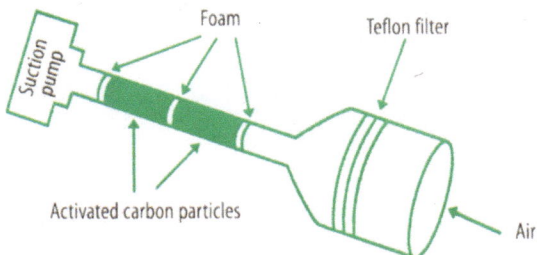

Gaseous samples. The atmosphere contains a wide variety of organic and inorganic analytes (pollutants) including gases, particulates, droplets (aerosols), *etc*. The sampling system to be used in each case will thus depend on the type of information required. By way of example, consider the sampler above, used for the analysis of particulates and the determination of organic gases. The tube is connected to a suction pump through which a given volume of air controlled via the flow-rate of the pump is passed. Particulates are retained on the Teflon filter and gases (*e.g.* SO_2) by a bed of carbon impregnated with a reagent which interacts with the analyte to form a product (*e.g.* a sulphate) that is adsorbed by the active support.

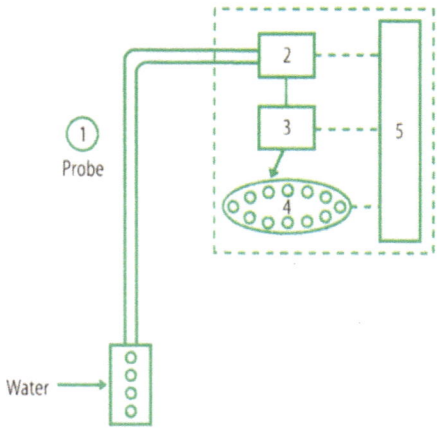

Liquid samples. Liquid environmental samples are easier to collect, particularly when readily available. The sole caution to be exercised is to use an appropriate container in order to avoid losses or gains of unwanted traces during transfer or storage. Automatic modules for sampling water (from the sea, a lake, a river) at variable depths and times as scheduled in the sampling plan have recently been made available. Typically, one such module consists of

(1) a probe of variable length through which the sample is aspirated,
(2) a suction pump,
(3) an automatic dispenser of aspirated samples,
(4) an autosampler holding 20 – 100 containers and
(5) a microprocessor governing the whole process.

Solid samples. Soil pollution can be studied by collecting surface samples at variable depths or determining the average composition of a zone in between the soil surface and a given depth. There is a wide range of available samplers for this purpose the most frequently used of which are screw drills. These are vertically screwed into the ground, whether mechanically or by hand, and piloted during the process. After the drill is drawn, the soil retained along its thread is removed for analysis.

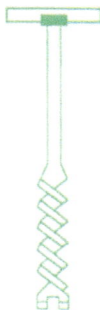

Sampling errors proper have a direct influence on the precision (uncertainty) in the final result. They originate from material heterogeneity. If a material is assumed to possess a Gaussian distribution, then sampling errors can be characterized by the standard deviation (σ_s), a measure of sample dispersion (heterogeneity). Sampling errors are different from and add to random errors made in other steps of a CMP. Because variances (σ^2) are additive, the total random error in a result can be calculated from its total variance (σ_t^2, which is the sum of the variances arising from sampling, other preliminary operations, signal measurement and data processing.

$$\sigma_t^2 = \sigma_s^2 + \sigma_{\text{other PO}}^2 + \sigma_{\text{signal meas}}^2 + \sigma_{\text{data proc}}^2$$

The variance due to sampling (σ_s^2) has a high statistical weight in the total variance. In fact, σ_s^2 is typically 5–10 times greater than the other variances in the previous equation. This confirms the importance of the sampling step and the need to improve it as far as possible.

The amount of mass to be sampled depends strongly on the particle size and heterogeneity of the material, as well as on the precision level required.

4.3.4 Sample Treatment

The term "sample treatment" encompasses a series of sub-steps following sampling that are intended to make the test sample ready for the second step of the CMP. Occasionally, some such sub-steps (e.g. storage, analyte preconcentration) are carried out simultaneously with sampling.

Figure 4.9 illustrates some of the more common sample treatment operations. The specific conditioning operations to which a sample must be subjected are dictated by the particular CMP, the choice of which is determined by the analytical problem concerned (see Fig. 4.3). The most influential factors in this respect are variability in the sample and its analytes (Fig. 4.7).

Mass or **volume measurements** may be required at different times during the preliminary operations of a CMP, namely:

(a) immediately after sampling, to determine the amount of test sample with which the CMP is started and to which the final result will be referred;

(b) between preliminary operations (e.g. when making to volume, withdrawing aliquots, etc.); and

(c) when a treated sample aliquot is incorporated into the analytical process. These operations use measuring instruments such as balances, pipettes, volumetric flasks, flow-meters, etc.

When the test sample is a solid, or a liquid containing suspended solids, it normally requires **dissolution**. If the sample has an organic matrix (e.g. pitch), one or more organic solvents are used in a pre-established sequence. If the sample is of inorganic nature (e.g. a soil), it is usually treated with an acid mixture or with several acids (e.g. HCl, HNO_3, $HClO_4$, aqua regia) in sequence. These operations can be expedited by increasing the temperature or pressure; as a result, most conditioning procedures rely on the use of (a) heated open containers, (b) airtight digesters (heating of which raises the temperature within) or (c) microwave ovens using open or closed systems.

Inorganic analytes present in a matrix of organic (*e.g.* metal traces in petroleum) or biological nature (*e.g.* iron in serum) require **destruction of organic matter** – an alternative to sample dissolution. This operation involves the oxidation of organic matter (*viz.* that of carbon, hydrogen, nitrogen and sulphur to CO_2, H_2O, N_xO_y and SO_2, respectively) by using appropriate reagents in different possible ways. So-called "dry procedures" heat the sample at 550–600 °C in an appropriate container (a crucible); the oxidant is obviously atmospheric oxygen. This procedure has the disadvantage that it results in partial or even complete losses of volatile elements such as arsenic or mercury. "Wet procedures" heat the sample with an oxidizing acid (HNO_3, $HClO_4$) or an acid mixture in a appropriate vessel (see Box 4.7). There are alternative, special procedures for the same purpose such as those based on the Shöninger oxygen flask or the Kjeldahl process for the determination of organic and ammonia nitrogen, respectively.

_____   Box 4.7

The determination of metal traces in organic or biological samples by wet methods entails the use of an oxidizing acid (*e.g.* HNO_3, $HClO_4$) in the presence or absence of an additional oxidant (*e.g.* H_2O_2) to decompose the sample into a residue of metal oxides readily dissolved in acids or a concentrated solution of the oxides. This treatment can be performed in a variety of ways.

Those samples that contain abundant moisture (*e.g.* biological materials) are best freeze-dried and then subjected to high vacuum in order to remove the water and break the matrix structure (*e.g.* any macromolecules in it). The resulting residue can be readily dissolved in an organic acid under mild conditions.

Alternatively, organic and biological samples can be attacked with an oxidizing acid in a flask furnished with a condenser, with heating for an interval of half an hour to several hours. This procedure is also slow and hazardous (it involves the use of sizeable volumes of toxic, polluting solvents).

In order to expedite the process and reduce the volume of oxidizing acid needed, samples can be digested in an air-tight enclosure known as a "pressurized reactor". This is usually a Teflon container lined with the same polymer material that is accommodated inside a steel cylinder which is opened and closed via threaded parts. Once the sample and acids have been inserted, the reactor is shut and heated. The combined effect of temperature and the pressure inside the vessel is thus exploited.

Microwave ovens have recently proved highly efficient for digestion inside pressurized reactors. In fact, the effect of concentrated microwave energy is a more drastic attack in a shorter time. The reactors must be made of Teflon or quartz and the steel housing removed to facilitate irradiation of the sample and avoid reflection of waves.

Solid samples that cannot be dissolved in an acid require **disaggregation**, usually by fusion with a solid alkali (*e.g.* Na_2CO_3), alkali-oxidant (*e.g.* Na_2O_2), alkali-reductant (*e.g.* KCN + C) or acid ($KHSO_4$), depending on the nature of the sample matrix, in a platinum or nickel crucible. Once cool, the molten mass is dissolved in an appropriate solvent.

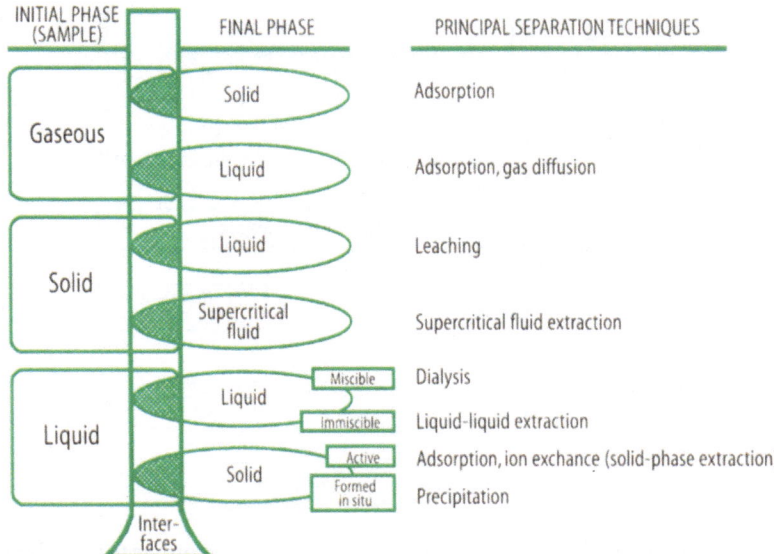

Fig. 4.13. Principal analytical separation techniques, classified according to state of aggregation of the sample (first phase) and nature of second phase. Mass transfer takes place across the interface between the two phases

Analytical separation techniques play a prominent role in preliminary operations. In fact, few CMPs use none. These techniques are based on mass transfer between two phases, which takes place in a static or dynamic manner. The material to be transferred may be either the analytes or any species potentially interfering with their determination. Figure 4.13 shows the more common interfaces in this context and the associated separation techniques. One of the phases may be the starting sample or the product of a preliminary step (*e.g.* a liquid resulting from dissolution of a solid or a filter used to collect atmospheric particulates), and hence solid, liquid or gaseous. The second phase, which is added or produced *in situ*, is usually a solid, liquid or supercritical fluid (or, less often, a gas).

Figure 4.14 classifies analytical separation techniques in terms of operational dynamics, efficiency and the way they are linked to the measuring instrument – the agent of the second step of the CMP as noted earlier.

In *discrete separation techniques*, the two phases are simply brought into mutual contact; at some time, however, they may require stirring to expand their interface (*e.g.* in manual or mechanical liquid-liquid extractions). The aim is usually to isolate the analytes, in groups, from their interferences. As a rule, the separation is carried out off-line, *i.e.* with no direct linkage to the detector. In *continuous separation techniques*, at least one phase is in continuous motion, so the process is fully dynamic. These techniques can be chromatographic or non-chromatographic. The former, which are the more efficient, are used to separate analytes in mixtures from one another and from their interferents; they are

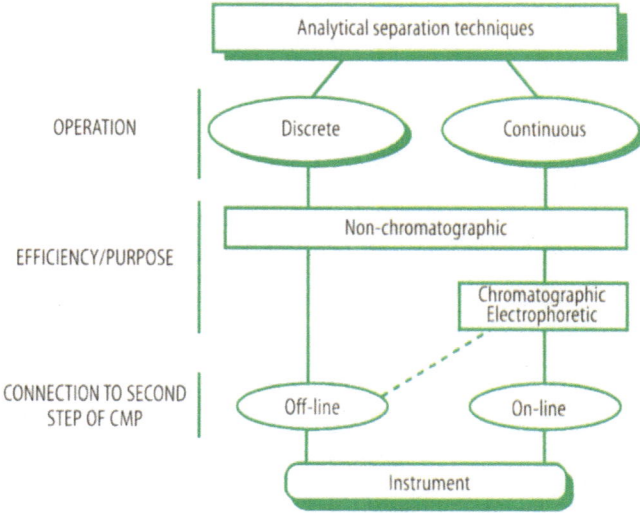

Fig. 4.14. Operational modes of separation techniques, arranged according to procedure, efficiency and/or purpose, and relationship to the instrument

usually coupled (on-line) to detection. Box 4.8 describes various ways of implementing liquid-liquid extraction.

_____ Box 4.8

Mass transfer between two immiscible liquids

The explanation below illustrates the alternatives of Fig. 4.14 as applied to a direct liquid-liquid interface. Because, unlike dialysis, the two liquids are not miscible, no separatory "barrier" is needed.

Technically, **liquid-liquid extraction** can be implemented in a number of ways. Only the three most typical options with an aqueous phase (A) and an organic phase (O) are discussed here, however.

(1) In the batch manual process, the two phases are used to fill a separatory funnel that is shaken until partitioning equilibrium is reached (once). After the emulsion is broken and the two phases have separated, an aliquot of the organic phase – now containing the analyte or its reaction product – is transferred, by hand, to the measuring instrument (operating off-line). This is a non-chromatographic approach.

(2) In doubly dynamic, continuous processes, streams of both phases are merged to form a succession of alternate segments of the two. Mass transfer takes place along a coil. A continuous phase separator allows a stream of organic phase, containing the analyte, to be connected on-line with the detector. This is a non-chromatographic continuous approach.

(2)

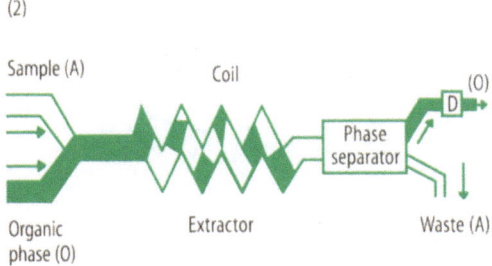

Sample (A)   Coil   (O)
Phase separator
Organic phase (O)   Extractor   Waste (A)

(3) The chromatographic liquid-liquid extraction system below is also of the dynamic type. However, one of the phases (O) is immobilized on a support located in the chromatographic column, through which the mobile (aqueous or aqueous-organic) phase is passed following insertion of the sample via an injection valve (IV). Mass transfer takes place in the column and equilibrium is reached many times (10^2-10^6), which significantly boosts the separation efficiency. In this way, one can not only isolate one component but also separate several from one another (see Fig. 4.15, 2b). Following separation, the analytes are continuously detected since the column output is coupled on-line to a suitable measuring instrument.

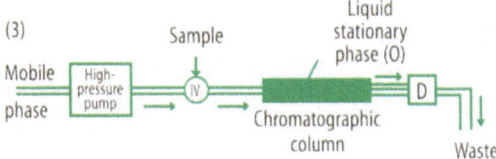

(3)   Sample   Liquid stationary phase (O)
Mobile phase   High-pressure pump   IV   Chromatographic column   D   Waste

The significance of separation techniques to the analytical process lies in the fact that they support the capital property accuracy (Fig. 2.1) by indirectly increasing sensitivity through preconcentration and selectivity through removal of interferences (clean-up). Figure 4.15 illustrates both contributions to quality in a CMP and hence to the results it produces.

The ability to use a low volume of the second phase (V_2) to separate an analyte A contained at a low concentration in a large volume (V_1) of the first phase, considerably raises the final concentration of analyte, $[A_f]$,

$$[A_f] = [A_0] \frac{V_1}{V_2}$$

thereby facilitating preconcentration.

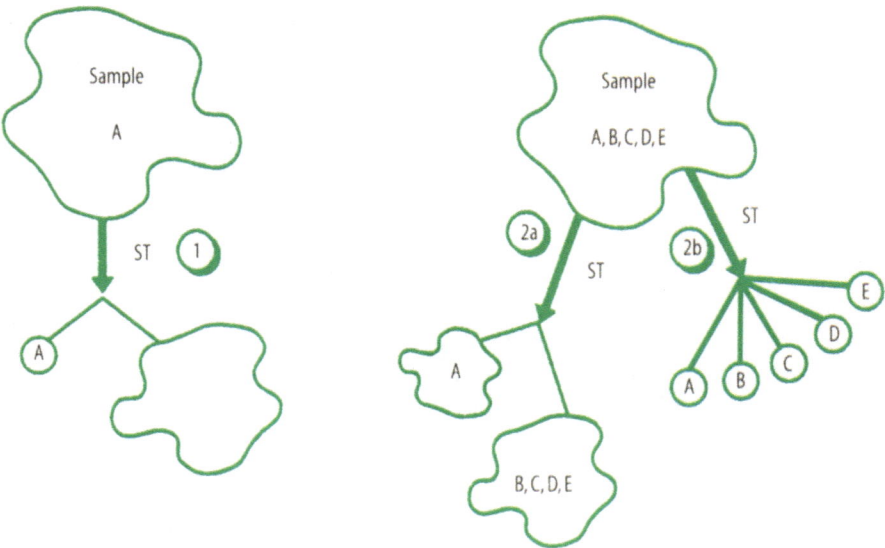

Fig. 4.15. Principal goals of analytical separation techniques (ST). (1) Preconcentrating an analyte (A). (2) Removal (clean-up) of interferents to isolate the analyte (A) from the other sample components (2a) or separate the different components (considered to be the analytes) from one another (2b)

_____ Box 4.9

A flame atomic absorption spectrometer (FAAS) is available to determine lead in water from wells near a mining area. The limit of detection achieved by direct insertion of aqueous standards of Pb^{2+} into the instrument is 1.0 mg/L (*i.e.* in the parts-per-million range). However, when real water samples are inserted, no signal is obtained; this suggests that their lead contents are below the limit of the detection. The sensitivity of an FAAS is thus inadequate for the purpose as the legal threshold for lead in water for animal or human consumption is 0.5 μg/L (*i.e.* in the parts-per-billion range). One must therefore increase the sensitivity of the available technique by preconcentrating the lead.

To this end, a volume of 4 litres of water is passed through a chelating column (Chelex-100), which selectively retains Pb^{2+}. Subsequently, the column is eluted with 2 mL of 0.1 N HNO_3 and the eluate is inserted into the instrument. In this way, a preconcentration factor of 4000 : 2 = 2000 is achieved. The new limit of detection is thus 2000 times lower, *i.e.* it has been reduced to 0.5 μg/L, which, coincidentally, is the same as the legal limit. Based on the underlying criterion of the analytical problem, if a preconcentrated real sample now provides a signal, then the water can be assumed to be contaminated with lead.

One could also have used a more sensitive technique [*e.g.* anodic stripping voltammetry (ASV), electrothermal-atomization atomic absorption spectrometry (ETAAS)] to avoid the need for such a high concentration factor and the time-consuming manipulations involved – in fact, passing 4 litres of water through the column takes a long time and using a high flow-rate to expedite the process may decrease the analyte retention efficiency below the minimum 90–95% required to ensure quantitative results.

_____ Box 4.10

The direct identification by chemical reaction and the human senses (Classical Qualitative Analysis as described in Chap. 1) of Fe^{3+}, Bi^{3+}, Cr^{3+}, Ca^{2+}, Ni^{2+}, Zn^{2+}, Cd^{2+}, Hg^{2+} and Cu^{2+} in aliquots of the same solution is rendered impossible by unsurmountable mutual interferences. A separation by precipitation (liquid–solid interface, second phase formed *in situ*, Fig. 4.13), using a non-chromatographic batch procedure (filtration, Fig. 4.14) is thus required. For this purpose, a buffered, highly concentrated solution of ammonium chloride/ammonia is used as precipitant. This solution precipitates the hydroxides (basic salts) of the tervalent cations and leaves the ammine complexes of the divalent ones in solution. After the hydroxide precipitate is dissolved in hydrochloric acid, its constituent cations can be identified through chemical reactions involving changes perceptible by human senses, using aliquots of each of the two resulting solutions. This is a straightforward example of interference removal.

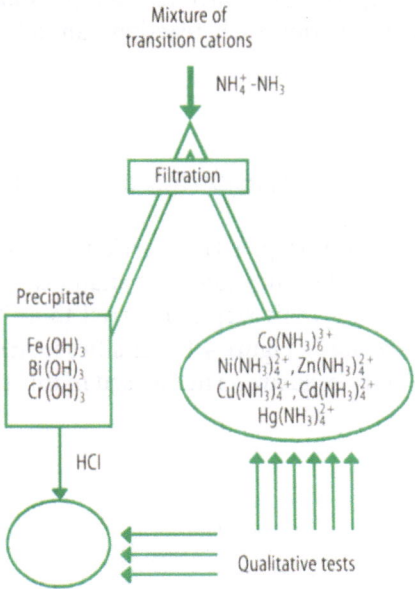

The extent to which the initial concentration is thus indirectly increased depends directly on the volume factor (see Box 4.9). While this goal is shared by both operating modes, it is more frequent in discrete and non-chromatographic continuous techniques (Fig. 4.15.1).

Removing interferents potentially disturbing the determination of the analytes is one other essential goal of analytical separation techniques (see Box 4.10). As shown in Fig. 4.15.2, the problem can be approached in two different ways. One involves allowing the analyte (A) to accumulate, free of interferents (B–E), or *vice versa*, in the new phase; this is normally accomplished by using a non-chromatographic technique. The analytical goal, however, may be more ambitious. For example, determining several analytes, A–E, in the same sample entails the physical separation of portions containing the analytes in isolation, which can only be realized by using a chromatographic or electrophoretic technique coupled on-line (column chromatography, capillary electrophoresis) or off-line (classical electrophoresis, planar chromatography) to detection.

Some of the more common sub-steps of preliminary operations (Fig. 4.9) can also be considered sample treatments. Such is the case with **preliminary non-analytical reactions**, which are used to facilitate development of the CMP. Typical examples include the use of fluoride ion to avoid the interference of ferric ion by formation of a stable, soluble, colourless complex (FeF_6^{3-}) and that of Jones reductor to ensure that all dissolved iron is converted into Fe^{2+} for titration with a standard solution of potassium dichromate. On the other hand, the use of **analytical chemical reactions** in preliminary operations is usually intended to convert the analyte into products to which the available instrument will respond [*e.g.* the coloured complex FeL_3^{2+}, obtained by addition of 1,10-phenanthroline (the ligand, L) to facilitate the photometric determination of iron traces]; alternatively, one such reaction can act as the main titrimetric reaction (see Chap. 6).

4.4 Measurement and Transducing of the Analytical Signal

The second step of a CMP (Fig. 4.4) is carried out by using a measuring instrument, which is the materialization of an analytical technique (see Fig. 1.16). Frequently, but improperly, this step is referred to as "detection" (and the instrument used as "detector"). Figure 4.16 classifies instruments according to the six complementary, non-exclusive, criteria used in this section.

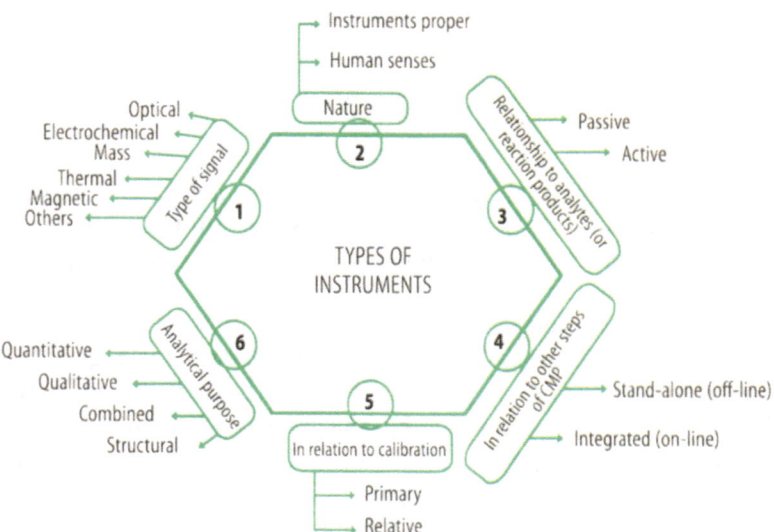

Fig. 4.16. Classification of instruments used in the second step of the CMP according to six complementary criteria

_____ Box 4.11

The following scheme places some analytical instruments typically used in CMPs in the different categories of Fig. 4.16.

Instrument	Category of Fig. 4.16					
	1	2	3	4	5	6
Balance	Mass	Instrument proper	Passive	Off-line	Primary	Quantitative
Manual burette	Mass volume	Instrument proper Human senses	Passive	Off-line	Primary	Quantitative
Autoburette	Mass volume	Instrument proper	Active	Off-line	Primary	Quantitative
UV-Visible photometer	Optical	Instrument proper	Active	Off-line On-line (chromato-graph)	Relative	Quantitative
Atomic emission spectrometer	Optical	Instrument proper	Active	Off-line	Relative	Qualitative and quantitative
Potentiometer (pH, pM measurements)	Electro-ana-lytical	Instrument proper	Passive	Off-line	Relative (special)	Quantitative

Note: Only the most usual options are considered.

The measured signal can be a response to a chemical (reactivity) or physico-chemical property of the analytes or their reaction products. As such, signals can be of various types (optical, electrochemical, mass, thermal, magnetic) and require instruments of widely variable design for measurement. The primary signal is transduced (and amplified) into another – usually electrical – signal which is that usually measured in practice (in volts or millivolts). The output can be in analog, digital or printed form, but is normally given in units related to those of the initial signal (e.g. absorbance in photometry, fluorescence intensity in luminescence techniques). Computers are being increasingly used to acquire and process signals, and also to deliver the ensuing analytical data or results.

The **human body** (specifically, senses such as sight and smell) can serve as an instrument in applications such as classical qualitative analysis and titrimetries (*e.g.* when reading the volume of titrant used from a graduated scale). More often, however, signals are measured by using an **instrument** proper (*e.g.* an atomic absorption spectrometer for optical signals, a balance or a mass spectrometer for mass signals, a differential thermal analyser for thermal signals, a magnetic resonance spectrometer for magnetic signals, *etc.*).

Depending on their relationship to the analytes (or their reaction products), instruments can be of the active or passive type (Fig. 4.16, 3). In **passive instruments**, the signal is not induced; it arises from physico-chemical properties of the analytes or their products (*e.g.* mass in balances, the luminous energy of chemiluminescence produced by photomultiplier tubes) or from the reactivity of the former (*e.g.* in the visual location of the liquid meniscus on the volumetric scale of a burette once the end-point of the main titrimetric reaction has been reached). **Active instruments** induce a signal by imposing some type of energy (*e.g.* luminous, electric) on the analytes or their products (*e.g.* in photometry, fluorimetry, polarography).

Depending on their relationship to the previous and following step of the CMP (Fig. 4.16, 4), measuring instruments can be of the **stand-alone** (off-line) or **integrated** type (on-line). Thus, with respect to preliminary operations (Sect. 4.3), an instrument can be (*a*) stand-alone when it receives an aliquot of a collected, treated sample that is manually or automatically (discretely) inserted into it; and (*b*) integrated or coupled on-line (*e.g.* a chromatograph or a capillary electrophoresis system including a detector connected to the output of the capillary separation column). With respect to the third step of the CMP (Sect. 4.5), the instrument is most often interfaced on-line to a computer; some instruments, however, provide digital or, less often, analog readings that are processed by the operator or input into a computer for calculation of the results.

The relationship of an instrument to chemical and SI base standards (Chap. 3) dictates the type of quantitative calibration to be applied (Fig. 4.16, 5). So-called **primary instruments** (and the balance) require "equipment calibration" only (see Fig. 4.5, I1); on the other hand, **relative instruments**, which operate by comparing the signals for standards and the sample, call for "method calibration" in addition to equipment calibration (Fig. 4.5, MC1).

Depending on the objective of the analysis (Fig. 1.20), instruments can provide responses for – preferentially – **qualitative, quantitative, structural** or **combined** purposes (Fig. 4.16, 6). Instruments for preferentially qualitative purposes provide a wealth of multivariate information that allows the reliable identification of the analytes; typical examples include Fourier transform infrared (FTIR) spectrophotometers and mass spectrometers. This, however, does not exclude their use to quantify amounts or concentrations of analyte. Instruments for essentially quantitative processes provide a reliable signal-concentration relation but cannot deliver – by themselves – dependable qualitative information; such is the case with molecular absorption and emission spectrometers (photometers and fluorimeters, respectively). Some instruments (*e.g.* atomic absorption and emission spectrometers) are used for both purposes. Others, such as nuclear magnetic resonance spectrometers, are preferred for structural analysis.

4.5 Signal Acquisition and Data Processing

The third step of a CMP can be considered the link between the measuring instrument (second step) and the results, expressed in the required format (see Fig. 4.4). The subsequent materialization in a report (Fig. 1.15) is within the scope of the analytical problem (Chap. 7).

This step involves two sequential sub-steps, namely:

Acquisition of transduced signals, which are the raw (absorbance, mass, current intensity) data obtained from the measurement process (see Fig. 1.15). Data can be acquired **manually** (*e.g.* by inspecting the colour of a precipitate, reading from a burette scale, locating the position of a gauge on an analog scale, reading the figures of a digital display), **semi-automatically** (when the instrument provides an informative output such as a spectrum, chromatogram, graph, signal-time plot, *etc.*, from which data can be extracted for purposes such as identifying a molecule from its IR spectral bands, measuring the height of a chromatographic peak or the slope of a kinetic curve) and **automatically** (when a computer is used to process one-, two- or three-dimensional information such as a mass or volume, signal-wavelength or signal-time relations, and signal-wavelength-time relations, respectively).

Processing of data, through computations based on chemometric methods, to express the results in the required format. A distinction should be made among data types and the treatments used. As can be seen from Fig. 4.17, the data involved in this sub-step can be of widely variable type and origin.

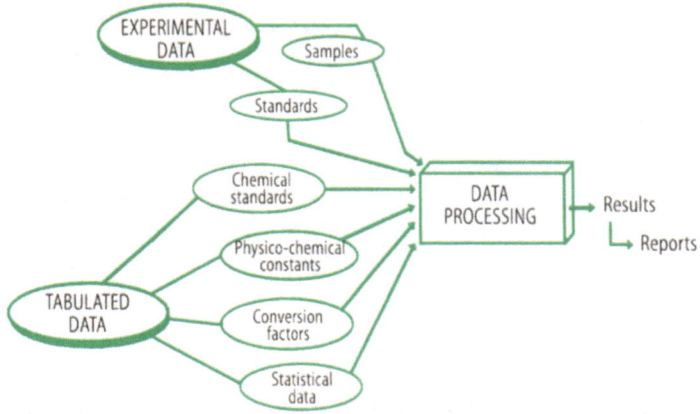

Fig. 4.17. Data sources and types that can be efficiently processed to deliver results in the required format and produce reports. For details, see text

Thus, there are experimental data (signals) provided by the measuring instrument (second step) for the samples or measurement standards used. In addition, a laboratory can use tabulated or stated data not experimentally produced in the CMP concerned; such is the case with values assigned to chemical standards (*e.g.* atomic weights, the Faraday), physico-chemical constants (*e.g.* ionic equilibrium constants, partition coefficients in separation techniques), conversion factors (which are dimensionless numbers such as that used to convert protein nitrogen contents when determined in various types of matrix) by which some experimental data must be multiplied to be made consistent with the known constraints of a CMP. This type of datum also includes statistical values such as Student's *t* (see Box 2.2) that allow one to express the specific uncertainty of the results when a number of aliquots are individually subjected to the CMP in question (see Fig. 2.10). This step can also be conducted by hand (with or without a pocket calculator, for example) or automatically (with a computer). The way experimental data are to be treated is also strongly dependent on the type of analytical method (primary, relative) used in the CMP, as well as on the static or dynamic nature of the primary data it produces.

The growing availability and power of computers has brought about dramatic changes in the third step of the CMP. Computers play a central role in this context and will foreseeably continue to do so. The relationship between an instrument and a computer is schematically depicted in Fig. 4.18. When coupled on-line, the computer can govern the instrument's operation via an active electronic interface; in addition, the computer can acquire data via a passive interface and, with simple, commercially available, software, also process them. In the off-line mode, the human interface acquires the data, which are manually input into the computer for processing. Current instruments are increasingly being equipped with dedicated microcomputers for this purpose.

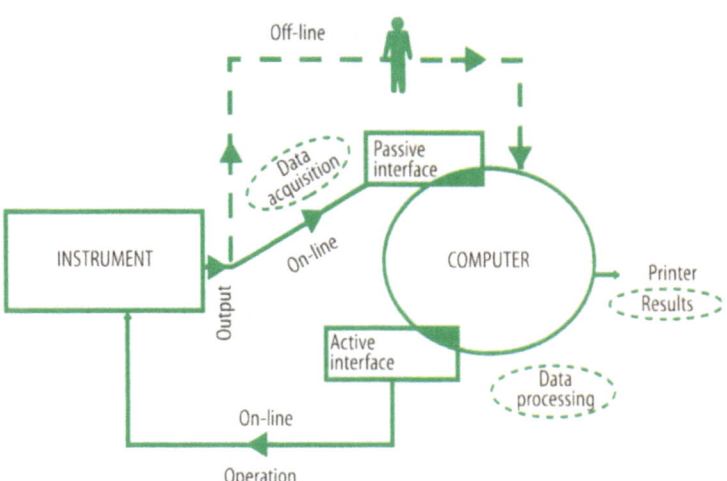

Fig. 4.18. Most common types of relationship between an instrument and a computer in the third step of the analytical process: off-line and on-line. Control of the instrument operation by the computer is also considered. For details, see text

4.6 Validation of a Chemical Measurement Process

To validate is to demonstrate, in experimental and formal terms, that a general (CMP) or particular **system** (*e.g.* the sampling or data acquisition and processing operations) has performed as expected and continues to do so. Also, to validate is to experimentally and formally demonstrate that an **object** (an instrument, apparatus, device) possesses specific properties and retains them.

The validation of a CMP is of the exclusive concern of analytical quality assessment systems (Chap. 8). A CMP can be validated for two essential purposes, namely:

(a) To identify and assure quality (and constancy) in the analytical information (results, reports) produced. This involves so-called **internal validation**, by which a CMP is characterized via such properties as accuracy (traceability), precision (uncertainty), selectivity, determination range, expeditiousness, cost-effectiveness, robustness, *etc.*

(b) To ensure consistency between the analytical signal produced and the information demanded by society, industry, trade, science or technology. This **external validation** (Chap. 8) additionally allows one to categorize (characterize) CMPs and analytical tools, facilitates quality audits, assures transferability of CMPs and tools among laboratories, and encourages harmonization and mutual acceptance of results among laboratories.

Figure 4.19 shows the different types of analytical systems and objects amenable to validation, and ranks their obvious mutual relationships. At the first level of the ranking is **method validation**; this rests on the second level, which comprises validation of

(a) sampling processes;

(b) analytical tools (objects such as reagents, reference materials, instruments, apparatus and devices); and

(c) data (hardware and software validation). Computer hardware, at the fourth level of the ranking, can be an apparatus (*e.g.* a computer) or a device (*e.g.* an interface).

A CMP can be internally validated in two different ways, namely:

(a) In a stepwise process involving validation of the analytical aspects at the second, third and fourth level in Fig. 4.19; while this is labour-intensive and time-consuming, it occasionally provides valuable specific information about the particular instrument, sampling schedule, data, *etc.*

(b) By checking the CMP as a whole.

On the other hand, a CMP is externally validated by checking for consistency between the results it produces and the information initially required.

A detailed description of validation procedures, of increasing use in the analytical chemical field, is beyond the scope of this introductory book on Analytical Chemistry. Note, however, that all validation processes share the need for

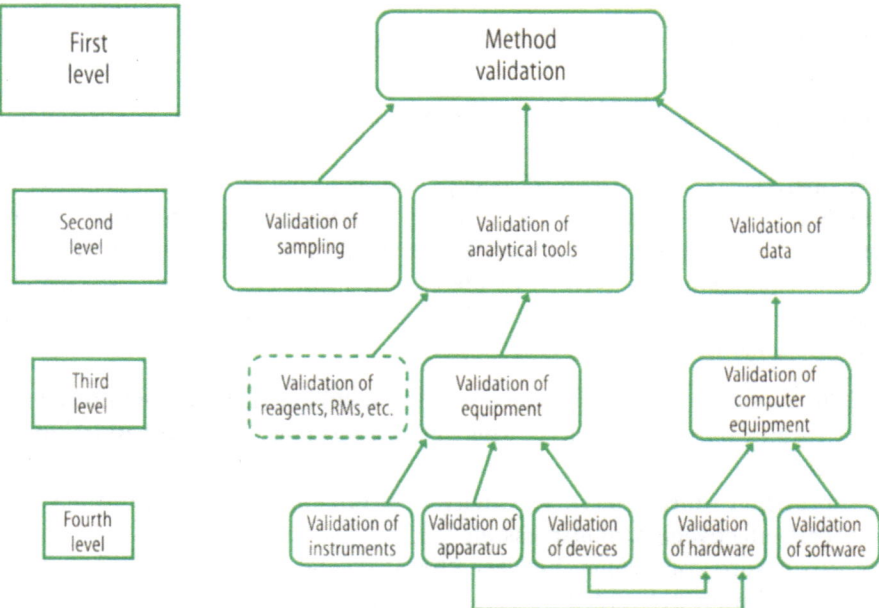

Fig. 4.19. Ranking of analytical aspects amenable to validation and mutual relations among them. For details, see text

references (to validate essentially involves to compare). Thus, the global validation of a CMP can be accomplished by comparing the results it provides in analyses of separate aliquots of a certified reference material (CRM) with the material's stated values (see Sect. 3.5.4.2). Alternatively, one can compare the results obtained by subjecting aliquots of the same sample to the CMP in question and to a primary method (gravimetry, coulometry). Spiked samples, which are real samples that have been supplied with a given amount of analyte, provide a less powerful means for the global validation of CMPs.

The validation of analytical tools requires the use of a variety of procedures dependent on their nature. As a rule, such procedures use reference materials, whether certified or otherwise. It should be noted that instrument validation is virtually identical with calibration of both equipment (primary methods) and analytical methods (relative methods) (see Sect. 3.5.4.1).

4.7 Salient Current Trends

Chemical measurement processes require continual refining in response to the increasingly varied, heavy demand for chemical information; this has fostered intensive research and development work, which constitutes the first fundamental element of Analytical Chemistry (Fig. 1.8). Future developments in this context will inevitably rely on scientific and technological innovations inspired by the three major current trends shown in Fig. 4.20.

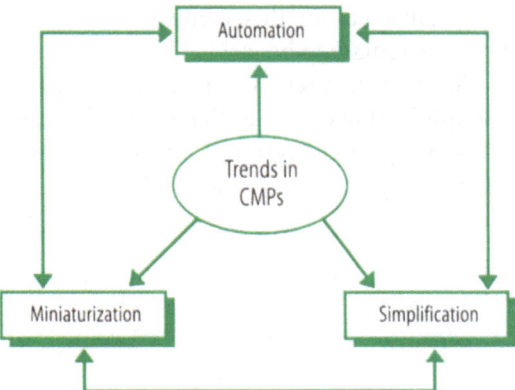

Fig. 4.20. Incorporation of major scientific and technological trends at the turn of the XXI century into CMPs

The complete or partial **automation** of a CMP entails reducing or even fully avoiding human intervention in it. There have been substantial advances in this direction over the past twenty years; as a result, a wide range of automated apparatus and instruments for use in CMPs have been made commercially available. At the top of integrated equipment for this purpose are automatic analysers, frequently called "autoanalysers" (see Fig. 1.17). The most salient advantage of using automated equipment in Analytical Chemistry is – occasionally very substantial – improvements in basic (*e.g.* precision) and, especially, accessory analytical properties (expeditiousness, cost-effectiveness and personnel safety). The different steps of the CMP have been automated to rather different degrees. Thus, equipment, and the data acquisition and processing operations, are now highly automated thanks to late advances in micromechanics, microelectronics and computer science. On the other hand, preliminary operations – which, as shown above, are decisively influential on the final results– continue to be conducted largely in a non-automated manner, mainly because of difficulties such as their high diversity and technical complexity (see Fig. 4.6). Reducing human intervention in them continues to be a priority R&D topic.

Miniaturization also has a strong impact on the ability of CMPs to satisfy information needs. This trend has materialized in a dramatic reduction in the size of material tools (*e.g.* capillary electrophoretic systems embedded in silicon chips that are only a few millimeters in size) and in the integration of modules that perform the steps and/or sub-steps of a CMP. So-called "micro total analytical systems" (µTAS) have a highly promising future in this context as they will help break the traditional barriers with which Analytical Chemistry has been confronted (Fig. 1.13); however, dedicated R&D work is still required with a view to their consolidation.

Simplification of some CMPs by use of novel analytical tools such as responsive sensors and portable NIR or RMS analysers has brought about a revolution in the traditional approach to the three steps of the CMP, which, however, are somehow implicitly developed during the process. These systems meet the requirements imposed by the need to process large numbers of samples or to obtain a rapid response, a global measure or a binary (yes/no) response. There have been substantial advances in this context, even though a number of problems in various fields remain to be tackled.

As can be seen in Fig. 4.20, the previous three trends are inevitably related to one another. Thus, simplification entails both reducing human intervention and miniaturizing – particularly for integration purposes –, and miniaturization leads to automation and simplification of the CMP.

_____ Box 4.12

The following example illustrates the practical implications of automating a CMP. The analytical method used involves the photometric determination of aluminium in a solution obtained by pretreating a rock specimen. The procedure comprises the following five steps: (1) acidification with 0.1 N HCl; (2) addition of ascorbic acid to reduce Fe^{3+} to Fe^{2+} in order to avoid its interference; (3) addition of a ligand (Xylenol Orange, XO) to form a soluble, strongly coloured chelate with the analyte; (4) use of a masking ligand (EDTA) to minimize or avoid interferences from divalent and tervalent metal ions accompanying Al^{3+} in the sample (EDTA does not destroy the Al–XO coloured complex under the working conditions used); and (5) measuring the absorbance of a sample aliquot.

The procedure can be performed in a manual or automated manner.

Manual procedure. Its conduct requires ordinary glassware (volumetric flasks, pipettes) and equipment (photometers, measuring cells). An aliquot of dissolved sample is collected by means of a high-quality pipette – one with a calibration certificate – and transferred to a calibrated flask – one of lower accuracy and precision than the pipette is acceptable for this purpose – containing 0.1 N HCl, 2% w/w aqueous ascorbic acid, 0.1% XO in ethanol/water and 0.2 M aqueous EDTA. After 30 min, the flask is made to the mark and its contents thoroughly mixed. Then, an aliquot is withdrawn and transferred to a glass cuvette of 1 cm light path that is placed in the photometer compartment to measure the solution absorbance at 560 nm. The same procedure is applied to standard solutions containing 0.5, 1, 2, 4 and 6 μg Al^{3+}/mL in order to construct a calibration curve as a relative method is being used. The aluminium concentration in the sample is obtained by interpolation.

Automatic procedure. This is performed in an continuous-flow system into which the sample is injected, *i. e.* in a flow injection (FI) assembly. A peristaltic pump propels four streams carrying the chemical ingredients of the CMP at a flow-rate of 1 mL/min each. The streams are sequentially merged as shown in the drawing below. The combined stream is passed at a rate of 4 mL/min through a flow-cell of 1 cm light path permanently housed in the photometer. The sample (100 μL) is injected into the acid stream via an injection valve and meets the three reagents at different merging points (1, 2, 3). Finally, the sample zone, brightly coloured, passes through the flow-cell and yields a peak-shaped signal. The method is calibrated by injecting standard solutions. The procedure is much more rapid, simple and precise than its manual counterpart.

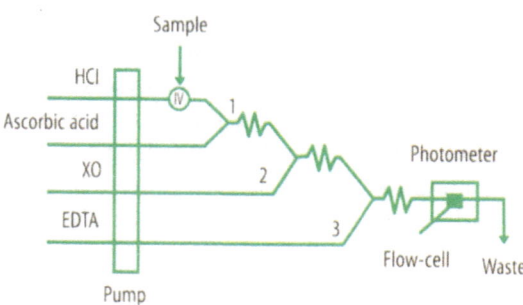

Annex I
Determination of metals in water

Learning objective:
To illustrate an analytical process used in the determination of a wide variety of metals in different types of water. Depending on the particular concentration level, a direct process or one that includes preconcentration by liquid–liquid extraction of the corresponding metal chelates is used.

Reference:
Standard Methods for Examination of Water and Wastewater
American Public Health Association, Washington, D.C., 1989, pp 152–157.

303 A. Determination of Antimony, Bismuth, Cadmium, Calcium, Cesium, Chromium, Cobalt, Copper, Gold, Iridium, Iron, Lead, Lithium, Magnesium, Nickel, Platinum, Potassium, Rhodium, Ruthenium, Silver, Sodium, Strontium, Thallium, Tin and Zinc by Direct Aspiration into an Air–Acetylene Flame

1. Apparatus
Atomic absorption spectrophotometer and associated equipment: See Sect. 303.2. Use burner head recommended by the manufacturer.

2. Reagents
a. *Air,* cleaned and dried through a suitable filter to remove oil, water and other foreign substances. The source may be a compressor or commercially bottled gas.

b. *Acetylene,* standard commercial grade. Acetone, which always is present in acetylene cylinders, can be prevented from entering and damaging the burner head by replacing a cylinder when its pressure has fallen to 689 kPa acetylene.

c. *Metal-free water:* Use metal-free water to prepare all reagents and calibration standards, and also as dilution water. Prepare metal-free water by de-ionizing tap water and/or by using one of the following processes, depending on the metal concentration in the sample: single distillation, redistillation or sub-boiling. Always check de-ionized or distilled water to determine whether the element of interest is present in trace amounts. (CAUTION: If the source water contains Hg or other volatile metals, de-ionized and single- or redistilled water may not be suitable for trace analysis because these metals distill over with the distilled water. In such

cases, use sub-boiling to prepare metal-free water).

d. *Calcium solution:* Dissolve 630 mg calcium carbonate, $CaCO_3$, in 50 mL of 1 + 5 HCl. If necessary, heat and boil gently to obtain complete solution. Cool and dilute to 1000 mL with water.

e. *Hydrochloric acid.* HCl, conc.

f. *Lanthanum solution:* Dissolve 58.65 g lanthanum oxide, La_2O_3, in 250 mL conc. HCl. Add acid slowly until the material is dissolved and dilute to 1000 mL with water.

g. *Hydrogen peroxide.* 30%.

h. *Nitric acid.* HNO_3, conc.

i. *Aqua regia:* Add 3 volumes conc. HCl to 1 volume conc HNO_3.

j. *Iodine solution,* 1 N: Dissolve 20 g potassium iodide, KI, in 50 mL water, add 12.7 g iodine and dilute to 100 mL.

k. *Cyanogen iodide* (CNI) solution: To 50 mL water add 6.5 g potassium cyanide, KCN, 5.0 mL 1 N iodine solution, and 4.0 mL conc. NH_4OH. Mix and dilute to 100 mL with water. Prepare fresh solution every 2 weeks.

l. *Standard metal solutions:* Prepare a series of standard metal solutions in the optimum concentration range by appropriate dilution of the following stock metal solutions with water containing 1.5 mL conc HNO_3/L. Thoroughly dry reagents before use. In general, use reagents of the highest purity. For hydrates, use fresh reagents.

(1) *Antimony:* Dissolve 2.7426 g antimony potassium tartrate hemihydrate (analytical reagent grade), $K(SbO)C_4H_4O_6 \cdot \frac{1}{2}H_2O$, in 1000 mL water: 1.00 mL = 1.00 mg Sb.

(2) *Bismuth:* Dissolve 1.000 g bismuth metal in a minimum volume of $1+1$ HNO_3. Dilute to 1.000 mL with 2% (v/v) HNO_3; 1.00 mL = 1.00 mg Bi.

(3) *Cadmium:* Dissolve 1.000 g cadmium metal in a minimum volume of $1+1$ HCl. Dilute to 1000 mL with water; 1.00 mL = 1.00 mg Cd.

(4) *Calcium:* To 2.4972 g $CaCO_3$ add 50 mL water and add dropwise a minimum volume of conc. HCl (about 10 mL) to complete solution. Dilute to 1000 mL with water: 1.00 mL = 1.00 mg Ca.

(5) *Cesium:* Dissolve 1.267 g cesium chloride, CsCl, in 1000 mL water: 1.00 mL = 1.00 mg Cs.

(6) *Chromium:* Dissolve 2.828 g anhydrous potassium dichromate, K_2CrO_4, in about 200 mL water, add 1.5 mL conc. HNO_3, and dilute to 1.000 mL with water: 1.00 mL = 1.00 mg Cr.

(7) *Cobalt:* Dissolve 1.407 g cobaltic oxide, Co_2O_3, in 20 mL hot conc. HCl. Cool and dilute to 1.000 mL with water: 1.00 mL = 1.00 mg Co.

(8) *Copper:* Dissolve 1.000 g copper metal in 15 mL of $1+1$ HNO_3, and dilute to 1000 mL with water: 1.00 mL = 1.00 mg Cu.

(9) *Gold:* Dissolve 0.1000 g gold metal in a minimum volume of aqua regia. Evaporate to dryness, dissolve residue in 5 mL conc. HCl, cool, and dilute to 100 mL with water; 1.00 mL = 1.00 mg Au.

(10) *Iridium:* Dissolve 1.147 g ammonium chloroiridate, $(NH_4)_2IrCl_6$, in a minimum volume of 1% (v/v) HCl and dilute to 100 mL with 1% (v/v) HCl; 1.00 mL = 5.00 mg Ir.

(11) *Iron:* Dissolve 1.000 g iron wire in 50 mL of $1+1$ HNO_3 and dilute to 1000 mL with water: 1.00 mL = 1.00 mg Fe.

(12) *Lead:* Dissolve 1.598 g lead nitrate, $Pb(NO3)_2$, in about 200 mL water, add 1.5 mL conc. HNO_3, and dilute to 1000 mL with water; 1.00 mL = 1.00 mg Pb.

(13) *Lithium:* Dissolve 5.324 g lithium carbonate, Li_2CO_3, in a minimum volume of $1+1$ HCl and dilute to 1000 mL with water; 1.00 mL = 1.00 mg Li.

(14) *Magnesium:* Dissolve 4.952 g magnesium sulphate, $MgSO_4$, in 200 mL water, add 1.5 mL conc. HNO_3, and dilute to 1000 mL with water; 1.00 mL = 1.00 mg Mg.

(15) *Manganese:* Dissolve 3.076 g manganous sulphate, $MnSO_4 \cdot H_2O$, in about 200 mL water, add 1.5 mL conc. HNO_3, and dilute to 1000 mL with water; 1.00 mL = 1.00 mg Mn.

(16) *Nickel:* Dissolve 1.273 g nickel oxide, NiO, in a minimum volume of 10% (v/v) HCl and dilute to 1000 mL with water; 1.00 mL = 1.00 mg Ni.

(17) *Platinum:* Dissolve 0.1000 g platinum metal in a minimum volume of aqua regia and evaporate just to dryness. Add 5 mL conc. HCl and 0.1 g NaCl, and again evaporate just to dryness. Dissolve residue in 20 mL of $1+1$ HCl and dilute to 100 mL with water; 1.00 mL = 1.00 mg Pt.

(18) *Potassium:* Dissolve 1.907 g potassium chloride, KCl, in water and make up to 1000 mL; 1.00 mL = 1.00 mg K.

(19) *Rhodium:* Dissolve 0.412 g ammonium hexachlororhodate, $(NH_4)_3RhCl_6 \cdot 1.5$ H_2O, in a minimum volume of 10% (v/v) HCl and dilute to 100 mL with 10% (v/v) HCl; 1.00 mL = 1.00 mg Rh.

(20) *Ruthenium:* Dissolve 0.2052 g ruthenium chloride, $RuCl_3$, in a minimum volume of 20% (v/v) HCl and dilute to 100 mL with 20% (v/v) HCl; 1.00 mL = 1.00 mg Ru.

(21) *Silver:* Dissolve 1.575 g silver nitrate, $AgNO_3$, in water, add 1.5 mL conc. HNO_3, and make up to 1000 mL; 1.00 mL = 1.00 mg Ag.

(22) *Sodium:* Dissolve 2.542 g sodium chloride, NaCl, dried at 140 °C, in water and make up to 1000 mL; 1.00 mL = 1.00 mg Na.

(23) *Strontium:* Dissolve 2.415 g strontium nitrate, $Sr(NO_3)_2$, in 1000 mL of 1% (v/v) HNO_3; 1.00 mL = 1.00 mg Sr.

(24) *Thallium:* Dissolve 1.303 g thallium nitrate, $TlNO_3$, in water. Add 10 mL conc. HNO_3 and dilute to 1000 mL with water; 1.00 mL = 1.00 mg Tl.

(25) *Tin:* Dissolve 1.000 g tin metal in 100 mL conc. HCl and dilute to 1000 mL with water; 1.00 mL = 1.00 mg Sn.

(26) *Zinc:* Dissolve 1.000 g zinc metal in 20 mL $1+1$ HCl and dilute to 1000 mL with water; 1.00 mL = 1.00 mg Zn.

3. Procedure

a. *Instrument operation:* Because of differences between makes and models of atomic absorption spectrophotometers, it is not possible to formulate instructions applicable to every instrument. See manufacturer's operating manual. In general, proceed according to the following: install a hollow cathode lamp for the desired metal in the instrument and roughly set the wavelength dial according to Table 303:I. Set slit width according to manufacturer's suggested setting for the element being measured. Turn on instrument, apply to the hollow cathode lamp the current suggested by the manufacturer and let instrument warm up until energy source stabilizes, generally about 10 to 20 min. Readjust current as necessary after warm-up. Optimize wavelength by adjusting wavelength dial until optimum energy gain is obtained. Align lamp in accordance with manufacturer's instructions.

Install suitable burner head and adjust burner head position. Turn on air and adjust flow-rate to that specified by manufacturer to give maximum sensitivity for the metal being measured. Turn on acetylene, adjust flow-rate to value specified and ignite flame. Aspirate a standard solution and adjust aspiration rate of the nebulizer to obtain maximum sensitivity. Atomize a standard (usually one near the middle of the linear working range) and adjust burner both up and down, and sideways, to obtain maximum response. Record absorbance of this standard when freshly prepared and with a new hollow cathode lamp. Refer to these data on subsequent determinations of the same element to check consistency of instrument set-up and aging of hollow cathode lamp and standard.

The instrument now is ready to operate. When analyses are finished, extinguish flame by turning off first acetylene and then air.

b. *Standardization:* Select at least three concentrations of each standard metal solution (prepared as in above) to bracket the expected metal concentration of a sample. Aspirate each in turn into flame and record absorbance. For calcium and magnesium calibration, mix 100 mL of standard with 10 mL lanthanum solution (see above) before aspirating. For chromium calibration mix 1 mL 30 % H_2O_2 with each 100 mL chromium solution before aspirating. For iron and manganese calibration, mix 100 mL of standard with 25 mL calcium solution before aspirating.

Prepare a calibration curve by plotting on linear graph paper absorbance of standards versus their concentrations. For instruments equipped with direct concentration readout, this step is unnecessary. With some instruments it may be necessary to convert percent absorption to absorbance by using a table generally provided by the manufacturer. Plot calibration curves for calcium and magnesium based on original concentration of standards before dilution with lanthanum solution. Plot calibration curves for iron and manganese based on original concentration of standards before dilution with calcium solution. Plot calibration curve for chromium based on original concentration of standard before addition of H_2O_2.

Check standards periodically during a run. Recheck calibration curve by aspirating at least one standard after completing analysis of a group of samples. For instruments with built-in memory, enter one to three standards to register a calibration curve for use in subsequent sample analyses.

c. *Analysis of samples:* Rinse nebulizer by aspirating water containing 1.5 mL conc HNO_3/L. Atomize blank and zero instrument. Atomize sample and determine its absorbance.

When determining calcium or magnesium, dilute and mix 100 mL sample with 10 mL lanthanum solution before atomization. When determining iron or manganese, mix 100 mL with 25 mL of calcium solution before aspirating. When determining chromium, mix 1 mL 30 % H_2O_2 with each 100 mL sample before aspirating.

Analyse standards at the beginning and end of a run, and at intervals during longer runs. Run a blank or solvent between each sample or standard to verify baseline stability. Determine metal concentration from calibration curve.

4. Calculations

Calculate concentration of each metal ion, in micrograms per litre, by referring to the appropriate calibration curve.

1. Apparatus

a. *Atomic absorption spectrophotometer and associated equipment:* See Section 303.2.

b. *Burner head,* conventional. Consult manufacturer's operating manual for suggested burner head.

c. *Separatory funnels,* 250 mL, with TFE stopcocks.

2. Reagents

a. *Air:* See 303A.2a.

b. *Acetylene:* See 303A.2b.

c. *Metal-free water:* See 303A.2c.

d. *Methyl isobutyl ketone (MIBK),* reagent grade. For trace analysis, purify MIBK by redistillation or by sub-boiling distillation.

e. *Ammonium pyrrolidine dithiocarbamate (APDC) solution:* Dissolve 4 g APDC in 100 mL water. If necessary, purify APDC with an equal volume of MIBK. Shake 30 s in a separatory funnel, let separate, and withdraw lower portion. Discard MIBK layer.

f. *Nitric acid,* HNO$_3$, conc., ultrapure.

g. *Standard metal solutions:* See 303A.21.

h. *Potassium permanganate solution,* KMnO$_4$, 5% aqueous.

i. *Sodium sulphate,* Na$_2$SO$_4$, anhydrous.

j. *Water-saturated MIBK:* Mix one part purified MIBK with one part water in a separatory funnel. Shake 30 s and let separate. Discard aqueous layer. Save MIBK layer.

3. Procedure

a. *Instrument operation:* See Section 303A.3a. After final adjusting of burner position, aspirate water-saturated MIBK into flame and gradually reduce fuel flow until flame is similar to that before aspiration of solvent.

b. *Standardization:* Select at least three standard metal solutions (prepared as in 303A.21) to bracket expected sample metal concentration and to be, after extraction, in the optimum concentration range of the instrument. Adjust 100 mL of each standard and 100 mL of a metal-free water blank to pH 3 by adding 1 N HNO$_3$ or 1 N NaOH. For individual element extraction, use the following pH ranges to obtain optimum extraction efficiency:

Element	pH Range for Optimum Extraction
Ag	3 – 5 (complex unstable)
Cd	1 – 6
Co	2 – 10
Cr	3 – 9
Cu	0.1 – 8
Fe	2 – 5
Mn	2 – 4 (complex unstable)
Ni	2 – 4
Pb	0.1 – 6
Zn	2 – 6

Transfer each standard solution and blank to an individual 250-mL separatory funnel, add 1 mL APDC solution and shake to mix. Add 10 mL MIBK and shake vigorously for 30 s. (The maximum volume ratio of sample to MIBK is 40). Let contents of each separatory funnel separate into aqueous and organic layers, drain off aqueous layer and discard. Make sure that none of the aqueous layer remains in funnel stem. Drain organic layer into a 10-mL glass-stoppered graduated cylinder or glass-stoppered centrifuge tube. If a centrifuge tube is used, the extract can be centrifuged to remove entrained water.

Aspirate organic extracts directly into the flame (zeroing instrument on a water saturated MIBK blank) and record absorbance. With some instruments it may be necessary to convert percent absorption to absorbance by using a table generally provided by the manufacturer.

Prepare a calibration curve by plotting on linear graph paper absorbances of extracted standards against their concentrations before extraction.

c. *Analysis of samples:* Rinse atomizer by aspirating water-saturated MIBK. Aspirate organic extracts treated as above directly into the flame and record absorbances.

With the above extraction procedure only hexavalent chromium is measured. To determine total chromium, oxidize trivalent chromium to hexavalent chromium by bringing sample to a boil and adding sufficient $KMnO_4$ solution dropwise to give a persistent pink colour while the solution is boiled for 10 min. Cool extract an aspirate.

During extraction, if an emulsion forms at the water-MIBK interface, add anhydrous Na_2SO_4 to obtain a homogeneous organic phase.

Extraction period: partitioning time for silver is critical (keep constant to within 30 s).

To avoid problems associated with instability of extracted metal complexes, determine metals immediately after extraction.

4. Calculations

Calculate the concentration of each metal ion in micrograms per litre by referring to the appropriate calibration curve.

Annex II
Determination of cyanide in water

Learning objective:
To illustrate the analytical process for the determination of total cyanide (*viz.* free cyanide and cyanide bound as metal complexes) in water. Following distillation to avoid matrix effects and preconcentrate the analyte, cyanide can be determined in two ways: titrimetrically and photometrically.

Reference:
Standard Methods for Examination of Water and Wastewater
American Public Health Association, Washington, D.C., 1989, pp 317–322.

412 B. Total Cyanide after Distillation

1. General Discussion
Hydrogen cyanide (HCN) is liberated from an acidified sample by distillation and purging with air. The HCN gas is collected by passing it through an NaOH scrubbing solution. Cyanide concentration in the scrubbing solution is determined by either titrimetric, colorimetric or potentiometric procedures.

2. Apparatus
The apparatus is shown in the figure below. It includes:
  a. *Boiling flask*, 1 L, with inlet tube and provision for water-cooled condenser.

  b. *Gas absorber*, with gas dispersion tube equipped with medium-porosity fritted outlet.
  c. *Heating element*, adjustable.
  d. *Ground glass ST joints*, TFE-sleeved or with an appropriate lubricant for the boiling flask and condenser. Rubber stopper joints also may be used.

3. Reagents
a. *Sodium hydroxide solution:* Dissolve 10 g NaOH in water and dilute to 1 L.
  b. *Magnesium chloride reagent:* Dissolve 510 g $MgCl_2 \cdot 6H_2O$ in water and dilute to 1 l.
  c. *Sulphuric acid,* H_2SO_4, 1 + 1.

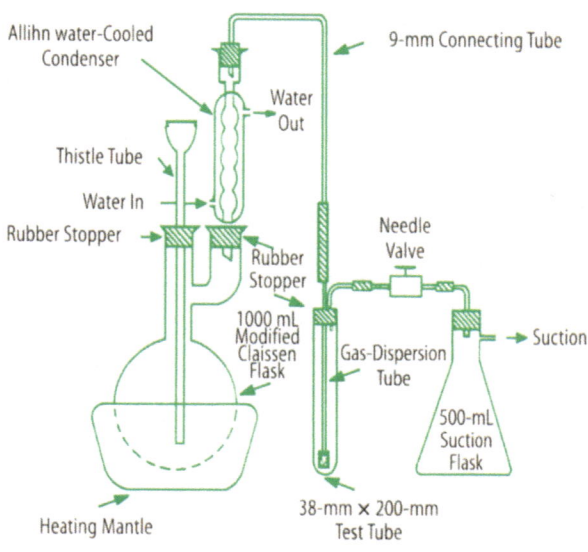

4. Procedure

a. Add 500 mL sample, containing no more than 100 mg CN/L (diluted, if necessary, with distilled water) to the boiling flask. Add 50 mL NaOH solution to the gas washer and dilute, if necessary, with distilled water to obtain an adequate depth of liquid in the absorber. Connect the train, consisting of boiling flask air inlet flask, condenser, gas washer, suction flask trap and aspirator. Adjust suction so that approximately 1 air bubble/s enters the boiling flask. This air rate will carry HCN gas from flask to absorber and usually will prevent a reverse flow of HCN through the air inlet. If this air rate does not prevent sample back-up in the delivery tube, increase air-flow rate to 2 air bubbles/s. Observe air purge rate in the absorber, where the liquid level should be raised not more than 6.5 to 10 mm. Maintain air flow throughout the reaction.

b. Add 50 mL 1 + 1 H_2SO_4 through the air inlet tube. Rinse tube with distilled water and let air mix flask contents for 3 min. Add 20 mL $MgCl_2$ reagent through air inlet and wash down with stream of water. A precipitate that may form redissolves on heating.

c. Heat with rapid boiling, but do not flood condenser inlet or permit vapours to rise more than halfway into condenser. Adequate refluxing is indicated by a reflux rate of 40 to 50 drops/min from the condenser lip. Reflux for at least 1 h. Discontinue heating but continue air flow. Cool for 15 min and drain gas washer contents into a separate container. Rinse connecting tube between condenser and gas washer with distilled water, add rinse water to drained liquid and dilute to 250 mL in a volumetric flask.

d. Determine cyanide content by titration method (C) if cyanide concentration exceeds 1 mg/L or by the colorimetric method (D) if the cyanide concentration is less. Use titration, the electrode probe method or the spot test to approximate CN content. Alternatively, use the cyanide-selective electrode in the concentration range 0.05 to 10 mg CN/L (Method E).

e. Distillation gives quantitative recovery of even refractory cyanides such as iron complexes. To obtain complete recovery of cobalticyanide use ultraviolet radiation pretreatment. If incomplete recovery is suspected, distill again by refilling the gas washer with a fresh charge of NaOH solution

and refluxing 1 h more. The cyanide from the second reflux, if any, will indicate completeness of recovery.

f. As a quality control measure, periodically test apparatus, reagents and other potential variables in the concentration range of interest. As an example, a minimum 98% recovery from 1 mg CN/L standard should be obtained.

412 C. Titrimetric Method

1. General Discussion

a. *Principle:* CN^- in the alkaline distillate from the preliminary treatment procedure is titrated with standard silver nitrate ($AgNO_3$) to form the soluble cyanide complex, $Ag(CN)\lvert O_2^+$. As soon as all CN^- has been complexed and a small excess of Ag^+ has been added, the excess Ag^+ is detected by the silver-sensitive indicator, p-dimethylaminobenzalrhodanine, which immediately turns from a yellow to a salmon colour. The distillation has provided a 2:1 concentration. The indicator is sensitive to about 0.1 mg Ag/L. If titration shows that CN^- is below 1 mg/L, examine another portion colorimetrically.

2. Apparatus

Koch microburette. 5-mL capacity.

3. Reagents

a. *Sodium hydroxide solution,* NaOH, 1 N.

b. *Sulphuric acid solution,* H_2SO_4, 1 N.

c. *Indicator solution:* Dissolve 20 mg p-dimethylaminobenzalrhodanine in 100 mL acetone.

d. *Standard silver nitrate titrant,* 0.0192 N: Dissolve 3.27 g $AgNO_3$ in 1 L distilled water. Standardize against standard NaCl solution, using the argentometric method with K_2CrO_4 indicator, as directed in Chloride, Section 407 A.

Dilute 500 mL $AgNO_3$ solution according to the titre found so that 1.00 mL is equivalent to 1.00 mg CN.

4. Procedure

a. From the absorption solution take a measured volume of sample so that the titration will require approximately 1 to 10 mL $AgNO_3$ titrant. Dilute to 250 mL or some other convenient volume to be used for all titrations. For samples with low cyanide concentration (< 5 mg/L) do not dilute. Add 0.5 mL indicator solution.

b. Titrate with standard $AgNO_3$ titrant to the first change in colour from a canary yellow to a salmon hue. Titrate a blank containing the same amount of alkali and water. As the analyst becomes accustomed to the end point, blank titrations decrease from the high values usually experienced in the first few trials to 1 drop or less. with a corresponding improvement in precision.

5. Calculation

$$mg\ CN/L = \frac{(A - B) \times 1000}{mL\ portion\ used}$$

$$\times \frac{250}{mL\ portion\ used}$$

where:

A = mL standard $AgNO_3$ for sample and
B = mL standard $AgNO_3$ for blank.

6. Precision and Accuracy

For samples containing more than 1 mg CN/L that have been distilled or for relatively clear samples without significant interference, the coefficient of variation is 2%. Extraction and removal of S^{2-} or oxidizing agents tend to increase the coefficient of variation to a degree determined by the amount of manipulation and the type of sample. The limit of sensitivity is approximately 0.1 mg CN/L, but at this concentration the end point is indistinct. At 0.4 mg/L, the coefficient of variation is four times that at CN concentration levels > 1.0 mg/L.

412 D. Colorimetric Method

1. General Discussion

a. *Principle:* CN^- in the alkaline distillate from preliminary treatment is converted to CNCl by reaction with chloramine-T at pH < 8 without hydrolysing to CNO^-. (CAUTION: CNCl is a toxic gas: avoid inhalation). After the reaction is complete, CNCl forms a red-blue dye on addition of a pyridine-barbituric acid reagent. If the dye is kept in an aqueous solution, the absorbance is read at 578 nm. To obtain colours of comparable intensity, have the same salt content in sample and standards.

b. *Interference:* All known interferences are eliminated or reduced to a minimum by distillation.

2. Apparatus

Colorimetric equipment: One of the following is required:

a. *Spectrophotometer,* for use at 578 nm, providing a light path of 10 mm or longer.

b. *Filter photometer* providing a light path of at least 10 mm and equipped with a red filter having maximum transmittance at 570 to 580 nm.

3. Reagents

a. *Chloramine-T solution:* Dissolve 1.0 g white, water-soluble powder in 100 mL water. Prepare weekly and store in refrigerator.

b. *Stock cyanide solution:* Dissolve approximately 2 g KOH and 2.51 g KCN in 1 L distilled water. (CAUTION: KCN is highly toxic: avoid contact or inhalation). Standardize against standard silver nitrate ($AgNO_3$) titrant as described in Section 412C.4, using 25 mL KCN solution. Check titre weekly because the solution gradually loses strength: 1 mL = 1 mg CN.

c. *Standard cyanide solution:* Based on the concentration determined for the KCN stock solution, calculate volume required (approximately 10 mL) to prepare 1 L of a 10 µg CN/mL solution. Dilute with the 2 g NaOH/L solution. Dilute 10 mL of the 10 µg CN/mL solution to 100 mL with the 2 g NaOH/L solution: 1.0 mL = 1.0 µg CN. Prepare fresh daily and keep in a glass-stoppered bottle. (CAUTION: Toxic –take care to avoid ingestion).

d. *Pyridine–barbituric acid reagent:* Place 15 g barbituric acid in a 250-mL volumetric flask and add just enough water to wash sides of flask and wet barbituric acid. Add 75 mL pyridine and mix. Add 15 mL conc. hydrochloric acid (HCl), mix and cool to room temperature. Dilute to mark with water and mix. This reagent is stable for up to 1 month: discard if a precipitate develops.

e. *Sodium dihydrogen phosphate,* 1 M: Dissolve 138 g $NaH_2PO_4 \cdot H_2O$ in 1 L distilled water. Refrigerate.

f. *Sodium hydroxide solution:* Dissolve 2 g NaOH in 1 L distilled water.

4. Procedure

a. Preparation of calibration curve: Prepare a blank of NaOH solution. From the standard

KCN solution prepare a series of standards containing from 0.2 to 6 µg CN in 20 mL solution using the 2 g NaOH/L solution for all dilutions. Treat standards in accordance with Section b below. Plot absorbance of standards against CN concentration (micrograms).

Recheck calibration curve periodically and each time a new reagent is prepared.

On the basis of the first calibration curve, prepare additional standards containing less than 0.2 and more than 6 µg CN to determine the limits measurable with the photometer being used.

b. Colour development: Adjust photometer to zero absorbance each time using a blank consisting of the NaOH dilution solution and all reagents. Take a portion of absorption liquid obtained in Method B, such that the CN concentration falls in the measurable range, and dilute to 20 mL with NaOH solution. Place the portion in a 50-mL volumetric flask. Add 4 mL phosphate buffer and mix thoroughly. Add 2.0 mL chloramine-T solution and swirl to mix. Immediately, add 5 mL pyridine-barbituric acid solution and swirl gently to mix. Dilute to mark with water; mix well by inversion.

Measure absorbance with the photometer at 578 nm after 8 min but within 15 min from the time of adding the pyridine-barbituric acid reagent. Even with the specified time of 8 to 15 min there is a slight change in absorbance. To minimize this, standardize time for all readings. Using the calibration curve and the formula in Section 5 below, determine CN concentration in original sample.

5. Calculations

$$\text{mg CN/L} = \frac{A \times B}{C \times D}$$

where:

A = µg CN read from calibration curve (50 mL final volume);

B = total volume of absorbing solution from the distillation, mL;

C = volume of original sample used in the distillation, mL; and

D = volume of absorbing solution used in colorimetric test, mL.

6. Precision

The analysis of a mixed cyanide solution containing sodium, zinc, copper and silver cyanides in tap water gave a precision within the designated range as follows:

$$S_T = 0.115X + 0.031$$

where:

S_T = overall precision and
X = CN concentration, mg/L.

Annex III
Determination methylmercury in fish and shellfish

Learning objective:
To exemplify the speciation of pollutants in samples used as environmental indicators.

Reference:
Official Methods of Analysis of AOAC. 15th Edition
AOAC, Arlington, MA, 1990, pp 268, 269.

988.11 Mercury (Methyl) in Fish and Shellfish Rapid Gas Chromatographic Method First Action 1988

A. Principle
Organic interferences are removed from homogenized seafood by acetone wash followed by toluene wash. Protein-bound Me Hg is released by addition of HCl and extracted into toluene. Toluene extract is analysed for CH_2HgCl by electron capture GC.

B. Reagents
Equivalent reagents may be used.

(a) *Solvents.* Acetone, toluene and isopropanol, all distilled in glass (Burdick and Jackson Laboratories, Inc., or EM Science OmniSolv® reagents). Caution: toluene is harmful if inhaled and is flammable; conduct all operations with toluene in laboratory hood.

(b) *Hydrochloric acid* solution (1 + 1). Add concentrated HCl to equal volume distilled or de-ionized H_2O and mix. Use 2 volumes toluene to extract potential interferences from 1 volume HCl solution by vigorously shaking mixture 15 s in separator. Discard toluene extract. Repeat extraction step 4 times. Solution may be mixed in advance. However, extraction must be performed immediately before HCl solution is used to avoid formation of electron-capturing compounds which produce extraneous peaks in chromatograms.

Before beginning analysis, check quality of reagents by chromatographing blank taken through method. Do not use HCl and solvents which produce extraneous peaks at retention time of Me Hg.

(c) *Carrier gas.* GC quality Ar – CH_4 (95 + 5).

(d) *Sodium sulphate. Anhydrous* reagent grade. Heat overnight in 600 °C furnace, let

cool, and store in capped bottle. Line cap with acetone-washed Al foil to prevent contamination from cap. Peaks appearing at 14 – 15 min may be eliminated by refiring Na_2SO_4 (600 °C overnight).

(e) *Methyl mercuric chloride* standard solutions. Keep tightly stoppered. Seal stopper with Teflon tape. *(1) Stock standard solution.* – 1000 µg Hg/mL. Weigh 0.1252 g CH_3HgCl (ICN-K&K Laboratories, Inc., PO Box 28050, Cleveland, OH 44128-0250) into 100 mL volumetric flask. Dilute to volume with toluene. *(2) High level intermediate standard solution.* – 40 µg Hg/mL. Dil. 10.0 mL stock standard solution to 250.0 mL with toluene. *(3) Low level intermediate standard solution.* – 2.0 µg Hg/mL. Dil. 10.0 mL high level intermediate standard solution to 200.0 mL with toluene. *(4) Working standard solutions.* – 0.005 – 0.10 µg Hg/mL. Prepare monthly by diluting with toluene in volumetric flasks as follows: Dilute 10.0 mL of 2.0 µg Hg/mL solution to 200.0 mL for 0.10 µg Hg/mL. Dilute 20.0 mL of 0.10 µg Hg/mL solution to 25.0 mL, 15.0 mL to 25.0 mL, 10.0 mL to 25.0 mL, 10.0 mL to 50.0 mL, 10.0 mL to 100.0 mL, and 10.0 mL to 200.0 mL for 0.080, 0.060, 0.040, 0.020, 0.010, and 0.005 µg Hg/mL, respectively.

(f) *Mercuric chloride column treatment solution.* – 1000 ppm $HgCl_2$. Dissolve 0.1 g $HgCl_2$ in 100 mL toluene.

(g) *Fortification solutions. (1) Stock solution.* – 1000 µg Hg/mL. Weigh 0.1252 g CH_3HgCl into 100 mL volumetric flask. Dil. to volume with H_2O. *(2) Working fortification solution.* – 15 µg Hg/mL. Dilute 1500 µL stock fortification solution to 100.0 mL with H_2O.

C. Apparatus

Wash all glassware with detergent (Micro Laboratory Cleaner, International Products, PO Box 118, Trenton, NJ 08601-0118) and rinse thoroughly with hot tap H_2O followed by distilled or de-ionized H_2O. Then rinse 3 times with acetone and 3 times with toluene. Dry in hood.

Equivalent apparatus may be used except use packed column specified.

(a) *Centrifuge.* Model IEC CRU-5000 or CR6OOO (International Equipment Co.).

(b) *Centrifuge tubes.* Glass, 50 mL capacity with Teflon-lined screw caps (Cat. No. 9212-K78, Thomas Scientific).

(c) *Graduated cylinders.* Glass, class A, 50 mL capacity, with ground-glass stoppers (Kimbie 20036).

(d) *Transfer pipettes.* Disposable glass, Pasteur-type.

(e) *Dropping pipettes.* Glass, 5 mL capacity (No. 13-710B, Fisher Scientific Co.).

(f) *Mechanical shaker.* Model S-500 shaker-in-the-round, with timer (Glas-Col Apparatus Co., 711 Hulman St, PO Box 2128, Terre Haute, IN 47802.).

(g) *Gas chromatograph.* Hewlett-Packard Model 5710A equipped with linear ^{63}Ni electron capture detector, Model 7131A recorder, and 6 ft × 2 mm id silanized glass column packed with 5% DEGS-PS on 100-120 mesh Supelcoport (Supelco, Inc.). Pack column no closer than 2.0 cm from injection and detector port nuts and hold packing in place with 2 cm high quality, silanized glass wool at both ends. Install oxygen scrubber and molecular sieve dryer (No. HGC-145, Analabs, Inc.) between carrier gas supply and column. Condition column according to manufacturer's instructions as follows: flush column 0.5 h with carrier gas flowing at 30 mL/min at room temperature. Then heat 1 h at 50 °C. Next, heat column to 200 °C at 4 °C/min and hold at 200 °C overnight. Do not connect column to detector during this conditioning process. Maintain 30 mL/min carrier gas flow at all times during conditioning, treatment and use. Operating conditions: column 155 °C, injector 200 °C, detector 300 °C; carrier gas flow 30 mL/min; recorder chart speed 0.5-1.0 cm/min. Under these conditions and with $HgCl_2$ column treatment procedure described below, CH_3HgCl peak appears 2-3 min after injection of extract.

D. Mercuric Chloride Column Treatment

Column of 5% DEGS-PS, conditioned according to manufacturer's instructions, can be used to determine CH_3HgCl only after treatment by $HgCl_2$ solution. Because column performance degrades with time, also treat column periodically during use. Perform appropriate $HgCl_2$ treatment procedures described below.

(a) Following 200 °C column conditioning and after every 2-3 days of analyses. If column has just been conditioned according to manufacturer's instructions or has been used 2-3 days to analyse extracts, proceed as follows: adjust column temperature to 200 °C and inject 20 µL $HgCl_2$ treatment solution 5 times at 5-10 min intervals. Maintain 200 °C temperature overnight. Chromatogram will contain large, broad peaks. Adjust column temperature to 155 °C next morning and inject 20 µL $HgCl_2$ treatment solution 2 more times. Large, broad chromatographic peaks appearing at *ca.* 1-2 h signal completion of treatment process and that column is ready for use.

(b) On day preceding analyses. If column has been treated by procedure (a) or used 1 day at 155 °C to analyse extracts, column may be treated at end of working day for next day's use as follows: Lower column temperature to 115 °C and inject 20 µL $HgCl_2$ treatment solution 1 time. After large, broad peaks appear in chromatogram (11-20 h), treatment process is complete. Next working day, increase column temperature to 155 °C operating temperature. When baseline is steady, column is ready for use.

(c) During extract analyses at 155 °C. If column has been used at 155 °C to analyse extracts or if column performance and peak height have degraded enough to require $HgCl_2$ treatment, inject two 20 µL aliquots of $HgCl_2$ treatment solution. Large, broad peaks will appear in chromatogram 1-2 h after $HgCl_2$ injection, signalling completion of treatment process. Wait for steady baseline; then column is ready for use.

E. Extraction of Methyl Mercuric Chloride

Perform all operations except weighing in laboratory hood. Take empty centrifuge tube through all steps for method blank determination. Accurately weigh 1 g homogenized test sample into 50 mL centrifuge tube. Add 25 mL acetone; tightly cap and vigorously shake tube by hand 15 s. Loosen cap and centrifuge 5 min at 2000 rpm. Carefully decant and discard acetone. (Use dropping pipette to remove acetone, if necessary.) Repeat 25 mL acetone wash step 2 more times. Break up tissue with glass stirring rod before shaking tube, if necessary. Add 20 µL toluene; tightly cap and vigorously shake tube by hand 30 s. Loosen cap and centrifuge 5 min at 2000 rpm. Carefully decant (or draw off with dropping pipette) and discard toluene. Extraneous peaks in final GC chromatogram may indicate that more vigorous shaking with acetone and toluene is required. In products for which Me Hg recoveries are to be determined, fortify tissue at this point by adding working fortification solution, to centrifuge tubes.

Add 2.5 mL HCl solution, to centrifuge tube containing acetone and toluene washed sample. Break up tissue with glass stirring rod, if necessary. Extract CH_3HgCl by adding 20 mL toluene and shaking tube gently but thoroughly 5 mm on mechanical shaker at setting 5 (2 min by hand). Loosen cap and centrifuge 5 min at 2000 rpm. If emulsion is present after centrifugation, add 1 mL isopropanol and gently stir into toluene layer with glass stirring rod to reduce emulsion. Do not mix isopropanol with aqueous phase. Add equal amounts of isopropanol to blank and test solutions. If emulsion is not present, do not add isopropanol to blank or test solutions. Vigorous mixing of isopropanol with HCl may produce interfering peaks in chromatograms. Recentrifuge. With dropping pipette, carefully transfer toluene to graduated cylinder. Rinse walls of centrifuge tube with 1–2 mL toluene and transfer rinse to graduated cylinder. Repeat extraction step 1 more time. Combine both extracts in graduated cylinder, dilute to 50 mL with toluene, stopper and mix well. Add 10 g Na_2SO_4 and mix again. Tightly stoppered extracts (sealed with Teflon tape) may be refrigerated and held overnight at this point. Analyse by GC.

F. Gas Chromatography

Verify that system is operating properly by injecting 5 µL standard solution containing 0.005 µg Hg/mL into GC system. Difference between CH_3HgCl peak heights for 2 injections should be ≤4%. Check detector linearity by chromatographing all working standard solutions.

Inject 5 µL standard solution with concentration approximately equal to or slightly greater than concentration of extract. Immediately after CH_3HgCl peak appears, inject another 5 µL extract. Immediately after CH_3HgCl and background peaks for extract appear, inject another 5 µL aliquot of standard solution. Because column performance and peak height slowly decrease with time, calculate Hg concentration in each test sample by comparing peak height for each test extract to average peak height for standard solutions injected immediately after test extract.

Correct height of CH_3HgCl peak for test extract by subtracting height of peak for method blank obtained at same attenuation and recorder sensitivity. Calculate Me-bound Hg content of test sample expressed as µg Hg/g (ppm Hg) by comparing height of peak from injection of test extract to average height of peak from duplicate injections of standard solution as follows:

$$\mu g\ Hg/g\ \text{fish} = (R/R') \times (C'/C) \times 50$$

where

R = corrected height of CH_3HgCl peak from injection of test extract;

R' = average height of CH_3HgCl peak from duplicate injections of standard solution;

C = weight (g) of test portion;

C' = concentration (µg/mL) of Hg in standard solution; and

50 = final volume (mL)

Annex IV
Determination of silica in rocks

Learning objective:
To illustrate an analytical process based on the gravimetric technique (specifically, on the weight difference between the sample before and after treatment with hydrofluoric acid). The method relies on the volatilization of the silica in the following reaction:

$$SiO_2 + 4\,HF \leftrightarrow F_4Si\uparrow + 2\,H_2O$$

Reference:
P. G. Jeffery
Chemical Methods of Rock Analysis
Pergamon Press, Oxford, 1975, pp 406, 407

Method

Reagents
Hydrofluoric acid, concentrated.
Sulphuric acid, 20 N.

Procedure
Ignite a clean platinum crucible of about 30-mL capacity over the full flame of a Meker burner, allow to cool and weigh empty. Transfer to it approximately 1 g of the finely ground quartzitic material and reweigh to give the weight of the sample.[1] Again ignite the crucible over the Meker burner, gently at first and then finally over the full flame of the burner for a period of 1 h. Allow to cool and reweigh the crucible and contents. Repeat the ignition until no further loss in weight occurs. Report the change in weight as the "loss on ignition". This loss is due to the oxidation of ferrous iron and organic matter, and to the volatilization of water, carbon dioxide and other gases.

Add 4 drops of 20 N sulphuric acid to the residue, followed by 20 mL of concentrated hydrofluoric acid. Transfer the crucible to a hot plate or sand bath and remove the silica and excess hydrofluoric acid by volatilization in the usual way. Gently heat the crucible on the hot plate to remove the excess sulphuric acid and then ignite over the full flame of the Meker burner for 10 min. Allow to cool and reweigh the crucible. Repeat the ignition until no further loss in weight occurs. Determine a reagent blank by volatilization in the same way, but omitting the sample material. The blank value obtained with analytical grade hydrofluoric acid will probably amount to between 0.3 and 1.0 mg. This should be added to the loss on volatilization to give the total silica content of the sample.

In a very few samples some silicate mineral grains may remain undecomposed. These can usually be attacked by a second evaporation with smaller amounts of hydrofluoric and sulphuric acid.

---

[1] Tsubaki describes a modification to this procedure in which 3 g of boric acid is added to a 0.5 g sample portion prior to igniting at 1000 °C. The determination is then completed as described above.

Annex V
Determination of silica in rocks

Learning objective:
To exemplify a determination routinely performed by clinical control laboratories.

Reference:
W. R. Faulkner and S. Meites
Selected Methods for the Small Clinical Chemistry Laboratory
American Association for Clinical Chemistry, Washington, D.C., 1982, pp 353–355.

Triglycerides in Serum, Colorimetric Method

Introduction
Measurement of serum triglycerides is useful for diagnosis of hyperlipidemia and hyperlipoproteinemia. The correct diagnosis and treatment of hyperlipoproteinemia depends on accurate and precise methods for triglyceride determination. Presented here is a manual colorimetric method described by Neri and Frings (1). This method is relatively simple and requires only equipment that is usually available even in small clinical chemistry laboratories. It is ideally suited for laboratories that do not have a large number of requests for triglyceride assays and for those wanting to make their own reagents.

Principle
Isopropanol extracts of serum are prepared and then treated with alumina to remove glycerol, phospholipids and glucose. Triglycerides are then saponified to yield glycerol, which is oxidized by sodium metaperiodate to formaldehyde. Formaldehyde reacts with acetylacetone to form a yellow dihydrolutine derivative. The absorbance of this product at 405 nm is proportional to the serum concentration of triglycerides.

Materials and Methods

Reagents
1. Isopropanol. "Aldehyde-free," analytical grade reagent.
2. Alumina (Al_2O_3), washed. Wash alumina (Woelm, neutral, activity grade 1 for chromatography; Waters Associates, Inc., Framingham, MA 01701) until fines are removed. This usually requires eight to 10 washings, with approximately four bed-volumes of de-ionized water for each wash. Dry the alumina in an oven for 15 to 18 h at 100–110 °C. Do not keep the alumina in the oven longer than 18 h. It can be used satisfactorily for at least two months when stored in a tightly stoppered container at room temperature.
3. Saponification reagent. Dissolve 10.0 g of KOH in 75 mL of water and 25 mL of isopropanol. This reagent is stable for at least two months when stored at room temperature in a brown glass bottle.
4. Sodium metaperiodate reagent. Dissolve 77 g of anhydrous ammonium acetate in 700 mL of de-ionized water; add 50 mL of glacial acetic acid and 650 mg of sodium metaperiodate ($NaIO_4$). Dissolve the sodium metaperiodate in this solution and dilute to 1 L with de-ionized water. Do not change the order of addition of chemicals. This reagent is stable for at least two months when stored at room temperature in a brown glass bottle.
5. Acetylacetone reagent. Add 0.40 mL of 2,4-pentanedione to 100 mL of isopropanol. This reagent is stable for at least two months when stored at room temperature in a brown glass bottle.
6. Standard. The stock standard contains 1.000 g of triolein (A grade; Calbiochem, San Diego, CA 92112) per 100 mL of isopropanol. Prepare a working standard of 200 mg/dL by diluting 2.0 mL of the stock standard to volume with isopropanol in a 10-mL volumetric flask. The

stock standard is stable for at least two months at 4-7 °C in a tightly sealed container. Prepare working standard freshly each week and store in a brown glass bottle at 4-7 °C.

Apparatus
The procedure requires 16 × 125 mm screw-capped culture tubes, a mechanical rotator, a vortex-type mixer, a water bath at 65-70 °C, and a spectrometer or spectrophotometer to measure absorbance at 405 nm.

Specimen Collection
Because serum triglyceride concentrations are increased after an intake of most foods, blood samples must be taken after a 10-12 h fast. This method requires 0.2 mL of serum, but plasma from blood collected with ethylenediaminetetraacetate (EDTA) as an anticoagulant is also satisfactory. Triglycerides in serum are stable at least three days when stored at room temperature (24 °C) or refrigerated at 4-7 °C.

Procedure
1. To a 16 × 125 mm screw-capped culture tube (blank) containing 0.7 (± 0.1) g of washed alumina, add 5.0 mL of isopropanol and 0.20 mL of water.
2. To another tube (standard) containing 0.7 (± 0.1) g of washed alumina, add 5.0 mL of isopropanol and 0.20 mL of working standard.
3. To another tube (unknown) containing 0.7 (± 0.1) g of washed alumina, add 5.0 mL of isopropanol and 0.20 mL of serum.

4. Place all tubes on a mechanical rotator for 15 min.
5. Centrifuge for approximately 10 min at high speed to pack the alumina.
6. Transfer 2.0 mL of the clear supernatant fluid to appropriately labelled test tubes.
7. Add 0.60 mL of saponification reagent to a» tubes and mix with a vortex-type mixer.
8. Let all tubes stand for at least 5 min, but no longer than 15 min, at room temperature.
9. Add 1.5 mL of periodate reagent to all tubes and mix with a vortex-type mixer.
10. Add 1.5 mL of acetylacetone reagent to all tubes and mix with a vortex-type mixer. Tightly cover each tube with Parafilm.
11. Place all tubes in a 65-70 °C water bath for approximately 15 min.
12. Remove all tubes from the water bath, allow to cool at room temperature for 10-15 min and vortex-mix for 5 s.
13. Within 1 h measure the absorbance of standard and unknowns against the blank at 405 min in 10- or 12-mm cuvettes.
14. Calculate values for unknowns by comparing the absorbances (A) of the unknown and standard as follows:

$$(A_{unknown}/A_{standard}) \times 200$$
$$= \text{triglycerides, mg/dL}$$

Clinical Interpretation
Normal concentrations of serum triglycerides for fasting (10-12 h) adults are approximately 30-150 mg/dL.

Annex VI
Determination of glucose in serum

Learning objective:
To illustrate a determination routinely performed by clinical control laboratories.

Reference:
J. D. Bauer
Clinical Laboratory Methods
C. V. Mosby Co., St Louis, Missouri, 1984.

Glucose oxidase method

Principle
Glucose is oxidized to glucuronic acid by the enzyme glucose oxidase in the presence of oxygen (air). The reaction releases hydrogen peroxide, which, in the presence of another enzyme, peroxidase, is decomposed, the oxygen thus formed oxidizing the chromogen, which can then be measured colorimetrically. The amine *o*-dianisidine was used for this purpose in earlier methods; however, the phenol-aminoantipyrine mixture is a better chromogen as it gives a dark red colour. The reaction is highly specific for glucose, so it provides glucose contents very close to the true values. High concentrations of reductants – particularly ascorbic acid and, to a lesser extent, uric acid – can interfere with the reaction by competing with the chromogen for the oxygen released, thus leading to underestimated results. Hemoglobin also interferes as it causes the hydrogen peroxide to decompose prematurely, thereby also giving rise to underestimated values. This interference is seemingly less serious with the phenol-antipyrine mixture than it used to be with the former chromogen (*o*-dianisidine). With strongly hemolysed sera or whole blood, a Somogyi filtrate must be used.

Reagents
1. *Phosphate buffer*, 0.1 mol/L, pH 7.0
   Dissolve 8.5 g anhydrous disodium hydrogen phosphate (Na_2HPO_4) and 5.3 g potassium dihydrogen phosphate (KH_2PO_4) in about 8 dL water. Check solution

pH and adjust to 7.0 ± 0.1 by addition of a small volume of 1 mol/L NaOH or HCl, as required. Finally, dilute to 1 litre.
2. *Peroxidase reagent*
   Dissolve 175 mg (0.75 mmol) 4-amino-antipyrine (4-aminophenazone, Sigma Chemical Co.) and 2 mg peroxidase (Sigma Type II) in 5 dL of the phosphate buffer. This solution is stable for about four weeks if stored refrigerated.
3. *Glucose oxidase reagent*
   Add 2 mL glucose oxidase stock solution (Sigma Type V, 1000 unit/mL) to the peroxidase reagent. This solution is stable for one week if stored in a refrigerator.
4. *Phenol solution*
   Dissolve 2.0 g (21.3 mmol) phenol and 9 of NaCl in water, and make to 1 litre. This solution is stable for several months at room temperature.
5. *Standards*
   The standards used are the same as those employed in the previous direct determination procedure.

Procedure and calculations
Add 50 µL serum or standard to 2 mL of the glucose oxidase reagent in a test tube and mix. Next, add 2 mL of the phenol reagent, stopper the tube and shake to facilitate aeration. Heat the tube in a bath a 37 °C for 15 min, allow to cool and read the absorbances of the sample and standards at 510 nm against a reagent blank subjected to the same treatment.

Calculate the results from the following expression:

$$\frac{\text{Sample absorbance}}{\text{Standard absorbance}} \times \text{Standard concentration}$$

$$= \text{Sample concentration}$$

Constraints

With hyperlipidemic or slightly hemolysed sera, as well as sera containing large amounts of bilirubin, a serum blank can be prepared by adding the same amount of serum in the sample to a reagent consisting of the phenol solution diluted with an identical volume of water. This serum blank is measured against the reagent alone and any absorbance obtained is subtracted from that provided by the serum treated with the whole reagent. With strongly hemolysed sera and sera containing large amounts of bilirubin, use of a Somogyi filtrate prepared by 10:1 dilution is to be preferred; the filtrate volume to be used is 10 times as high as that of serum employed in the direct method. Standards must be diluted to the same extent (10:1) and added in proportional amounts in order to ensure constancy in the calculations. Acid precipitants (*e.g.* trichloroacetic acid) are inappropriate for use in the glucose oxidase method.

Glucose oxidase is specific to β-D-glucose; however, dissolved glucose exists 36% as the α form and 64% as the β form. For the glucose to be thoroughly oxidized, the α form must be mutarotated to the β form. The rate of this conversion depends on pH and temperature. Some commercially available glucose oxidase preparations contain the enzyme glucomutarotase, which accelerates the reaction.

Enzymatic processes for determining glucose should never be calibrated with freeze-dried serum. In fact, freeze-drying can cause some glucose to bind to proteins and protein-bound glucose is unavailable for enzymatic analysis under moderate conditions. By contrast, the strongly acidic conditions of the *o*-toluidine method ensure recovery of the whole glucose.

The glucose oxidase method performs quite well; however, preparing the reagents can be tedious and laboratory personnel may opt for using commercially available reagent kits such as those from Biodynamics/Boehringer Mannheim Corp. (Indianapolis), Sclavo (Wayne, NJ) or Worthington Biochemical Corp. (Freehold, NJ).

Normal values and interpretation

The fasting concentration of glucose in serum or plasma as determined by the above-described methods typically ranges from 70 to 110 mg/dL (3.9–5.8 mmol/L). The levels provided by the earlier methods are about 5% higher. The glucose concentration in whole blood is somewhat different. Although most methods (enzymatic and *o*-toluidine) currently in use for this purpose analyse serum or plasma, glucose values thus determined are taken to be fasting blood sugar (FBS) values; unless otherwise stated, however, the values given in this and subsequent sections are glucose concentrations in serum.

High fasting glucose levels (up to 500 mg/dL or 28 mmol/L) are typically found in diabetic patients, depending on the severity of their condition.

Hypoglycemia (low sugar levels in blood) is most often the result of insulin overdose during antidiabetic treatments.

Questions

(1) Relate CMP to "technique", "method" and "procedure".

(2) Comment on the "black box" notion implicit in the generic definition of CMP.

(3) Why should equipment calibration also be considered in conducting preliminary operations?

(4) Give your opinion about the suitability of the designation CMP for the body of operations involved in obtaining an analytical chemical result (Box 4.1).

(5) How does the analytical problem influence the way a CMP should be designed and applied?

(6) Can calibration operations (Fig. 4.5) be considered steps of a CMP?

(7) Relate "calibration" to "chemical measurement process".

(8) Why does the type of analyte involved dictate the particular CMP to be used? Give an example illustrating the fact that the same sample may require different types of treatment if different analytes are to be determined.

(9) Give some examples of apparatus calibration in the analytical process.

(10) Explain the significance of preliminary operations in terms of their features (Fig. 4.6).

(11) Which group of analytical properties is most markedly influenced by the preliminary operations of the CMP?

(12) What are the most salient differences between physical and chemical measurement processes?

(13) Why are preliminary operations so varied in nature? Why is this a constraint rather than an asset?

(14) Give some examples of instrument calibration in the analytical process.

(15) Why do the specific analytical tools available dictate the way a CMP is to be designed?

(16) Why do preliminary operations affect representativeness of results?

(17) Define "computer (hardware and software) validation" as a part of a CMP.

(18) Give one example of each type of interface shown in the schemes of Fig. 4.7.

(19) The first example in Box 4.4 (the determination of metals in soil) involves two rather different types of treatment. Why?

(20) Give two examples of integration of two steps of the analytical process.

(21) What is the usual origin of the problems encountered during sampling?

(22) Can a random sample be representative? Can a selective sample? Why?

(23) Why do random errors accumulate in the course of a CMP?

(24) What is a bulk sample? What is an aliquot?

(25) In what situations are analytical separation methods redundant?

(26) Comment on the characteristics of a sampling plan.

(27) How can spatial and temporal heterogeneity in a sample be defined?

(28) What is the "clean-up" operation? What analytical property does it affect?

(29) What preliminary operations are conceivable for a solid (A), liquid (B), gaseous sample (C)?

(30) Why do analytical separation methods play such a prominent role in CMPs?

(31) Comment on the basic and technical aspects of preconcentration.

(32) Distinguish continuous and discrete separation methods.

(33) Comment on the validation-calibration binomial in a CMP.

(34) Distinguish chromatographic and non-chromatographic separation methods.

(35) Based on the example of Fig. 4.12, explain how the sampling plan is determined by the particular analytical problem.

(36) Distinguish on-line and off-line coupling of a computer to a CMP.

(37) Comment on the types of data used to compute the results of a CMP.

(38) Comment on human intervention in a CMP.

(39) What does transducing the analytical signal involve?

(40) Distinguish (A) passive and active, (B) primary and relative, and (C) stand-alone and integrated instruments. Give examples to illustrate the differences.

(41) How can the signals produced by instruments in the second step of a CMP be acquired?

(42) What is the difference between external and internal validation?

(43) Comment on the ranking of the different meanings of validation in the context of CMPs.

(44) Discuss the impact of automation on a CMP.

(45) How can a CMP be miniaturized?

(46) Give an example of simplification of a CMP.

(47) Why does the use of sensors encompass the first and second steps of the analytical process?

(48) Why can a chromatograph not span the entire first step of a CMP?

(49) Why are scientific and technical trends in CMPs related? Give some examples.

(50) Why do analytical separation techniques support the accuracy of the results produced by a CMP.

Seminar 4.1

Analytical Processes for the Determination of Inorganic Pollutants in Water

- **Learning objectives**
  - To examine different measurement procedures for the same type of sample (water) in environmental analysis.
  - To become acquainted with chemical measurement processes reported in the specialized literature.
- **Sources**
  Annexes I and II of this chapter, which reproduce procedures from a handbook containing detailed descriptions of the following:
  - the direct determination of metal traces;
  - the determination of metal traces requiring preconcentration; and
  - the determination of total cyanide.
- **Description of the procedure to be used in each case**
  - Identify and classify the analytical tools required.
  - Distinguish the three fundamental steps of the CMP.
  - Draw a detailed scheme for each process.
  - Describe the calibration procedures, discerning between method and equipment calibration.
  - Discuss the way the final results are computed and expressed.
- **Discussion**
  - Compare the CMPs considered in a critical manner and justify the differences among the methods employed.
  - Identify common and differential sources of error potentially affecting the quality of the final results.
- **Expanding the Seminar**
  - Use suitable handbooks to find one or more procedures for determining organic pollutants in water.
  - Describe the selected procedures.
  - Compare such procedures with one another and with those for the determination of inorganic pollutants described in Annexes I and II of this chapter.

Seminar 4.2

Analytical Processes for the Determination of Analytes in Solid Samples

- **Learning objectives**
  - To examine different CMPs suited to the type of solid sample to be analysed and analytes to be determined.

- To become acquainted with chemical measurement processes reported in the specialized literature.

■ **Sources**

Annexes III and IV of this chapter, which reproduce procedures contained in specialized handbooks for the determination of analytes of rather different nature and content in two disparate types of sample (biological and inorganic).

■ **Description of the processes**

- Identify and classify the analytical tools needed.
- Distinguish the three fundamental steps of the CMP.
- Draw a detailed scheme for each process.
- Comment on the calibration procedures and discern between method and equipment calibration.
- Discuss the way the final results are computed and expressed.

■ **Discussion**

- Compare the CMPs considered in a critical manner and justify the differences among the methods employed.
- Identify common and differential sources of error potentially affecting the quality of the final results.
- Comment on the preliminary operations involved when the sample is liquid (Seminar 4.1) and when it is solid.

■ **Expanding the Seminar**

- Use suitable handbooks to find CMPs for determining major components and traces in solid foods.
- Describe the selected procedures.
- Compare the preliminary operations with one another and with the procedures used in this Seminar in the light of the influence of the state of aggregation of the sample, the nature of the analyte and its concentration.

Seminar 4.3

Analytical Processes for the Determination of Clinical Parameters in Humans

■ **Learning objectives**

- To examine commonplace Clinical Analysis procedures.
- To become acquainted with the processing of clinical samples for the determination of various analytes.

■ **Sources**

Annexes V and VI of this chapter, which reproduce procedures contained in Clinical Analysis handbooks.

■ **Description of the processes**

- Identify and classify the analytical tools required.
- Distinguish the three fundamental steps of the CMP.

- Draw a detailed scheme for each process.
- Comment on the calibration procedures and discern between method and equipment calibration.
- Discuss the way the final results are computed and expressed.

■ **Discussion**
  - Compare the CMPs considered in a critical manner and justify the differences among the methods employed.
  - Identify common and differential sources of error potentially affecting the quality of the final results.
  - Comment on the significance of the results to diagnoses based on comparisons with normal values.

■ **Expanding the Seminar**
  - Find expeditious CMPs for determining the clinical parameters considered in this seminar (and also in others) developed by manufacturers and available at chemist's shops as test kits for urine and blood based on visual comparisons or photometric or electrochemical measurements.
  - Describe the selected procedures.
  - Compare conventional CMPs and those based on test kits.

Seminar 4.4

Selecting Analytical Procedures from the Literature

■ **Learning objectives**
  - To introduce the analytical literature concerned with the second fundamental component of Analytical Chemistry, and also with the first (Fig. 1.8).
  - To classify methods directly resulting from R&D (primary literature) on the one hand, and recommended, official and standard methods (secondary literature) on the other.
  - To help students develop critical awareness in selecting suitable methods, taking into account the factors of Fig. 4.3.

■ **Sources**
  - Handbooks, standards, *etc.*, containing descriptions of analytical methods.
  - Computer access to searches of methods published in the primary scientific literature (*e.g.* the Analytical Abstracts database of the Royal Society of Chemistry or the Chemical Abstracts database of the American Chemical Society).
  - Volumes of prime analytical journals (*e.g.* Analytical Chemistry, Analytica Chimica Acta, The Analyst, Fresenius' Journal of Analytical Chemistry).

■ **Suggested criteria for searching, by objective**
  Objective 1. Search handbooks for specific methods for each sample-analyte couple in a specific field. Examples:
  - Food analysis (determination of total protein in milk).

- Environmental analysis (determination of organic atmospheric pollutants).
- Clinical analysis (free and bound cholesterol in serum).
- Industrial analysis (moisture content in a petroleum product).

Objective 2. Compare analytical procedures for the same sample-analyte couple described in different handbooks. Note any peculiarities resulting in differences in analytical properties (*e.g.* those between manual and automatic procedures, which influence expeditiousness, cost-effectiveness, *etc.*).

Objective 3. Compare procedures for the same sample-analyte couple described in different handbooks and recently published papers (*e.g.* procedures based on the use of different equipment, which dictates CMP design, or CMPs developed on site and in the laboratory).

Suggested Readings

Preliminary Operations

(1) "Sample Pretreatment and Separation", R. Anderson, *Wiley*, Chichester, 1987.
   A highly pedagogical, straightforward book and an excellent complement to the contents of this chapter.
(2) "Analytical Supercritical Fluid Extraction", M.D. Luque de Castro, M.T. Tena and M. Valcárcel, *Springer-Verlag*, Heidelberg, 1994.
   Chap. 1 of this book presents a generic approach to preliminary operations of the analytical process that is consistent with the layout of this chapter.

Sampling

(3) "The Sampling of Bulk Materials", R. Smith and G.V. Hames, *Royal Society of Chemistry*, London, 1981.
   A general book that deals briefly with statistical theories and models, as well as with practical sample collection procedures for a variety of objects.

Analytical Separation Techniques

(4) "Técnicas Analíticas de Separación", M. Valcárcel and A. Gómez, *Reverté*, Barcelona, 1988.
   Provides a comprehensive description of a variety of analytical separation techniques.

Measurement and Transducing of the Analytical Signal

(5) "Instrumental Methods of Analysis", H. Willard, L. Herrit, J. Dean and F. Settle, Wadsworth, Inc., New York, 1988.

(6) "Instrumental Analysis", 4th edn, D. Skoog and J. Leary, Saunders College Pu., New York, 1992.

Trends: Automation

(7) "Automatic Methods of Analysis", M. Valcárcel and M. D. Luque de Castro, *Elsevier*, Amsterdam, 1988.
(8) "Automatic Chemical Analysis", P. B. Stockwell, *Taylor & Francis*, London, 1996.
(9) "An Introduction to Laboratory Automation", V. Cerdà and G. Ramis, *Wiley*, Chichester, 1990.

Commonly used Handbooks

(10) "Standard Methods for Examination of Water and Wastewater", *American Public Health Association*, Washington, D. C., 1989.
(11) "Official Methods of Analysis of AOAC", 15th edn, *Association of Official Analytical Chemists*, Arlignton, MA, 1990.
(12) "Selected Methods for the Small Clinical Chemistry Laboratory", *American Association for Clinical Chemistry*, Washington, D. C., 1982.

Objectives

- To introduce the essential features of qualitative analytical information.
- To define and characterize the YES/NO binary response and the errors (false positives and false negatives) potentially made in obtaining it.
- To outline the more commonplace qualitative analytical methodologies based on classical and instrumental techniques.
- To underscore the renewed significance of qualitative analysis.

Table of Contents

5.1 Introduction

Qualitative analysis is the first step in the hierarchy formed by the three primary goals of Analytical Chemistry according to the type of information required or delivered, *viz.* qualitative, quantitative and structural (see Fig. 1.20). In fact, quantifying an analyte entails previously checking that it is present in the sample concerned.

Samples can be classified into three broad categories depending on how deep available knowledge about them is – and hence on how difficult their qualitative analysis can be expected to be. **White samples** are samples with well-defined properties which, with some exceptions (*e.g.* those of spring water subjected to routine laboratory controls), remain virtually constant as a whole; as such, their qualitative analysis is straightforward and unequivocal. On the other hand, **grey samples** are only approximately known as regards composition; such is the case with samples analysed for environmental pollutants at a central laboratory receiving them from several factories. Obviously, grey samples are more complicated to analyse qualitatively than are white samples. Finally, **black samples** are samples of absolutely unknown composition (*e.g.* those involved in sporadic analyses for special purposes); this type of sample is the most difficult to analyse qualitatively of the three as it may in principle contain an indefinite number of species of widely variable nature and be subject to interferences among them.

_____ Box 5.1

Qualitative Analysis is designated in various ways in the analytical literature and in daily practice – with slightly different connotations, however. Two such designations are "detection" and "identification".

The word **detection** is usually employed to refer to a measurement process for qualitative purposes, as opposed to "determination", which is usually reserved for Quantitative Analysis. The distinction is not categorical, even though such combinations as "quantitative detection" and "qualitative determinations" are less frequently used.

The word **identification** is normally used to refer to a qualitative analytical process by which the analyte (or its reaction product) is recognized (identified) from some physical or physico-chemical property. Identifying entails the prior use of a standard to obtain one or more signals that are subsequently compared with those produced by the sample. This word is consistent with the comparison involved in metrological activity and hence more meaningful and appropriate than "detection" as an alternative to "Qualitative Analysis".

Figure 5.1 summarizes the essential features of qualitative (bio)chemical information. The most usual way of expressing it is a YES/NO binary response. As shown in this chapter, such a seemingly simple response possesses some quantitative connotations. In fact, the response can be qualified with some of the capital, basic and accessory properties described in Chap. 2 – with some adaptations, however. In clear contrast with chemical and biological metrology, qualitative information is of little significance to physical metrology.

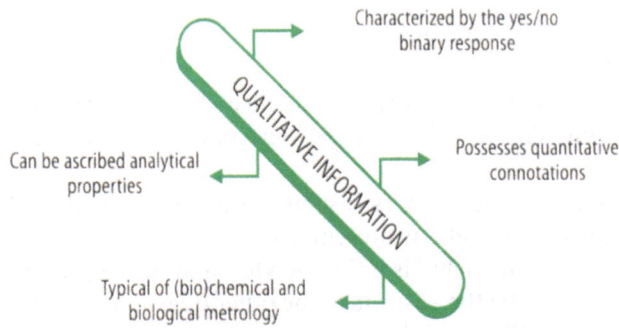

Fig. 5.1. General connotations of qualitative analytical chemical information

It is interesting to note that the qualitative slant of Classical Analysis, reflected in the substantial portion of Analytical Chemistry curricula it usually accounts for, has declined in the last few decades in favour of instrumental analytical techniques, which provide qualitative information with high sensitivity and selectivity. However, simple qualitative analyses have regained interest in recent years in response to an increasing demand for answers to "yes – no" questions such as "is the sample contaminated?", "does it contain a banned additive?", "did the athlete take any drugs?", "does this pepper batch contain any pesticides?", *etc.*

Box 5.2

A chemical measurement process (CMP) used to derive qualitative information is designated in practice with words such as "test" or "screening" rather than with "analysis".

A "test" is the action of testing, *i.e.* of examining and sensing the properties of things; as such, this concept is very close to that of Qualitative Analysis. In practice, the word "test" applies to simple, rapid analytical processes for essentially qualitative purposes (*e.g.* immunoassay tests, commercially available test kits for determining various parameters, *etc.*). Tests usually involve few or no preliminary operations.

"Screening" can be considered a part of Qualitative Analysis. In fact, it is more than a mere "analysis" as it entails making some selection and decision in relation to the analytical problem addressed. Although it is applied to both samples (sample screening) and analytes (analyte screening), there is no clear-cut difference between the two.

5.2 The Binary Response

The output of a measuring process used to derive qualitative chemical information is either YES or NO, *i.e.* binary information. This section describes the different types of binary responses one can obtain from qualitative analyses, their quantitative connotations and analytical attributes, and the errors potentially involved.

5.2.1 Types of Binary Response

The apparent simplicity of a YES/NO response can be misleading. In fact, as can be seen from Fig. 5.2, the information content of such an answer can vary widely.

The most simple possible binary response is the identification of an analyte and/or the confirmation that it is present in a sample as the answer to two complementary questions, namely: "is it the analyte?" and "is it in the sample?". This response is completed by the answers to the other questions asked in Fig. 5.2.

In deriving the qualitative information sought, one may need to consider a concentration limit or threshold imposed by the client or legislation. The actual question thus addressed will be something like "is the concentration of the analyte (*e.g.* a toxin, a pollutant) above the pre-established threshold?" The binary response to this question will obviously have more quantitative connotations than the response to the question stated in the most simple possible form.

There is a growing need to obtain discriminate qualitative information about the different forms in which an analyte may occur in a given type of sample. So-called "speciation" (Box 1.4) involves both the qualitative detection of different species and the determination of their individual concentrations. The question "in what form is the analyte?" must thus be answered with a multiple binary response (one per species potentially present) and hence involves chemical discrimination. This binary response relies on the presence of the analyte, *i.e.* on a YES answer to the first type of question: "is it present?". Box 5.3 gives several illustrative examples.

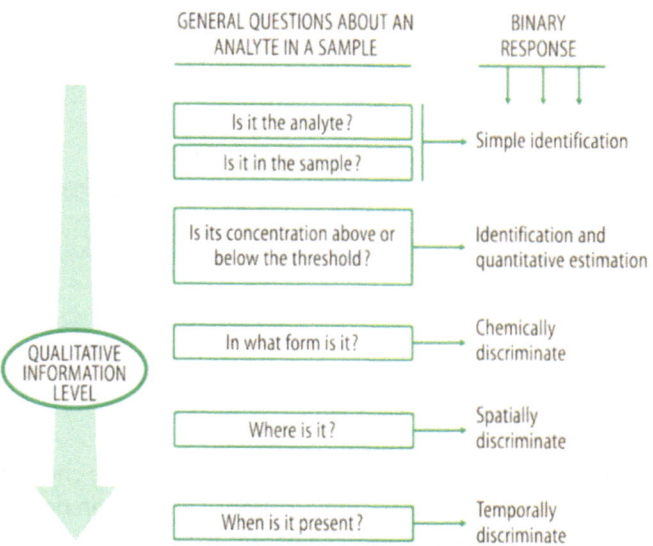

Fig. 5.2. Types of binary response, ranked in order of increasing qualitative information level, to questions that may arise in addressing an analytical problem. A YES answer to the most simple question is a prerequisite for the others to make sense

_____ Box 5.3

Below are some examples of possible binary responses (Fig. 5.2) to a demand of qualitative chemical information.

First-level information demands

- Is this beef of pork?
- Is there any cadmium in the paint coating of this toy?
- Are there any aflatoxins in this imported dried fruit batch?
- Has this meat product been adulterated with vegetable protein?

Second-level information demands

- Is this water contaminated with hydrocarbons as per the threshold global index set by the European Union?
- Is the aluminium concentration in this hemodialysis fluid acceptable?
- Is this soft drink fit for consumption as per the legally established limit for the concentration of the preservative it contains?

Third-level information demands

- Is this water, which may contain various mercury species (organic, organometallic), fit for drinking even if its total mercury content is near the tolerated limit?
- Is the price of this pharmaceutical preparation, where one enantiomer is the active principle and the other is therapeutically useless, acceptable?

Fourth-level information demands

- In what part of the surface of this object is the composition such that breakage or deterioration is most likely?
- At what point in this ore slag stack is the proportion of gold highest?
- At what time of year is the overall pollution index (*e.g.* chemical oxygen demand) highest at this point of the river?

Some analytical problems require expanding the most simple binary response with temporal (*e. g.* answering the question "when?" for dynamic objects from which the samples are withdrawn) or spatial discrimination (*i. e.* answering the question "where?" for heterogeneous objects such as those involved in surface analysis).

Consequently, even in its most simple possible form, the YES/NO binary response can differ chemically, spatially and temporally from case to case.

5.2.2 Quantitative Connotations of the Binary Response

As can be seen from Fig. 5.1, qualitative information invariably possesses some quantitative connotations; essentially, to obtain the former one compares data (signals) produced in response to given concentrations or amounts of analyte. To this end, one must use the references schematically depicted in Fig. 5.3, placed on an imaginary concentration scale. Such references are as follows:

(*a*) The **limit of detection**, C_{LOD}, which is the concentration yielding a signal X_{LOD} that can be statistically discriminated from a blank signal, $\bar{X}_B$, and

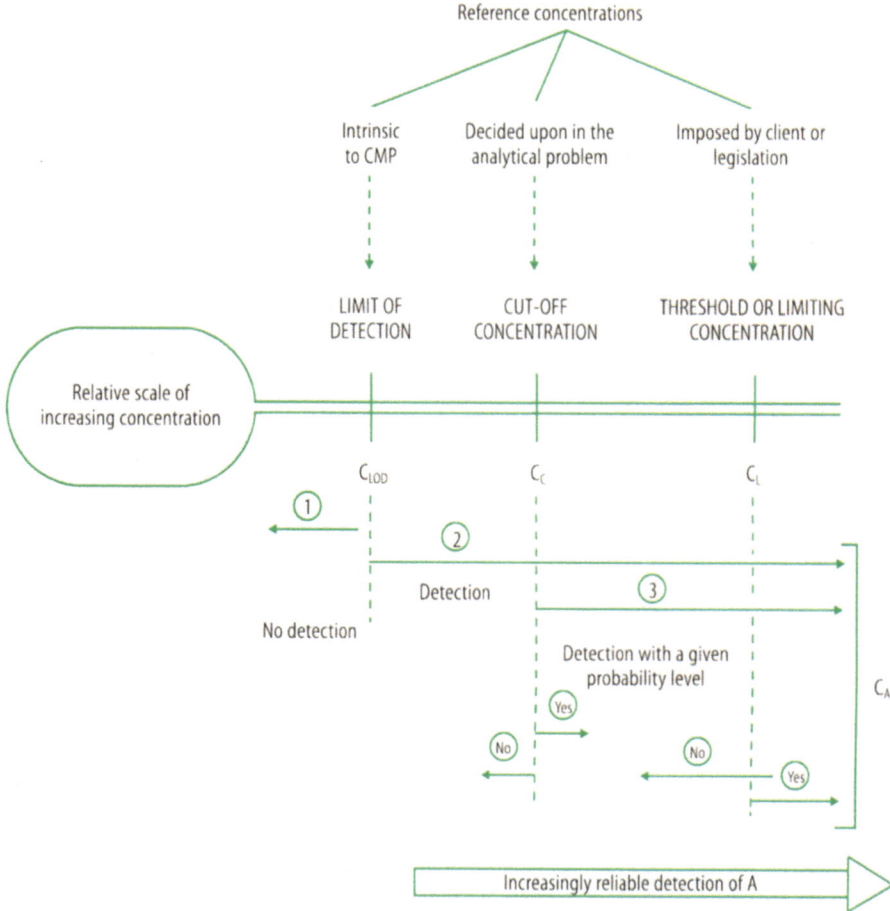

Fig. 5.3. Reference concentrations (C_{LOD}, C_C, C_L) used to establish the quantitative connotations of the detection–identification of an analyte at concentration C_A in a sample. For details, see text

is expressed as

$$X_{LOD} = \bar{X}_B + 3\sigma_B$$

where $\bar{X}_B$ is the mean for $n > 30$ blanks and σ_B its standard deviation (see Sect. 5.2). This concentration is an internal reference inherent in the chemical measurement process.

(b) The **cut-off concentration**, C_C, which is the concentration that produces a signal x_C and is established on an individual basis by the analyst in setting a given probability level to ensure the obtainment of a correct binary response. A detailed description of this concept is beyond the scope of this book.

(c) The **limiting concentration** or **threshold concentration**, C_L, which is the highest or lowest level, established by the client or legislation, to be used in deciding whether a sample (or the object it represents) warrants assignation of a given attribute (*e.g.* toxic, contaminated, nutritionally fit, fat-free, decaffeinated).

It is interesting to note the sequence and distance between the references of Fig. 5.3. Thus, the limit of detection, C_{LOD}, must always be lower than the other two parameters – otherwise, detection (identification) will be impossible. Also, the cut-off concentration, C_C, must exceed the limit of detection as it involves a higher probability level that detection will be error-free. Finally, any externally imposed limit or threshold, C_L, should be greater than C_C – and inevitably greater than C_{LOD} as well – if an error-free response is to be assured. The greater the difference $(C_C - C_{LOD})$ or $(C_L - C_C)$ is, the more reliable will be the qualitative identification achieved.

The concentration of the target analyte, C_A, may lie in different zones of the imaginary scale of Fig. 5.3. For the most simple binary question to have a reliable answer, C_A must be greater than the limit of detection $(C_A > C_{LOD})$, *i.e.* it must fall in zone 2. With $C_A > C_C$ (zone 3 of the scale), a given probability exists that the analyte will be detected. Derivation of the YES/NO binary response can rest on two different references, namely: an internal reference (the cut-off concentration, C_C) and an external one (the limiting concentration, C_L) produced outside the laboratory, with which the analyte concentration, C_A, is compared. The internal reference can always be established; the external one may or may not exist, or even differ with applicable legislation or the client's specific needs. The two coincide in some instances; also, C_C may exceed C_L, in which case the ensuing binary response will be scarcely reliable. The limiting concentration, C_L, is the most significant reference in practice as it allows one to determine whether the sample (or object) has the required quality, is fit for the intended purpose, *etc.*

These considerations are based on comparisons between concentrations. An alternative, more appropriate approach is to consider their respective specific uncertainties in order to compare concentration ranges rather than individual data points.

_____ Box 5.4

Providing the right answer to the question of whether the analyte concentration, C_A, is above or below the legally established or client-imposed threshold, C_L, entails considering not only C_A and C_L, but also their respective uncertainties. The figure below shows the limit and its uncertainty $(C_L \pm U_L)$, as well as various situations (1–6) that may arise in determining the analyte and obtaining a different value of C_A with the same uncertainty (U_A).

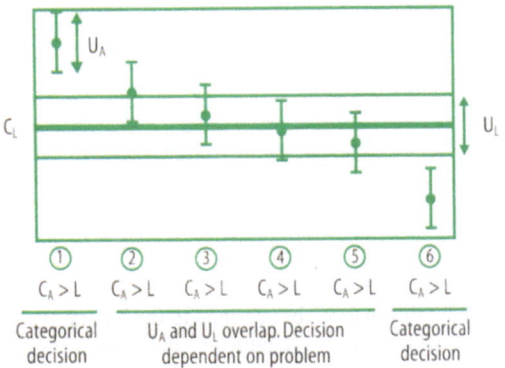

Only in the extreme situations (1 and 6) is the binary response absolutely reliable and can the analyte be said to be absent or present. In the intermediate situations (2–5), the uncertainty ranges overlap and the YES/NO decision depends on the particular connotations of the analytical problem concerned.

The figure above is purely theoretical as, in practice, the threshold concentration imposed by legislation or required by the client is rarely accompanied by its uncertainty.

5.2.3 Analytical Features of the Binary Response

The analytical process and the results of qualitative analyses can also be characterized by the analytical properties described in Chap. 2 –with some adaptation. Figure 5.4 illustrates the most salient relations, which are described below.

Representativeness. The YES/NO response, both as such and in its variably discriminating forms (Fig. 5.1), should be representative of the test sample, the object from which it is withdrawn, the way the analytical problem is approached and the information required, all in compliance with the scheme of Fig. 2.8, which is quite applicable in this context. This property is specially significant when the problem, and hence the CMP, are tackled with the imposed limit or threshold in mind.

Accuracy and precision. This capital and basic property, respectively, which are mutually linked in the analytical chemical field, are even more closely related in Qualitative Analysis. Their qualitative relation has produced a new property called **reliability**, which can be defined as *the proportion of right answers provided by individual tests carried out on aliquots of the same sample to identify an analyte*; such a proportion depends on two basic analytical properties: sensitivity and selectivity.

Sensitivity. The significance of this property was noted in describing the limit of detection in the previous section. In fact, an analyte present at a concentra-

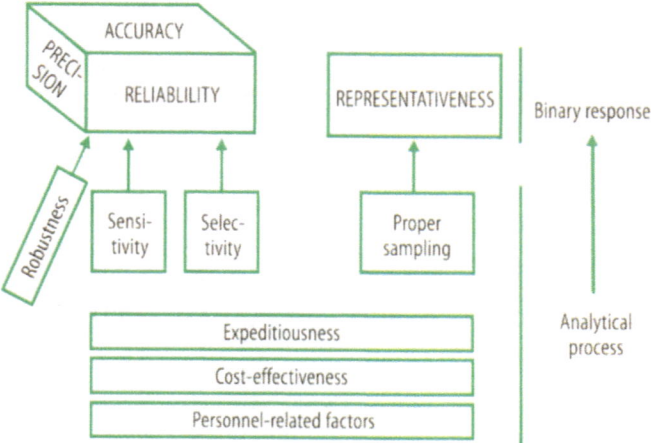

Fig. 5.4. Analytical properties that can be ascribed to the binary qualitative response and the analytical problem from which it arises (adapted from Fig. 2.1). For details, see text

tion (or in an amount) below the limit of detection of the CMP concerned will never be detected or identified. However, low and very low analyte levels can be made detectable by using an appropriate separation technique for preconcentration.

Box 5.5

Practical expressions of sensitivity in classical qualitative analysis

In Classical Qualitative Analysis, sensitivity is not expressed in terms of the limit of detection (C_{LOD}) described in Sect. 5.2.2; this would entail measuring n blanks, which makes no sense with visual (human) detection as the responses can never be the quantitative data required for statistical processing. Rather, the sensitivity is expressed in terms of other parameters such as the lowest amount or concentration of analyte that can be detected taking into account the results of the blank test. The following are the two most commonly used alternatives:

The limit of perceptibility, LP, which is the smallest amount of analyte, in nanograms, that can be identified in a qualitative test.

The limit of dilution or limiting dilution, D, which is the smallest amount of analyte, in nanograms, that can be identified per unit volume (in millilitres). Therefore,

$$D = \frac{LP\,(ng) \cdot 10^{-6}}{mL\ sample}$$

The limit of dilution is usually expressed as $pD = -\log D$ or $1:1/D$. Thus, for a limit of 1 ng in 1 mL of sample, the sensitivity can be expressed as

$$D = 10^{-6} \qquad pD = 6 \qquad 1:1000000$$

Selectivity. This property is defined differently depending on the particular goal of the analysis required to solve the analytical problem concerned. Thus, it may refer to a single analyte (*e.g.* the absence of interferences from other sample components). Also, it may refer to a compound "family" (*e.g.* organochlorine pesticides, hydrocarbons, heavy metals) and the absence of interferences from components other than the target congener(s). This property depends strongly on the relationship between the signal(s) produced by the analyte(s) and their properties; as such, it can differ markedly among analytes and may be zero in the presence of potential interferents. Analytical techniques for the separation of species, both individually and in groups, are commonplace in CMPs for qualitative purposes; in this context, selectivity refers to the separation proper.

Expeditiousness. As a rule, meeting the demand for (bio)chemical information in the industrial, ecological and clinical fields entails making timely decisions, which relies on a reasonably high sample throughput. The growing use of sensors and screening systems is one currently favoured trend with a view to achieving expeditiousness.

Cost-effectiveness. Obtaining the high sensitivity and selectivity required to assure reliable binary responses usually entails using analytical tools of high purchasing and maintenance costs. Such is the case with immunoassay reagents and sophisticated instruments (*e.g.* those used in chromatography-mass spectrometry combinations).

Robustness. The reliability of a binary response produced by a CMP occasionally (*e.g.* when using biochemical or biological tools) depends strongly on the stability of the reagents and on small changes in the operating conditions. Developing robust CMPs of high sensitivity and selectivity is one current challenge of analytical chemical R&D work.

The hierarchical foundation, contradictory and complementary relationships discussed in Sect. 2.7 are quite applicable in this context.

5.2.4 Errors: False Positives and False Negatives

The definition of the analytical property "reliability" in relation to the binary response contains an implicit statement of the errors made in Qualitative Analysis: the relative proportion (as a fraction of unity or percentage) of wrong YES/NO answers (*i.e.* of false positives and negatives that are obtained in subjecting n aliquots of the same sample to the same analytical process in order to produce qualitative analytical information).

As shown in Fig. 5.3, the reliability of the binary response increases with increasing concentration or amount of the target analyte (C_A). The proportion of errors will be greater in the vicinity of the limit of detection (C_{LOD}), consistent with the integration of the properties accuracy and precision. The notion of error usually employed in this context encompasses both systematic (determinate) and random (indeterminate) errors; however, the latter will obviously predominate when the analyte concentration is near the limit of detection.

As noted earlier, the errors contained in qualitative information are specifically called "false positives" and "false negatives", which, based on statistical principles, correspond to errors of the first kind (α, resulting from rejection of a true hypothesis) and errors of the second kind (β, made in holding a false hypothesis as true), respectively.

Figure 5.5 provides schematic definitions of errors in the binary response based on comparisons of the analyte concentration (C_A) with the reference

Concentration		Correct binary response	Errors	
Relative of analyte	Reference		Incorrect response	Designation
$C_A < C_{LOD}$	Limit of detection			
$C_A < C_C$	Cut-Off	No	Yes	False positive
$C_A < C_L$	Threshold or limiting			
$C_A > C_{LOD}$	Limit of detection			
$C_A > C_C$	Cut-Off	Yes	No	False negative
$C_A > C_L$	Threshold or limiting			

Fig. 5.5. Schematic definition of false positives and false negatives in Qualitative Analysis

concentrations of Fig. 5.3. A **false positive** arises when the signal for a sample containing the analyte at a level below the reference concentrations (false hypothesis) yields a YES response even though the sample is in fact a blank (*i.e.* it results from acceptance as positive of samples that are not such but blanks). On the other hand, a **false negative** results from the signal for a sample containing the analyte at a level above the reference concentrations (true hypothesis) being deemed a blank and a NO response thus being adopted; it is thus made in rejecting a signal for a sample containing the analyte in the mistaken belief that it is a blank (*i.e.* that it does not contain the analyte).

The precise implications of these errors depend on the particular analytical problem. As a rule, the results of screening tests and systems are confirmed by using a conventional CMP when any errors made may have a significant social or economic impact. False negatives are especially serious when detecting or identifying a toxic chemical or chemical family as no confirmation analyses are usually performed.

_____ Box 5.6

The specific example of this box is intended to illustrate the concepts of reliability, false positives and false negatives. It describes a CMP for screening benzodiazepines in human urine with a limit of detection of 0.5 μM. The proponents of the method chose the blank signal plus twice its standard deviation as the cut-off concentration. Such a concentration was thus below the limit of detection (see its definition in Sect. 2.5.2), in contrast with the situation of Fig. 5.3.

In order to determine the reliability of this screening method, 250 samples containing five different, increasing concentration levels (50 samples each), were processed. The analyte concentrations in them were (A) zero, (B) half the limit of detection ($0.5\,C_{LOD}$), (C) equal to C_{LOD}, (D) slightly higher than C_{LOD} ($1.2\,C_{LOD}$) or (E) twice the limit ($2\,C_{LOD}$). The following scheme shows the results obtained and their uncertainties, all on a (low) benzodiazepine concentration scale.

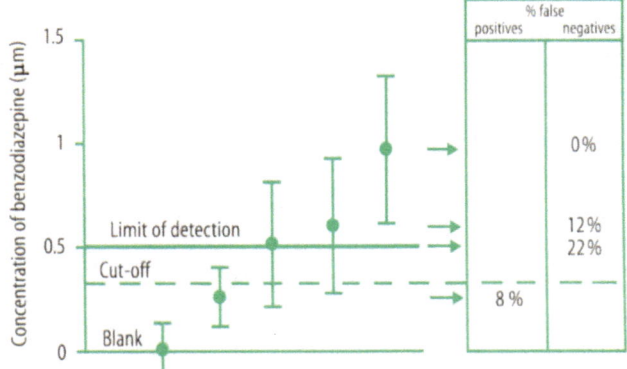

Notes:

• The reliability was calculated as the percentage of right answers. Thus, out of 50 samples containing a concentration equal to $1.2\,C_{LOD}$, 44 were deemed positive, so the proportion of false negatives was 12%.

- Note that the concentrations are accompanied by their uncertainties. This supports the choice of a cut-off concentration below C_{LOD} as it lies within the range $C_{LOD} \pm U_{LOD}$.
- Interestingly, reliability increased with increasing concentration; in fact, at an analyte concentration of $2C_{LOD}$, identifications were absolutely reliable (0% of false negatives).

5.3 Types of Qualitative Identification

Figure 5.6 shows several possible classifications of qualitative identification based on different criteria. It provides a comprehensive picture of Qualitative Analysis.

The target of a qualitative analysis (Fig. 5.6, 1) can be a single species, which will produce individual binary responses (*e. g.* in the identification of clenbuterol in beef), or several species belonging to the same chemical group or family, to which a global index or response will be assigned (*e. g.* in the identification of hydrocarbons in water).

Qualitative Analysis can also be classified according to various aspects of the chemical measurement process (CMP) used to derive the binary response. Thus:

(*a*) The analytical techniques involved in qualitative CMPs use either the human senses (Classical Analysis) or instruments (Instrumental Analysis). This is the criterion used in this chapter to discuss the more commonplace, representative approaches in this context (Fig. 5.6, 2).

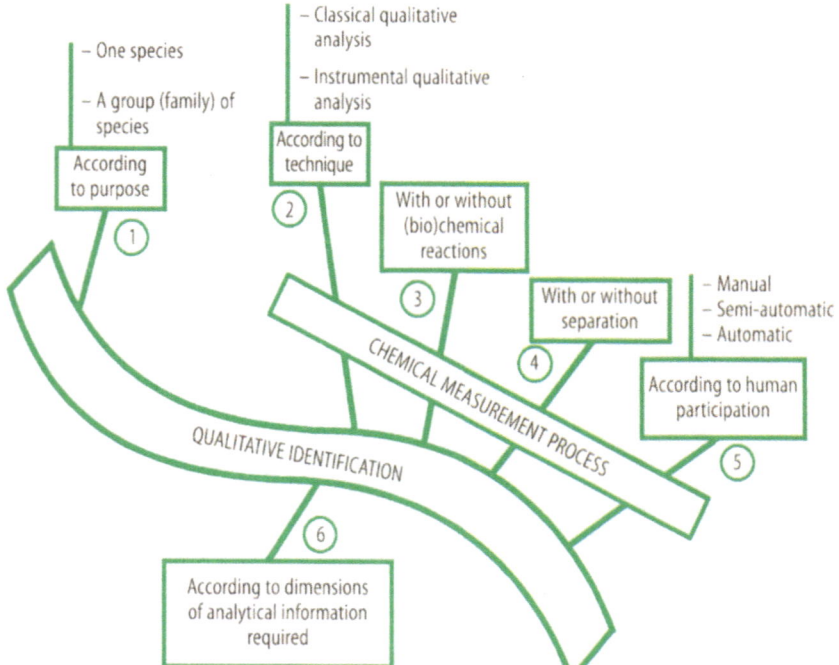

Fig. 5.6. Principal ways of identifying individual analytes in a mixture

(b) The signals provided by the analytical techniques used in a qualitative CMP can be produced by the analyte itself or the product of a chemical, enzymatic or immune reaction, for example (Fig. 5.6, 3). This substep of the CMP boosts sensitivity and selectivity in the identification. For example, the bluish colour of the cupric ammine, $Cu(NH_3)_4^{2+}$, is too light to be easily detected and can be masked by other, stronger colours. The addition of a ligand (cuproine) to form a bright-red, soluble chelate (CuL_2^+), dramatically increases the sensitivity and selectivity of the identification, which is thus made much more reliable.

(c) The need to obtain an acceptably reliable binary response has led to including a separation method in many qualitative CMPs. The characteristics and uses of separation methods are discussed in Sect. 4.3.4. Classical Qualitative Analysis uses non-chromatographic methodologies such as precipitation, liquid-liquid extraction or ion exchange. On the other hand, Instrumental Qualitative Analysis usually employs chromatographic approaches (gas, liquid and supercritical fluid chromatography) involving the use of a detector (instrument) to continuously monitor the signal produced by the eluent that emerges from the separation column (Fig. 5.6, 4).

(d) Qualitative CMPs can be conducted manually (e.g. by visually inspecting the formation of a precipitate and/or color, by comparing IR spectra), semi-automatically (e.g. when locating the retention time for an analyte in a chromatogram with the aid of an electronic integrator or computer) or fully automatically (e.g. in spectral searches of computer libraries) (see Fig. 5.6, 5).

(e) The reliability of the binary response additionally depends on the dimension of the information supplied by the CMP (Fig. 5.6, 6). Some instruments provide the values of one, two, three or even more parameters; also, signals can be single or multiple in number. The more raw data (parameter values and signals) are available and the better discriminated they are, the more reliable will be the identification. Thus, Classical Qualitative Analysis uses a multidetection system (the human senses and brain) and one, two or more signals for identification (e.g. the formation of a precipitate, this and the precipitate color, the previous two and the precipitate texture, etc.). On the other hand, the identification of a metal ion present in trace amounts in water entails using atomic absorption spectrometry: a signal at a preset wavelength (i.e. an instrumental parameter). A fluorescent analyte (or its reaction product) is identified fluorimetrically, i.e. from a fluorescence intensity signal obtained at two different instrumental parameter values (the excitation and emission wavelengths). The body of spectra provided by instrumental techniques (IR spectroscopy, mass spectrometry, nuclear magnetic resonance) for a given species – or a suitable combination of the information thus gathered – makes a powerful, highly reliable identification tool on account of the many signals and instrumental parameters it encompasses.

5.4 Standards and Calibration in Qualitative Analysis

The two types of calibration depicted in Fig. 3.10 are quite applicable to Qualitative Analysis.

Equipment calibration, based on standards not containing the analyte, is done to ensure correct functioning of the identification instrument. Thus, polyethylene film (a reference material, RM) is typically used to calibrate IR absorption spectrophotometers. The film spectrum should coincide with that supplied by the manufacturer; otherwise, the instrument should be recalibrated. A solid fluorophore (*e.g.* rhodamine) dispersed in a transparent plastic block is employed to check that the two monochromators of a fluorimeter operate to specification.

Method calibration uses standards containing the analyte to establish an unambiguous relationship between some physico-chemical property of the analyte (or its reaction product) and the signals provided by a pre-calibrated instrument under specific operating conditions. The identification involved in qualitative analyses relies on standards that are used in two different ways for method calibration, namely (see Box 3.13):

(*a*) Calibration based on the relationship between the signal and properties of the analyte (or its product). Such is the case with identification in Classical Qualitative Analysis: the product of a chemical reaction (a colourless species, gas or solid) is previously identified by the operator using a standard of pure analyte. Once the features (colour, odour, texture) of the product have been recorded by the brain, the identification test is performed on a sample aliquot; whether the new product is identical with that previously obtained from the standard is decided upon by the analyst (see Box 1.5). In Instrumental Qualitative Analysis, the standard provides two data sets (instrument parameters and signals) for comparison (see Box 3.9). The greater the number of data used and the more similar they are, the more reliable will be the determination. Comparisons can be made visually or with the aid of a computer; the latter allows the sample spectrum to be automatically checked against those for a wide range of standards included in a so-called "spectral library". The identification can be made even more reliable by using a derivative (first, second, *etc.*) of the original signal; this expands the initially available information with new bands and diminishes background noise.

(*b*) Calibration based on the behaviour of the analyte (or standard) in a dynamic instrumental system that produces signals of temporal or spatial dimension. Thus, in a column chromatographic system (capillary electrophoresis and liquid, gas or supercritical fluid chromatography), the analyte is identified from its retention time; by contrast, in planar chromatography and classical electrophoresis, identification relies on the distance travelled (migrated) by the analyte. The use of standards in this context is described in detail in Sect. 5.6.3.

5.5 Classical Qualitative Analysis

5.5.1 General Notions

The distinction between Classical and Instrumental Analysis made in Sect. 1.7 rests on the type of instrument used to produce or acquire the analyte signal. Classical Qualitative Analysis uses the human senses (mainly sight and smell) to detect the presence of an analyte; the analyte is subjected to a chemical (acid-base, complex-formation, precipitation, redox, condensation), biochemical or immune reaction which yields a product that is identified from a well-defined change (*e. g.* the formation of a gas, colour or precipitate).

Identification in Classical Qualitative Analysis thus relies on comparisons of the behaviour of an analyte standard, a blank and the unknown sample – which may or may not contain the analyte – towards a CMP. Comparisons can also rely on the use of a reference scale from which the operator will read semi-quantitative information. Such is the case with pH, active chlorine and glucose measure-

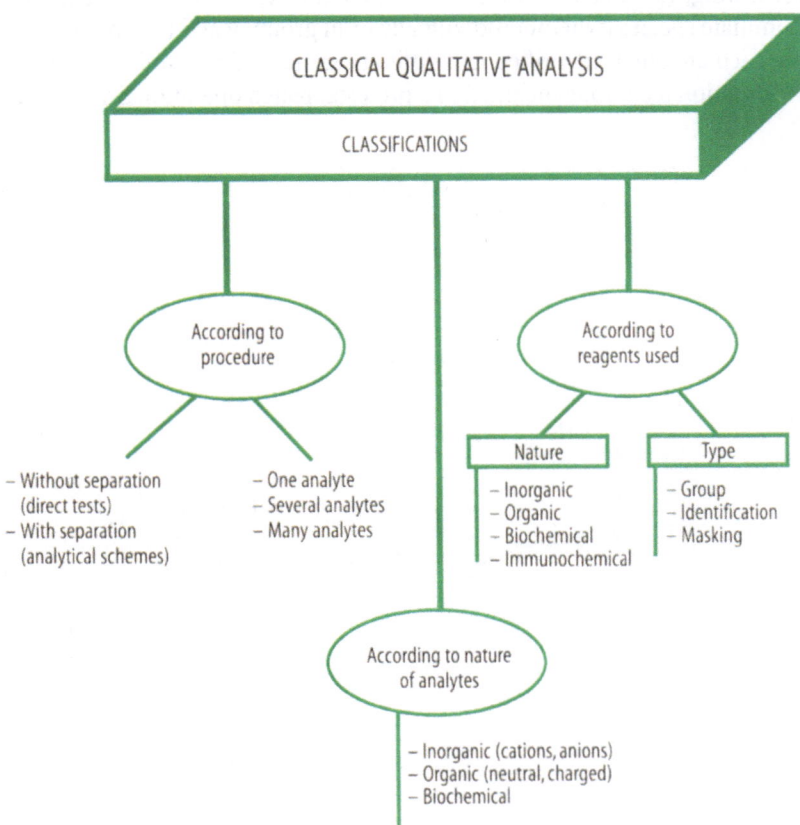

Fig. 5.7. General types of detection-identification processes used in Analytical Chemistry. For details, see text

ments in fluids, pool water and urine, respectively, provided by reagent strips; the colour taken by the strip allows one not only to identify the presence of the analyte but also to obtain an estimate of its concentration.

The limited capacity of the human senses and brain to detect small changes, low discrimination among signals and the scarce variety of information that can be derived severely restrict the identification scope of Classical Qualitative Analysis relative to Instrumental Analysis. As a result, reliability in this context depends strongly on the sensitivity and selectivity of the particular CMP used and, ultimately, on the chemical reaction(s) it involves. Boosting these two analytical properties – primarily with the aid of separation techniques – has been a permanent challenge to Classical Qualitative Analysis.

Figure 5.7 shows complementary classifications of Qualitative Classical Analysis according to various criteria. One is inherent in the CMP, which can be a straightforward, direct test involving no complicated operations or a systematic process (a "scheme") using separations to indirectly raise sensitivity and selectivity through preconcentration and interference removal, respectively (see Fig. 4.15). There are three generic approaches to the identification of individual analytes in mixtures (Fig. 5.8), namely: (*a*) the ideal situation, where every analyte can be identified by direct testing; (*b*) that which calls for systematic separations to obtain spatially discriminate species (whether individually or in groups); and (*c*) a mixed situation where each analyte is identified in a different sample aliquot following subjection to a separation technique included in a pre-established operational sequence.

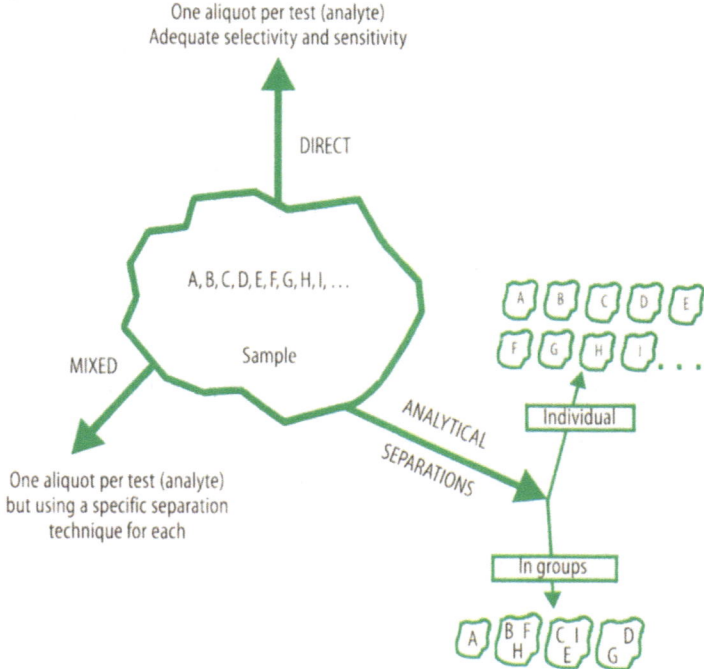

Fig. 5.8. Classification of Classical Qualitative Analysis according to various criteria. Letters A to I denote analytes. For details, see text

The qualitative CMP of choice differs widely depending on whether a single analyte (*e.g.* clenbuterol in beef for human consumption), a compound family (*e.g.* atmospheric aromatic hydrocarbons), a small group of species (*e.g.* pesticides in oranges) or a wide range of analytes (*e.g.* metals and non-metals in rocks) is to be identified. The process grows in complexity from white to grey to black samples (see Sect. 5.1). In fact, some black samples (*i.e.* samples of absolutely unknown composition) may contain scores of analytes.

5.5.2 Types of Reagents

The CMP of choice for a classical qualitative analysis is also dictated by the nature of the analytes involved (inorganic, organic, biochemical), which is thus one other possible classification criterion for Classical Qualitative Analysis (see Fig. 5.7).

Analytes can be identified by using reagents of varied nature. Selectivity in a qualitative CMP increases and the ideal situation of Fig. 5.8 is approached in the following sequence: inorganic < organic < biochemical < immune reagents. The many possible combinations of analyte and reagent types result in a wide range of situations of variable complexity (the more similar the analytes in a mixture are the more complicated will be their identification). In fact, such combinations dictate the type of identification reaction to be used in each case. Thus, inorganic analyte-inorganic reagent combinations typically involve a precipitation, redox or complex-formation reaction; inorganic analyte-organic reagent couples usually require a coloured or fluorescent chelate formation reaction; in the organic analyte-organic reagent and biochemical analyte-biochemical reagent combinations, the identification reaction is invariably of the organic and biochemical type, respectively.

The specific purpose of a reagent in Classical Qualitative Analysis can be widely variable. Figure 5.7 shows the three principal types of reagents used in this context.

Group reagents are intended to effect the separations described in Fig. 5.8. Most often, they rely on separation by precipitation (the reagent causes the formation of insoluble products), liquid-liquid extraction (the reagent gives hydrophobic products that can be readily extracted into organic solvents) or ion exchange (the reagent itself acts as the exchange material and selectively retains one or more analytes under specific experimental conditions). The purpose of group reagents is to isolate species in groups in such a way that each individual analyte can be ultimately identified with a high reliability.

Group reagents must meet some general requirements, namely:

(*a*) They should isolate every single analyte belonging to each group. Also, the reaction used to identify each individual analyte within the group should be adequately sensitive and selective.

(*b*) The separation should be selective: species not belonging to the particular group should remain in the original phase.

(*c*) Excess reagent and the second (added) phase should not interfere with subsequent identification reactions.

_____ Box 5.7

Examples of commonplace combinations between analytes and reagents of variable nature used in classical qualitative analysis and identification reactions

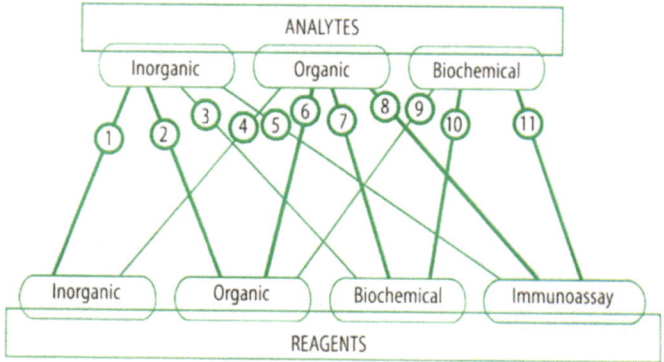

Note: The more usual relations are depicted with bold lines.

Combination	Examples		
	Analyte	Reagent	Reaction
1	Fe^{3+}	SCN^-	Formation of a red complex ($SCNFe^{2+}$)
2	Al^{3+}	Morine (L)	Formation of a fluorescent chelate (AlL_2)
6	Aldehydes	Dinitrophenylhydrazine	Formation of a yellowish orange product
	Analyte	Reagent	Reaction
7	Glucose	1) Glucose oxidase 2) Reagent	Formation of H_2O_2, which oxidizes the reagent to a coloured product
8	Aflatoxins	1) Solid-phase antibody 2) Enzyme-bound antibody 3) Substrate	ELISA (see Box 5.8)

The analytical schemes discussed below use one or more of these types of reagents in a sequential manner to gather species in different groups by precipitation-based separations. Cations are usually separated with S^{2-}, CO_3^{2-}, Cl^-, SO_4^{2-} or NH_3, and anions with Ca^{2+}, Ba^{2+} or Ag^+.

Identification reagents are reagents that give a reaction the externally apparent effect of which (*e.g.* formation of a precipitate or gas, a colour change) can be readily detected by the human senses. Their primary use is in the identification of individual analytes, whether in the starting sample (ideal situation) or in a group. Obviously, these reagents should provide tests of adequate sensitivity and selectivity; both properties, however, are indirectly improved by the separations included in an analytical scheme. Identification reagents can be of inorganic, organic, biochemical or immune nature (see Box 5.8). Table 5.1 gives selected examples of reagents used for the identification of inorganic analytes. Each example in the table is amenable to supplementary identification. In the first, heating the solution after the precipitate is formed causes its dissolution; subsequent abrupt cooling of the resulting solution produces microcrystals that scatter light in a dramatic effect. In the second example, the addition of a masking agent such as F^- ion, which forms complexes of increased stability with the analyte, displaces the formation of the red complex to that of FeF^{2+}, which is colourless. In the third, bubbling the gas over a Ca^{2+} solution leads to the formation of a white precipitate of calcium carbonate.

Table 5.2 gives several examples of identification of organic analytes in families, using organic reagents (alone or in combination with inorganic ones).

Table 5.1. Selected examples of identification of inorganic analytes

Effect	Analyte	Reagent	Product	Additional confirmation
Precipitate formation	Pb^{2+}	I^-	$PbI_2\downarrow$	Gold rain effect
Colour change	Fe^{3+}	SCN^-	$FeSCN^{2+}$ (red)	Discolored on addition of F^-
Gas formation	CO_3^{2-}	HCl	$CO_2\uparrow$	Bubbling over Ca^{2+} solution produces $CaCO_3\downarrow$

Table 5.2. Selected examples of identification of organic analytes

Effect	Analyte	Reagent	Product
Red colour or precipitate	Aromatic hydrocarbons	Azoxybenzene $AlCl_3$	*p*-Phenylazobenzene
Purple colour	Aldehydes	Fuchsin	Quinoid dye
Precipitate	Amines	5-Nitrosalicylaldehyde $NiCl_2$	*In situ* formed nickel chelate

The high selectivity of immunoreagents has opened up previously unimaginable avenues for Classical and Instrumental Qualitative Analysis. Box 5.8 illustrates selected qualitative identifications based on immunoassays; the analytical signal can be detected either by the human senses or by a straightforward instrument (*e. g.* a photometer or fluorimeter).

Masking reagents are added to the reaction medium in order to prevent sample components other than the analyte from interfering with the qualitative test – whether by exhibiting the same effect as the analyte or by hindering it. Masking reagents should never significantly disturb the main analytical reaction (*i. e.* that between the analyte and its identification reagent), so they should interact with neither its ingredients nor its product. The masking reagents used in classical qualitative analyses for inorganic species are normally ligands that form stable, soluble or colourless chelates with the potential interferents accompanying the analyte in the sample. Box 5.9 presents a typical example.

_____ Box 5.8

Immunoassay methods, particularly heterogeneous immunoassay (which is commonly known as "enzyme-linked immunosorbent assay", ELISA), are widely used in food analysis. In addition to quantitative immunoassay methods, a number of commercially available kits are used for qualitative screening with a view to the rapid detection of food contaminants and adulterants. Typical examples include the following:

● The detection of toxic contaminants such as aflatoxins (fungal metabolites) in dried fruits, cereals, spices, coffee, *etc.*
● The identification of meat types (beef, pork, ewe, horse, chicken, rabbit) as a means of fighting the fraudulent replacement of some meat products with others of poorer quality.
● The detection of soy protein, which is frequently added to meat products to improve texture, increase water and fat retention, and replace meat protein.
● Microbiological analyses for pathogens such as salmonella and listeria. Available immunoassays provide results within a few hours and thus save substantial time relative to classical culturing methods, which take no less than 3 – 4 days to complete.

Sandwich ELISA, one of the more widely used modes of this technique, comprises the following steps:

(1) Adsorption of the analyte-specific antibody (**Y**) onto a solid phase (a plastic tube, latex particles, nylon mesh, nitrocellulose paper, fibre glass).

(2) Addition of the sample to an active solid phase. The analyte (antigen, ⬭) is immobilized (bound) by the antibody; after an appropriate incubation time, the solid phase is rinsed to remove excess sample.

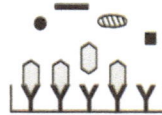

(3) Addition of the second specific antibody, bound to an enzyme (E, usually peroxidase, ), which serves as label. The analyte is thus sandwiched between both specific antibodies. Then, the solid phase is rinsed to remove excess label. The presence of the enzyme in the solid phase confirms that of the analyte in the original sample.

(4) Addition of the enzyme substrate (○), which is converted into a coloured product (●) under the catalytic effect of the enzyme.

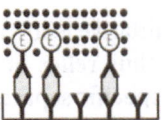

_____ Box 5.9

Use of masking reagents in classical qualitative analyses for inorganic species

The addition of a solution of sodium sulphide to a Cd^{2+} standard in ammonia (or alternatively bubbling gaseous H_2S through the standard) produces a bright yellow CdS precipitate.

If the sample contains additional cations such as Cu^{2+}, Hg^{2+}, Ni^{2+} or Co^{2+}, the reagent also precipitates CdS (yellow), CuS (black), HgS (black), NiS (black) and CoS (black), which will conceal the yellow colour of the analyte's sulphide.

In order to avoid this interference, the analyte solution is previously supplied with one of sodium cyanide (a masking reagent). This produces the cyanide complexes $Cd(CN)_4^{2-}$, $Cu(CN)_3^{2-}$, $Hg(CN)_6^{4-}$, $Ni(CN)_4^{2-}$ and $Co(CN)_6^{3-}$, of variable stability – the first is the most labile. The subsequent addition of the identification reagent precipitates yellow CdS alone. In fact, the masking reagent prevents all the black sulphides from the precipitating, thereby facilitating identification of the analyte (cadmium).

5.5.3 Analytical Schemes

Analytical schemes are sequential processes by which a variable number of analytes in a black or grey sample are identified (see Sect. 5.1). Their complexity and structure depend on a number of factors including

(a) how deep available knowledge about the sample is;
(b) the number of species potentially present;
(c) that of species to be identified, which will vary with the particular analytical problem; and
(d) what selective, sensitive reagents are available for the intended purpose.

Provided the nature of the sample is known (e. g. steel, vegetables), the analytical scheme can be quite simple; this is particularly the case when the sole information required is purely qualitative (*e.g.* the presence or absence of certain pesticides in lemons). On the other hand, the most complicated situation is that where the nature and origin of the sample are unknown and a comprehensive qualitative report of its composition is required.

The primary purpose of analytical schemes is to ensure maximal reliability in the qualitative test conducted on each analyte, which entails assuring adequate sensitivity and selectivity. No doubt, separation techniques play a central role in these processes. Below are briefly discussed the two most widely used types of analytical schemes.

5.5.3.1 Schemes without Group Separation

In this type of scheme, each qualitative test is applied to a separate aliquot of the original sample. Identification thus relies in the sequential use of direct tests – some, however, may include a specific separation if needed. They lie in between the direct and mixed options depicted in Fig. 5.8.

This type of analytical scheme has three salient technical features, namely:

(*a*) It uses highly sensitive and selective reagents.
(*b*) It is conducted in a strict operational sequence: from tests involving reagents of a high sensitivity and/or selectivity (a near-ideal situation) to others based on moderately sensitive and selective reagents.
(*c*) It applies separation techniques to individual sample aliquots when the information provided by previous qualitative tests suggests the presence of interferences with that being carried out. Obviously, previously identified analytes should not interfere with those to be tackled later in the sequence.

Figure 5.9 illustrates the operating procedure for an analytical scheme without group separation that possesses the above technical features. The first two tests are based on two ideal reagents (R_1 and R_2) that allow the first two analytes to be reliably identified in a direct manner. The third test is interfered by one or both of the previous analytes, so it requires the addition of masking ligand (ML_3) prior to the identification reagent proper (R_3). The fourth test entails the prior separation of the first three analytes by using an appropriate reagent (SR_4) before the new identification reagent (R_4) is added to detect the fourth analyte. Finally, the fifth test requires both separation (SR_5) and the use of a masking reagent (ML_5) in order to avoid disturbances to the identification reaction with reagent R_5.

As can be seen, the qualitative tests performed grow in complexity as the scheme progresses, to an extent that depends on the number of species present in the sample. Thus, the separation in the fourth test will be unnecessary if the first, second and third tests are negative.

These schemes are labour-intensive, time-consuming and complex when the number of analytes is relatively large. However, they are technically efficient and affordable for white and grey samples containing a limited number of analytes, and also in those cases where highly sensitive and selective reagents are available.

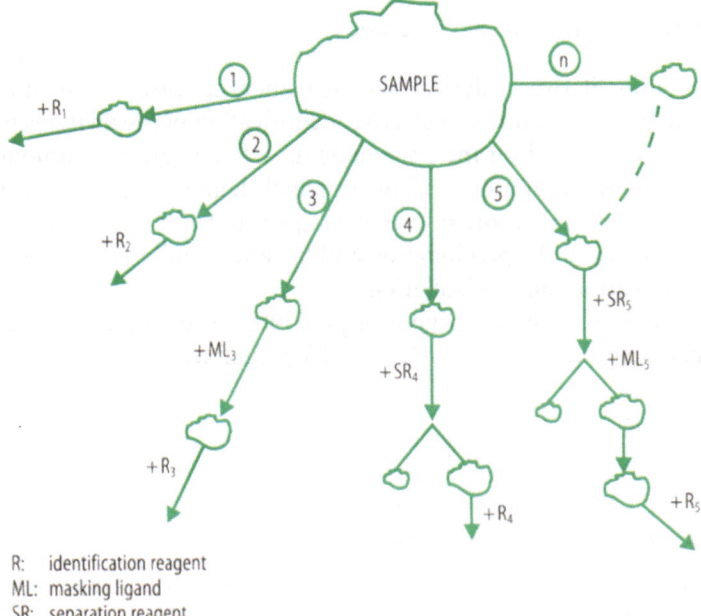

R: identification reagent
ML: masking ligand
SR: separation reagent

Fig. 5.9. Analytical scheme without group separation

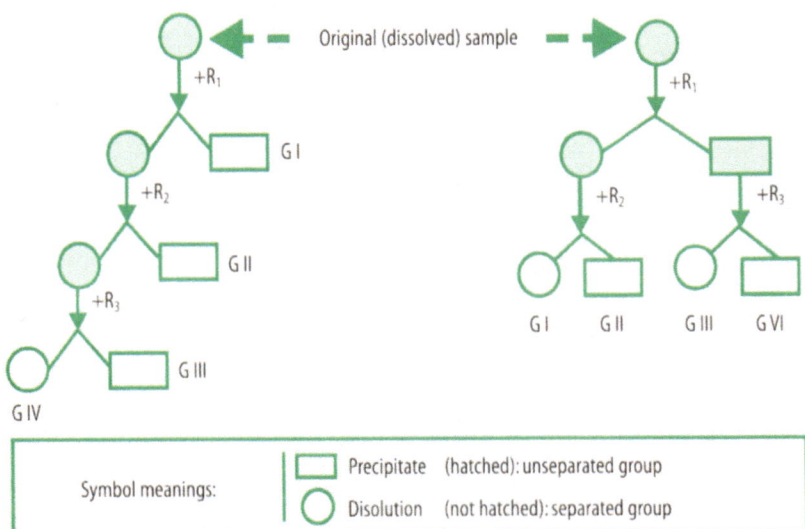

Fig. 5.10. Analytical schemes based on group separation by precipitation. Three precipitating group reagents (R_1–R_3) are used to split the initial analyte mixture into four groups in two different sequences. For details, see text

5.5.3.2 Schemes with Group Separation

This type of qualitative analytical scheme gathers species into small groups in order to facilitate the sensitive, selective identification of the analytes in each.

The process is based on the systematic use of separation techniques under appropriate operating conditions. The principal element of the scheme is the so-called "group reagent". Inorganic analyte species in the sample are normally separated into groups by precipitation and filtration, and organic ones by either liquid–liquid extraction or distillation.

Figure 5.10 depicts generic types of precipitation-based analytical schemes for inorganic species. In the first case (left) analytes are grouped in a fully

---

Box 5.10

Analytical scheme for the identification of inorganic cationic species based on group precipitation without the use of hydrogen sulphide

The present scheme allows 26 different species to be identified, but can be extended to another 20.

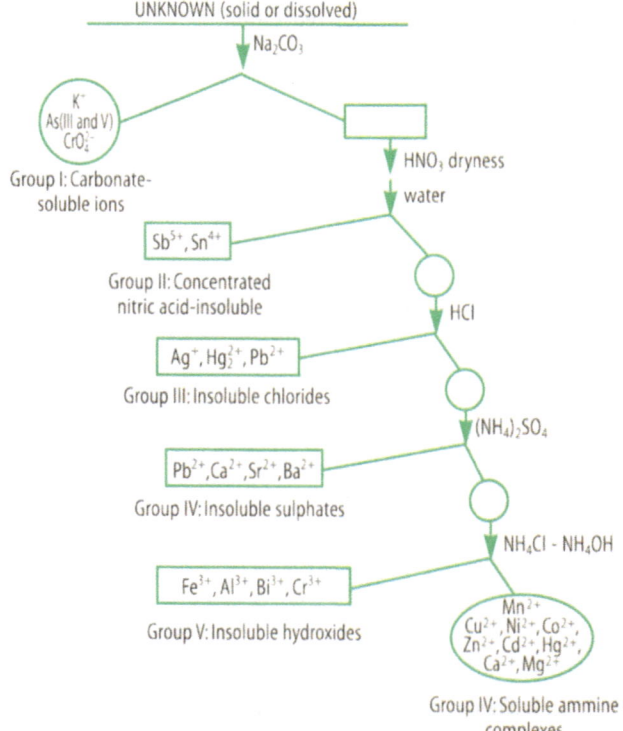

Notes:

- For experimental details, see the book by Arribas (suggested reading no. 4).
- The metals in the different groups are shown in cationic form, even though they are present as salts in the corresponding precipitates (e.g. $AgCl$, Hg_2Cl_2 and $PbCl_2$ in Group III).

sequential manner: a new precipitating reagent is added after the precipitate containing the analytes in the previous group is filtered off. The first three groups are initially precipitates, so they must be selectively dissolved to obtain aliquots for the identification of the species in each group. The fourth group contains the so-called "soluble analytes". The scheme on the right of Fig. 5.10 uses two types of separation reagents; one (R_1) produces global groups that require further processing for identification whereas the other (R_2 and R_3) yields groups where the analytes can be identified in a direct manner.

Boxes 5.10, 5.11 and 5.12 show different precipitation-based analytical schemes for the qualitative analysis of inorganic species. Analytical schemes for

---

Box 5.11

Analytical scheme for the identification of inorganic cationic species based on group precipitation

Fresenius' Hydrogen Sulphide Scheme

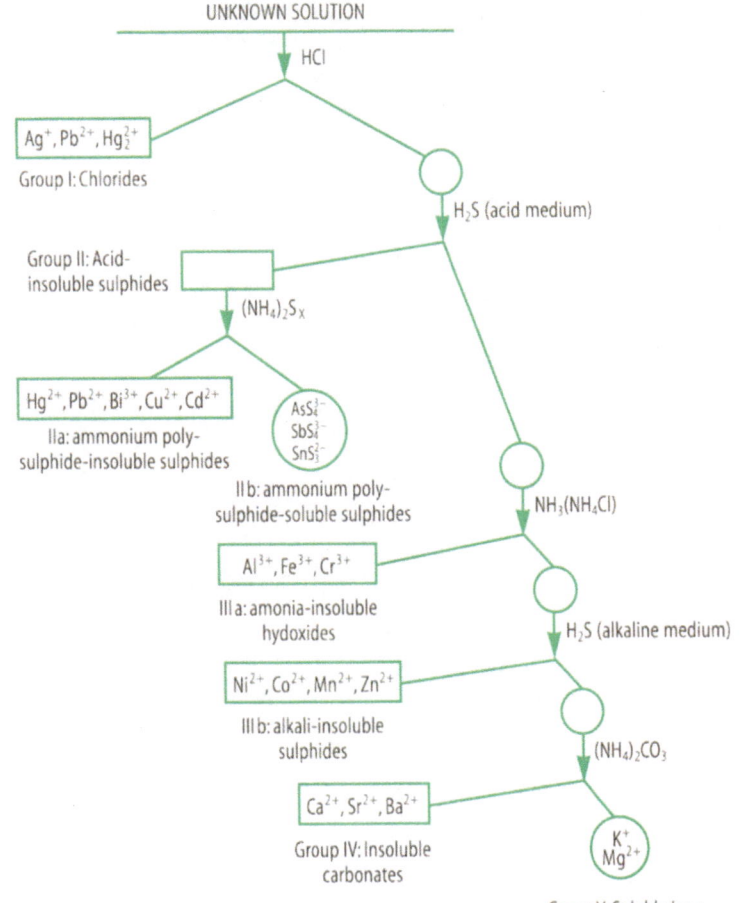

Notes:

- For experimental details, see the book by Burriel *et al.* (suggested reading no. 2).
- The metals in the different groups are shown in cationic form, even though they are present as salts in the corresponding precipitates (*e.g.* $AgCl$, Hg_2Cl_2 and $PbCl_2$ in Group I).

Box 5.12

Analytical scheme for the identification of inorganic anionic species based on group precipitation

Inorganic Anion Scheme

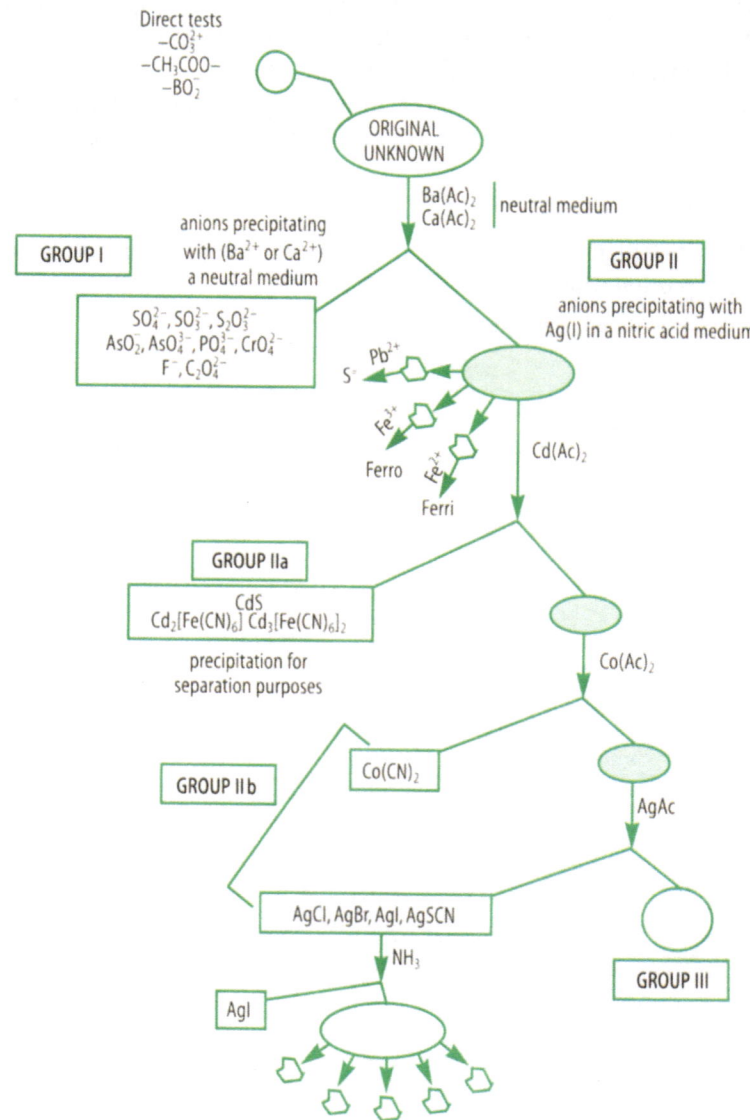

Box 5.13

Analytical scheme for the identification of families of organic compounds based on distillation and liquid-liquid extraction

Water-soluble Mixtures (some may contain esters)

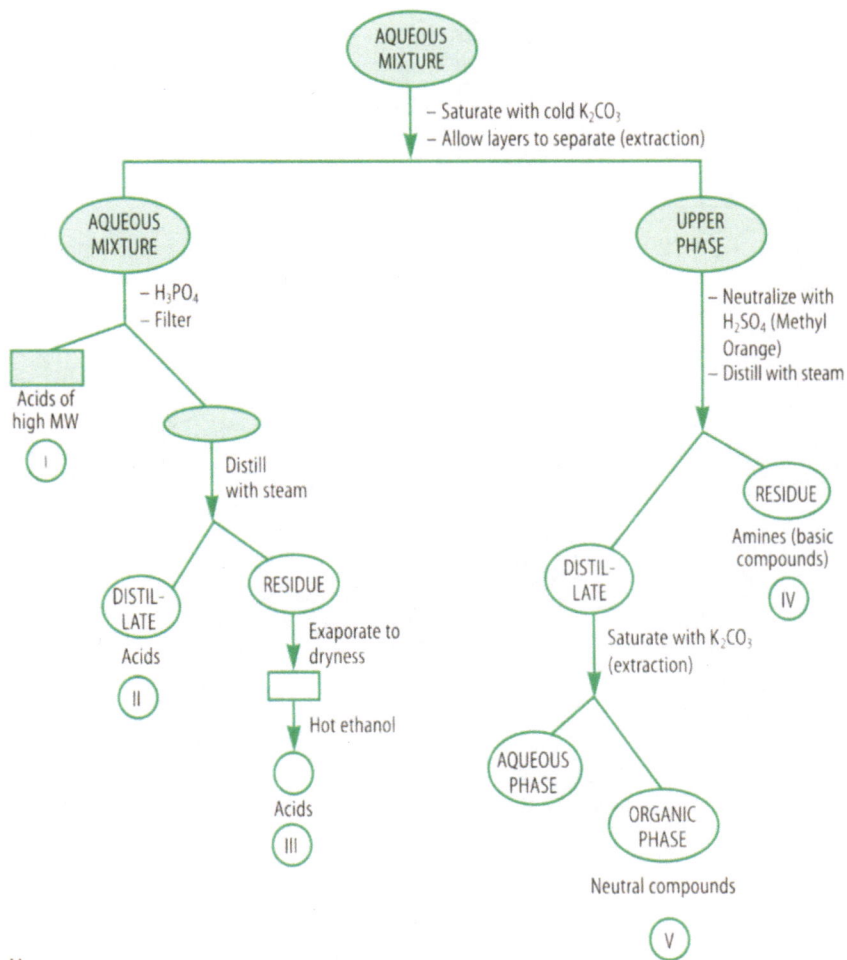

Notes:

● For experimental details, see the book by Shriner *et al.* (suggested reading no. 7).

● Separations produce only three large groups (I, II and III) of organic acids and one each of basic (IV) and neutral compounds (V).

● Groups I and II are separated by precipitation.

_____ Box 5.14

Analytical scheme for the identification of families of organic compounds based on distillation and liquid-liquid extraction

Mixtures of Water-insoluble Compounds

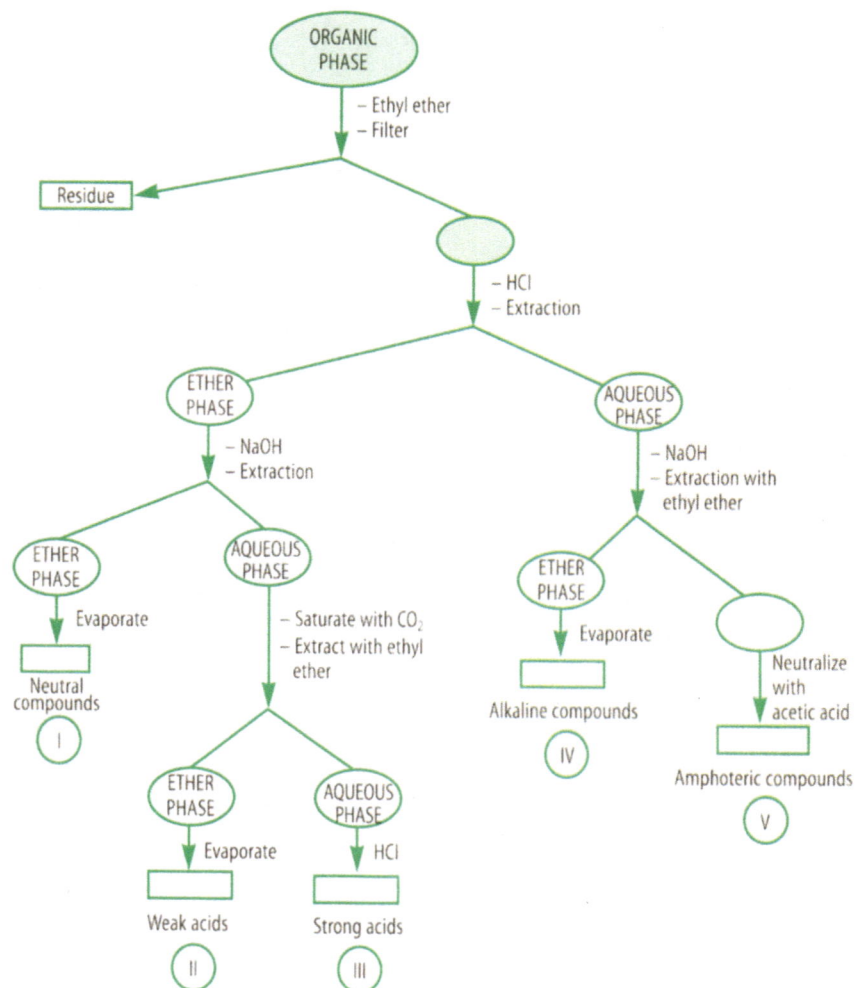

Notes:

- For experimental details, see the book by Shriner *et al.* (suggested reading no. 7).
- As in Box 5.13, separations produce large groups only.

the identification of organic analytes rely largely on separations by distillation (with or without steam) or liquid–liquid extraction – a few use precipitation-filtration. As a rule, they are based on polarity differences among the analytes. Species are grouped by their acid, basic, neutral or amphoteric character, and occasionally further split according to additional criteria. Boxes 5.13 and 5.14 show two separation schemes for samples of organic substances that are soluble and insoluble, respectively, in water. The analytes in each group are subjected to the required identification tests.

5.6 Instrumental Qualitative Analysis

5.6.1 General Notions

In Instrumental Qualitative Analysis, physico-chemical properties of the analyte or its reaction product are converted into signals that can be measured by optical, electroanalytical, thermal, magnetic or radiochemical instruments with

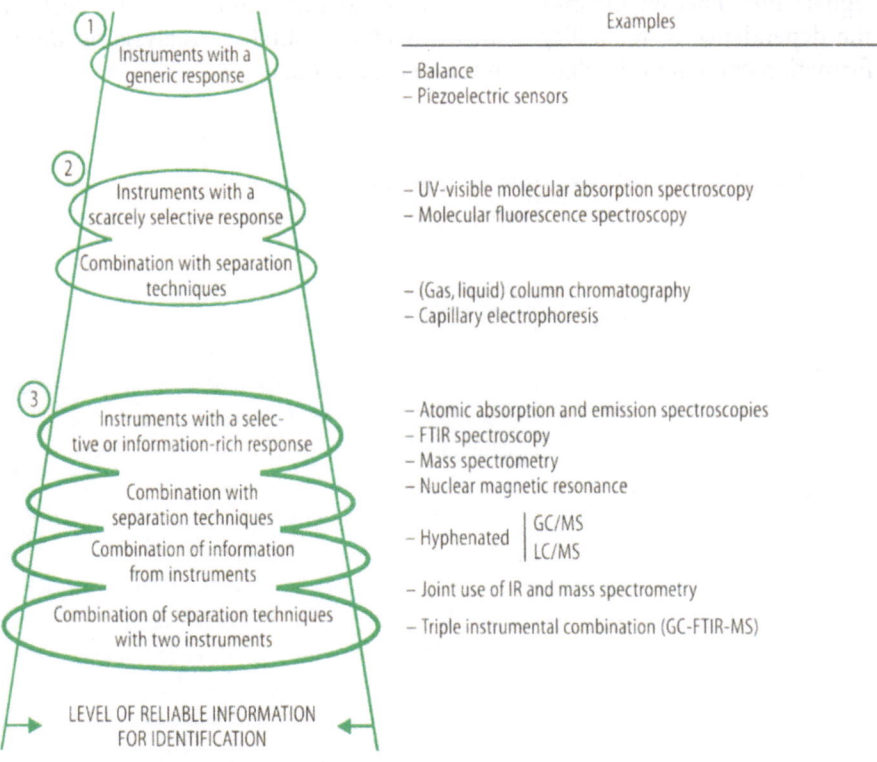

Fig. 5.11. General classification of instruments according to reliability in their response for use in Qualitative Analysis. Three different types of instruments, and the role of separation techniques in this context, are considered. For details, see text

a view to their identification. As noted above, the difference between Instrumental and Classical Qualitative analysis is that the latter uses the human body as "instrument". The substantially increased identification power of available equipment obviously makes Instrumental Qualitative Analysis much more reliable and widely applicable.

Identifying in this context involves comparing one or more signals obtained at a single or many instrumental parameter values for three different aliquots subjected to the analytical process, namely:

(*a*) A standard containing the analyte;
(*b*) a blank resembling the sample in composition but containing no analyte; and
(*c*) a sample, which may or may not contain the analyte.

The probability of obtaining YES or NO as the binary response will depend on the degree of consistency between the signals produced, their number and how well-resolved they are. The use of a single signal (*e.g.* absorbance) obtained at a single instrumental parameter value (*e.g.* wavelength) will obviously provide less reliable qualitative identifications than will multiple signals recorded at many different instrumental parameter values (*e.g.* the bands in an IR absorption spectrum). Molecular absorption spectra are much broader and general than are atomic absorption spectra; the latter contain much better resolved signals and hence enable much more reliable identification. Box 5.15 illustrates the dependence of reliability in Instrumental Qualitative Analysis on the information content of the data used for identification.

_____ Box 5.15

Information level of qualitative instrumental analyses

Number of instrumental parameters	Number of signals involved	Examples
ONE	ONE	• Atomic emission spectrophotometric signal at a characteristic wavelength for a specific element. • Half-wave polarographic potential for a species.
TWO	ONE	• Fluorescence intensity at an excitation wavelength (λ_{ex}) and an emission wavelength (λ_{em})
ONE	TWO	• Retention time (signal) for a peak (absorbance signal) at a single wavelength (λ_{max}) in liquid chromatography or capillary electrophoresis with UV-visible detection
MANY	MANY	• Identification based on whole spectra (IR, mass, NMR)

From Box 5.15 it follows that not all instruments are equally capable of providing reliable identifications; in fact, reliability in this context depends on the selectivity and information content of the responses they produce. Instrumental analytical techniques can be ranked in the three broad categories of Fig. 5.11 in this respect – the ranking, however, can obviously have some exceptions.

The instruments in **Group 1** provide a very general response (*i. e.* one shared by many analytes). Such is the case with the mass measured by a balance or a piezoelectric sensor. This type of information is useless for qualitative analyses unless altered in some way to increase its selectivity (*e. g.* in thermogravimetries, which provide mass–temperature two-dimensional information, and in using piezoelectric sensors coated with selective sorbent materials).

The instruments in **Group 2** give a scarcely selective response (*i. e.* one that can be subject to many interferences from sample components other than the analyte). One case in point is UV-visible absorption spectroscopy (photometry), where virtually every species exhibits absorption of incident light. The selectivity of this technique can be improved by derivatizing the analytes to products with differential spectral features. Alternatively, more selective techniques such as fluorimetry ensure higher reliability (fluorescence is a much more uncommon property than is molecular absorption). Even greater selectivity can be achieved from on-line combinations of instruments in this group with continuous column chromatographic separation techniques (see Fig. 4.14); the poor selectivity of the instrument is offset by the high selectivity of the continuous separation, which isolates the analytes within dynamic zones.

Finally, the instruments in **Group 3** provide highly selective information (*e. g.* that of atomic absorption and emission spectroscopies) or information containing many well-resolved signals obtained at multiple instrumental parameter values (*e. g.* those of IR spectroscopy or mass spectrometry). The use of this type of instrument as detector in chromatographic techniques has given rise to so-called "hyphenated techniques", which provide significantly increased reliability in the identification. Reliability can also be substantially improved by the joint use of information provided by one instrument each from Groups 2 and 3 (or two from Group 3). One typical example is the highly reliable (>99%) identification of analytes in a sample from its combined IR and mass spectra. However, the highest level of reliability is provided by the combination of a separation technique and two or more instruments from Group 2 or, better, Group 3 (*e. g.* the gas chromatography-mass spectrometry-IR spectroscopy tandem).

The detailed description of each individual technique used for qualitative purposes that would be required to provide readers with a comprehensive picture of the subject is beyond the scope of this introductory textbook. Instead, the following sections discuss specific examples involving time-dependent and time-independent signals that are intended to provide the reader with a clear view of the situation.

5.6.2 Static Systems

In static systems, the blank and standard are inserted into the instrument to obtain a time-independent analytical signal. This information is usually two-dimensional (a signal as a function of a single instrumental parameter) but can occasionally be three-dimensional (a signal as a function of two instrumental parameters). How reliable the identification is will depend directly on how well the information profile of the analyte is resolved from those for other species in the sample (*i.e.* on the group to which the instrumental technique belongs).

Figures 5.12–5.14 illustrate the use of optical molecular absorption techniques with differential (increasing) identification power. UV-visible molecular spectroscopy (Fig. 5.12) is based on a rather general property; however, it possesses some discriminating power – it is thus a Group 2 technique. The spectra for the three species (A–C) are strongly overlapped, which hinders identification. One way to facilitate it is by using a selective reagent for A to obtain a product AR with a very different (shifted) spectrum enabling the reliable identification of this analyte. If the signals in the spectral region where A, B and C absorb (200–500 nm) were compared, mutual interferences would pose an unsurmountable hindrance to identification (and also to quantitation).

Molecular emission spectroscopy (fluorimetry) relies on a much more uncommon property than is absorption; in fact, relatively few molecules emit fluorescence on being excited with light. Also, fluorescence signals are normally better resolved than are absorption signals. Fluorimetry is a Group 2 technique but close in performance to a Group 3 one. Figure 5.14 shows the excitation spectra (equivalent to absorption spectra) and emission spectra for two substances A and B. As can be seen, the portion of the emission spectrum at $\lambda_{em} > 550$ nm

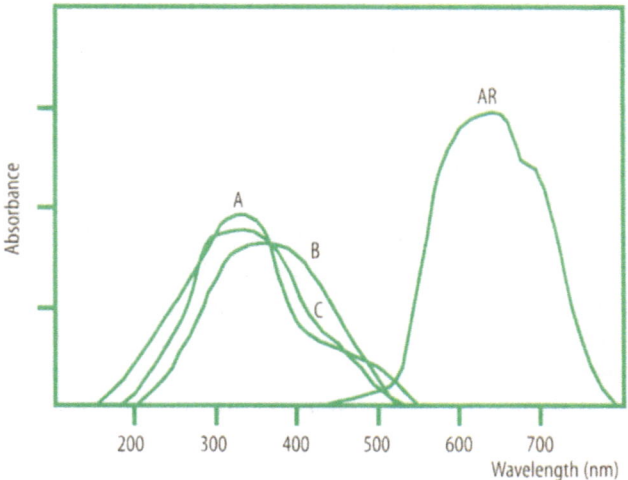

Fig. 5.12. UV-visible absorption spectra for three imaginary species *A*, *B* and *C*, and for the reaction product of *A*, used as a model to assess the potential of the photometric technique for identifying these species. For details, see text

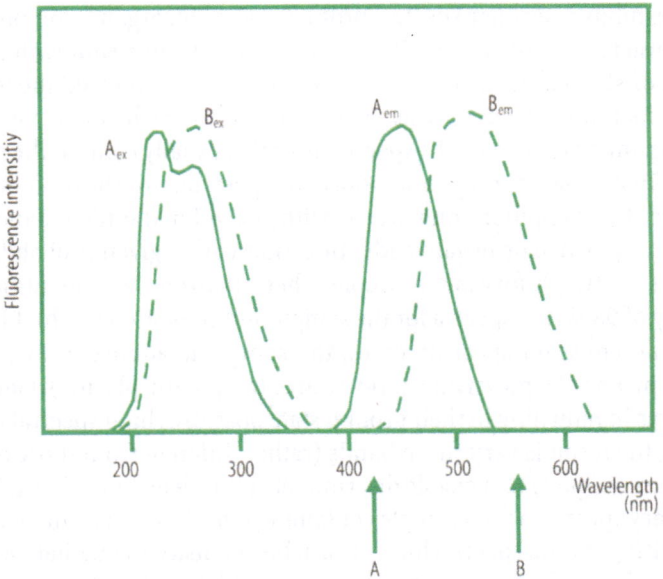

Fig. 5.13. Fluorescence excitation and emission spectra for two species A and B, which allow both to be identified by measuring the light emitted at 420 (A) and 565 nm (B), respectively. For details, see text

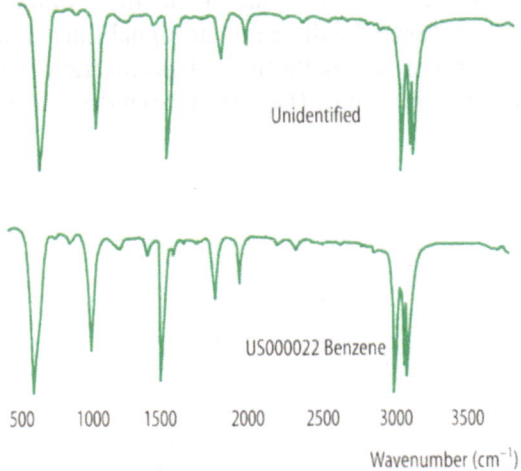

Fig. 5.14. Transmittance (molecular absorption) spectra in the infrared region for an unknown sample and the benzene standard included in a spectral library. For details, see text

allows the identification of B; also, if any signal (fluorescence intensity) is observed at $\lambda_{em} < 430$ nm, then the presence of analyte A can be assumed.

Those techniques that provide wealthier, more specific spectral profiles belong to Group 3 (that with the highest identification power). One such technique is molecular infrared (IR) absorption spectroscopy, where each organic function exhibits a peculiar spectral profile; as a result, organic compounds have highly characteristic IR spectra. By comparing an IR spectrum with those for a collection of standards, the unknown compound that produced the former can be readily identified in some cases. The comparison can be made visually or by using chemometric software. IR spectral libraries usually contain the spectra for 5000 to 100 000 organic compounds; once the spectrum for the unknown species is acquired, the computer compares it with those for the filed standards. The response is a positive or negative identification with a given probability. Figure 5.14 shows a straightforward example where benzene was identified with a probability of 98 % (the spectra for the sample and standard were highly similar).

In atomic emission spectroscopies, the sample is atomized and atoms are excited by means of a powerful energy source (arc, spark, plasma). Light emitted by the atoms in returning to their ground state possesses high spectral purity and sharpness; the result is very sharp bands (rather different from those of molecular optical techniques) that enable discriminate multidetection. Figure 5.15 shows an imaginary spectrum for a sample containing four different elements ($M_1 - M_4$) each exhibiting several spectral lines of variable intensity. In qualitative terms, the important thing is the scarce spectral overlap, which makes this technique suitable for multi-element identification (in addition to quantitation). This is thus a Group 3 technique with a high potential in Qualitative Analysis.

In Electroanalytical Chemistry, ion-selective electrodes (*e.g.* pH, pNa and pF sensors) can be used for qualitative purposes as well. These sensors, in combination with a reference electrode, are sequentially immersed in standard, blank and sample solutions; the discriminate signal thus obtained is normally used for quantitation but is equally fit for identification. The greatest shortcoming of the these sensors – the pH electrode excluded – is that their measure-

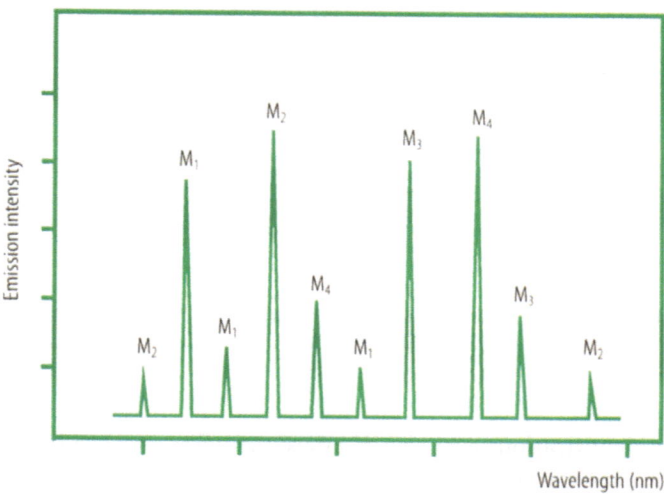

Fig. 5.15. Assessment of the potential of atomic emission spectroscopy for the identification of four elements ($M_1 - M_4$). The graph shows the emission spectrum for a sample containing all four. For details, see text

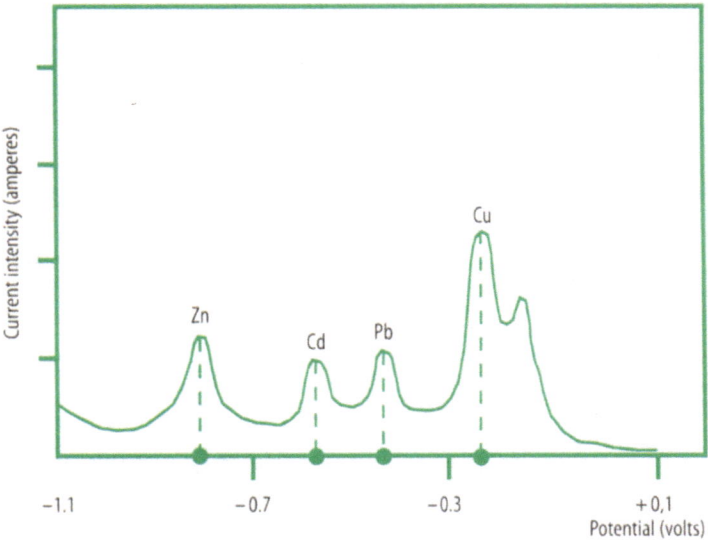

Fig. 5.16. Current-potential curves obtained by anodic stripping voltammetry for a sample containing zinc, cadmium, lead and copper. For details, see text

ments are subject to potential interferences from extraneous ions; based on their modest selectivity, they can be placed at an intermediate position in Group 2 (Fig. 5.11). Some electroanalytical techniques are selective enough for inclusion in Group 3, however. One such technique is anodic stripping voltammetry (Fig. 5.16), which involves two operational steps, namely:

(a) Concentration, by which metals or other species are deposited by reduction at the electrode ($M^{z+} + ze^- \rightarrow M$).

(b) Anodic stripping, where the electrode serves as anode (oxidation). As the applied potential is raised, previously deposited metals are sequentially oxidized at a characteristic potential each. The resulting microcurrent is suitably recorded and the redox potential at which a peak is obtained is used for identification.

As can be seen from the examples of Figs 5.12–5.16, identification relies on the information provided by a signal obtained at a given instrumental parameter value (on the x-axis of a two-dimensional plot).

5.6.3 Dynamic Systems

Dynamic systems provide time-dependent signals. Discrimination among species in the same sample, and hence their reliable individual identification, rest on time as the parameter. As a rule, the qualitative information required is derived from the x-axis of a signal-time two-dimensional plot. The dynamics of the signal may arise from (a) the instrumental technique itself (**instrumental**

_____ Box 5.16

General principles of column chromatography

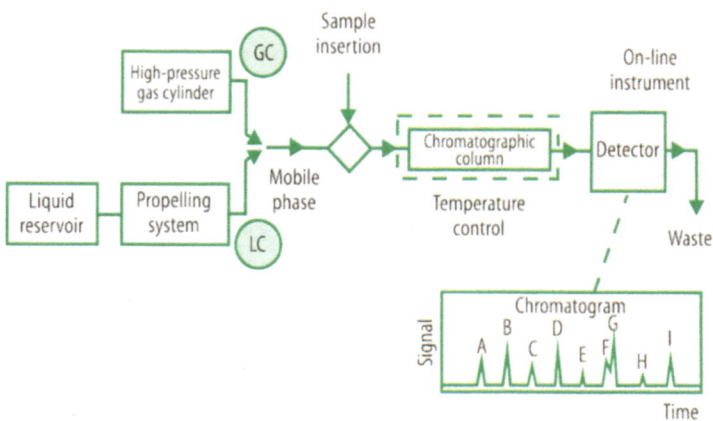

Gas chromatographs (GC) and liquid chromatographs (LC) comprise the basic elements shown in the figure above. The mobile phase is either withdrawn from a high-pressure gas reservoir (GC) or propelled by a high-pressure pump (LC). A microvolume of liquid or gaseous sample is inserted into it. The analytes are separated on the chromatographic column, which differs for GC and LC. The column contains a solid or liquid stationary phase. Separation is based on the differential affinity of the analytes for the two phases involved: elution of the more similar in nature to the stationary phase and the mobile phase is delayed and advanced, respectively, in relation to the others. These phenomena result in separation of the analytes in isolated or virtually isolated portions of the mobile phase. By continuously monitoring the eluent that leaves the column using a Group 2 or Group 3 instrument, a signal-time profile called a "chromatogram" (where each analyte exhibits a chromatographic "peak") is obtained.

kinetics), (*b*) a chemical reaction the development of which is monitored (**chemical kinetics**) or (*c*) the use of a separation technique (*e.g.* chromatography) coupled on-line to a Group 2 or Group 3 instrument (**physical kinetics**). Only the last is considered here, on account of its strong practical implications.

The ability of separation techniques to physically isolate the components of a mixture (sample) is used for group separation in Classical Qualitative Analysis. This ability can be substantially enhanced by on-line coupling to an instrument. Such is the case with column chromatography and capillary electrophoresis. As can be seen from Fig. 5.11, the information content of use for reliable identifications with Group 2 and 3 instruments can be increased by using a separation technique.

Box 5.16 outlines the different possible column chromatographic modes. Capillary electrophoresis is based on a different principle but uses a similar configuration (see Box 5.17). Both techniques are frequently used for identification purposes. Their high identification power frequently enables multidetection in the same sample.

Box 5.17

General principles of capillary electrophoresis

Capillary electrophoresis (CE) is a separation technique based on the differential rate of migration of the analytes along a capillary under a strong electrical field.

A typical CE system uses two buffers into which two electrodes connected to a high-voltage source are immersed; this creates a strong electric field within the intervening capillary. The capillary uses an on-line detector on one end to continuously monitor the analytes that circulate through it. The sample is inserted through one end in a discrete manner. Positively charged analytes migrate to the cathode, anionic ones to the cathode and uncharged species to neither. This approach is often altered in practice by effect of the electroosmotic flow from the anode to the cathode; this flow facilitates passage of the analytes through the detector in a sequential manner: cations first, neutral species then and anions last. Mobility differences among the analytes in these groups arise from differences in charge and mass: the more highly charged cations and anions migrate the faster and slower, respectively; also, the heavier ones migrate faster than the lighter ones.

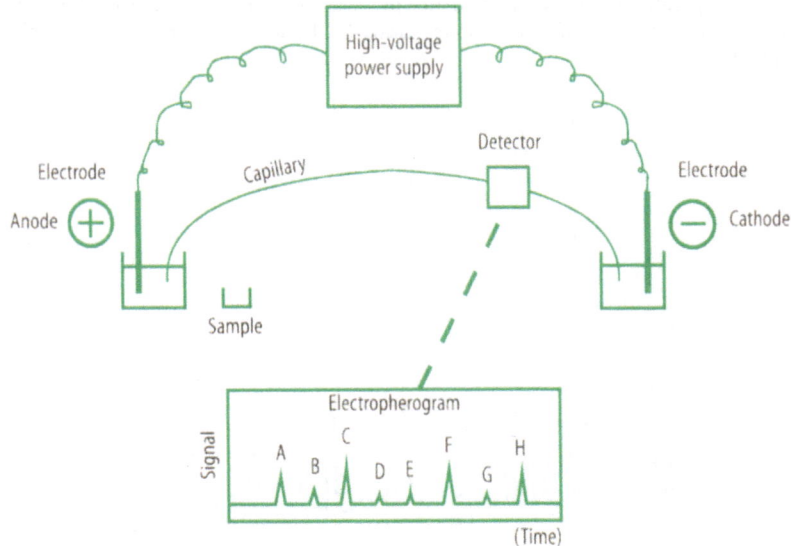

Figures 5.17 and 5.18 illustrate the high capacity of chromatography and capillary electrophoresis to discriminate among different species present in the same sample. Note that the electrophoretic separation of 30 species takes roughly 3 min – an extremely short time compared to identification by the classical procedure.

With these instrumental assemblies, each analyte is identified in terms of its "retention time", t_R, which is the time (on the x-axis of the chromatogram or electropherogram) corresponding to the top of the analyte peak. This process can be performed in various ways depending on the type of reference (standard) used, namely:

(a) By comparison with the t_R value for an analyte standard previously inserted into the system instead of the sample. However, the value provided by the standard is usually scarcely reliable as it depends strongly on the operating

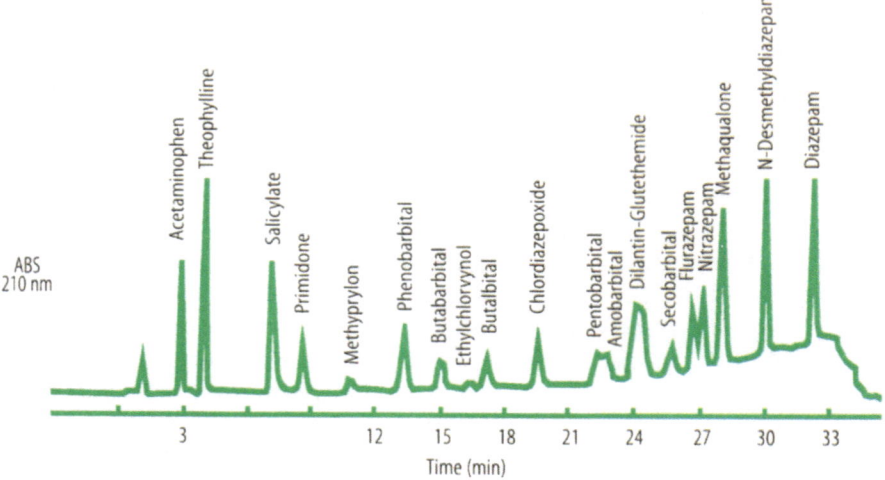

Fig. 5.17. Separation-detection of 19 drugs by reversed-phase liquid chromatography, using acetonitrile-phosphate buffer as mobile phase, C_{18} bonded silica as stationary phase and photometric detection

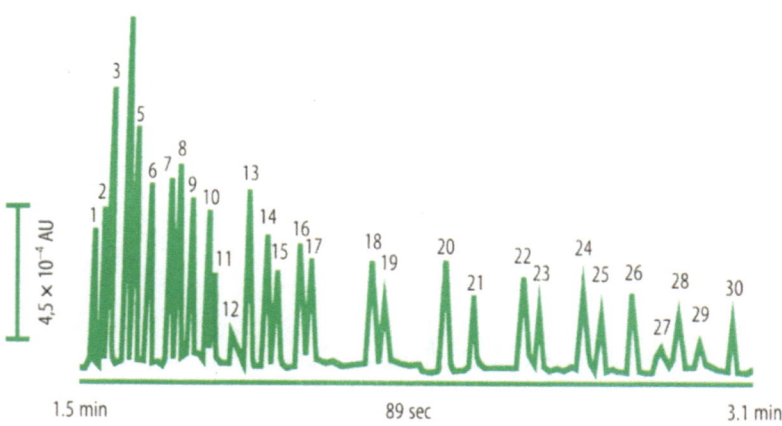

Fig. 5.18. Separation-detection of 30 inorganic and organic anions by capillary electrophoresis with photometric detection

conditions (pressure, temperature, flow-rate of the mobile phase, voltage); in fact, small changes in the conditions may considerably alter the dynamic behaviour of a given species in the standard and sample. For this reason, direct comparisons are inadvisable.

(b) By using the internal standard procedure described in Box 3.14, which involves adding a standard of a species other than the analyte to both the samples and the ordinary analyte standards used. The retention times thus obtained are either normalized or referred to the internal standard. In this way, the distorting influence of isolated fluctuations in the experimental conditions (*e.g.* temperature, pressure, flow-rate), which have crucial effects on the measured parameter (retention time), is minimized.

(*c*) By adding a standard of the analyte to the sample in order to facilitate its identification. Two chromatographic runs are performed, using the sample in one and the sample plus standard in the other. If any peak increases in height as a result, then the analyte will be the species contained in the added standard. Box 5.18 shows a typical example.

(*d*) By using one of the techniques in Group 3 (*e.g.* IR spectroscopy, mass spectrometry), which provide whole spectra for each chromatographic peak and facilitate highly reliable identification (as can be seen from Fig. 5.11, the qualitative information produced by these techniques is of a high level).

___ Box 5.18

Chromatographic identification of a sample component by addition of a standard of the analyte

Reversed-phase liquid chromatography (*viz.* with water as the mobile phase and a non-polar material such as C_{18} bonded silica as the stationary phase) enables the discrimination of various ingredients of soft drinks (*e.g.* ascorbic acid, saccharine and caffeine). If the chromatogram obtained by inserting a sample aliquot into the chromatograph (Box 5.16) contains a peak for an unknown substance, then one can suspect the presence of an additive. Because sodium benzoate is a frequent occurrence in this context, a new sample aliquot is supplied with an appropriate amount of the pure additive to make a standard for injection into the chromatographic system. If the previous, unknown peak grows as a result, then the binary answer to the question "is sodium benzoate present in the drink?" is YES. On the other hand, if the peak appears in a different zone (*i.e.* at a different time) in the chromatogram, then a different standard should be tested.

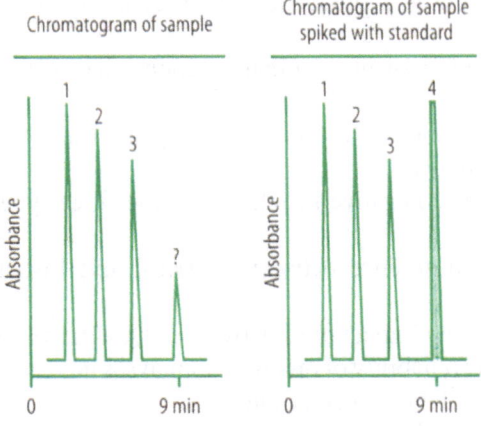

1: ascorbic acid; 2: saccharin; 3: caffeine; 4: benzoic acid (confirmed)

Questions

(1) What is a binary response?

(2) Define "test" and "screening" in connection with Qualitative Analysis.

(3) Why are physical measurements rarely qualitative in nature? Ask several questions with a binary answer in this context.

(4) Distinguish "detection" and "identification" in Qualitative Analysis.

(5) Why is Qualitative Analysis the first step in a quantitative analysis?

(6) What is a white sample? Give some examples.

(7) Give several examples of qualitative information requirements at different levels based on the scheme of Fig. 5.2.

(8) Elaborate on the differences between the first-level questions "is it the analyte?" and "is it in the sample?" through applicable examples.

(9) What is a grey sample? Give some examples.

(10) Distinguish "cut-off concentration" and "threshold concentration".

(11) Who imposes (a) the limit of detection, (b) the cut-off concentration and (c) the threshold concentration?

(12) Which is more important for the purpose of assuring quality in the binary response:

(a) that the limit of detection be high or low?;

(b) that the cut-off and threshold concentration be high or low?

(13) What is a black sample? Give several examples.

(14) Define the analytical property "reliability" and state the way it is expressed.

(15) Describe a false positive in relation to the three types of reference concentration.

(16) Give four examples of Qualitative Analysis and place them in each of the six classifications of Fig. 5.6.

(17) On what analytical properties does reliability in Qualitative Analysis depend?

(18) Which type of error in the binary response has a stronger impact on the analytical problem? Why?

(19) Comment on the impact of selectivity on Classical and Instrumental Qualitative Analysis.

(20) Describe a false negative in relation to the three types of reference concentration.

(21) What is the difference between Classical and Instrumental Qualitative Analysis?

(22) Comment on the impact of sensitivity on Qualitative Analysis.

(23) Why does the reliability of qualitative analyses increase with increase in the amount of available information?

(24) Give some examples of the use of standards in Classical Qualitative Analysis.

(25) What is an analytical scheme? What are the main types of scheme? What are their differences?

(26) Give several examples of equipment calibration in Instrumental Qualitative Analysis.

(27) What are the greatest shortcomings of Classical Qualitative Analysis?

(28) Give some examples of method calibration in Instrumental Qualitative Analysis and establish differences from the equipment calibration implicit in each.

(29) Compare the role of separations in Classical and Instrumental Qualitative Analysis, with emphasis on the more common choices in each.

(30) What advantages does Instrumental Qualitative Analysis have over Classical Qualitative Analysis?

(31) Give some examples of each type of reagent used in Classical Qualitative Analysis.

(32) Discuss the requirements a substance must meet for use as a group separation reagent in Classical Qualitative Analysis.

(33) Distinguish, through examples, the most usual types of reactions used to identify inorganic and organic species in Classical Qualitative Analysis.

(34) What advantages and disadvantages does the use of immunoreagents in Qualitative Analysis have?

(35) In what way are Group 2 and Group 3 instruments different as regards reliability in the qualitative information they provide?

(36) Why do separation techniques boost the reliability of identifications in Instrumental Qualitative Analysis?

(37) What is a mixed analytical scheme?

(38) Why is fluorimetry a more advantageous choice than photometry for identification purposes?

(39) Why is IR spectroscopy superior as an identification technique in relation to other molecular spectroscopies such as photometry and fluorimetry?

(40) In what way is an electropherogram (Fig. 5.18) different from a voltammogram (Fig. 5.16) and a chromatogram (Fig. 5.17)?

(41) Why must the use of a blank be considered in Classical and Instrumental Qualitative Analysis?

(42) What problems does it pose to use retention times as such for identification in electrophoretic and chromatographic systems?

(43) Distinguish "internal standard" and "analyte standard" in Instrumental Qualitative Analysis.

(44) In what way can a computer system be used for identification purposes?

(45) Give some examples of identification by use of instrumental systems.

(46) Why do instruments differ in identification power?

(47) How could the presence of copper traces in a wine be confirmed using anodic stripping voltammetry?

(48) How can a chromatographic peak for a given sample be identified? How can one in an electropherogram?

(49) What advantages does the use of an internal standard in Instrumental Qualitative Analysis have?

(50) Why has Qualitative Analysis been revitalized in recent times?

Seminar 5.1

The Hydrogen Sulphide Scheme

- **Learning objective**
  - To describe in detail the experimental procedure followed to identify metal cations in a mixture by separation into groups using precipitation as sulphides as the more relevant chemical tool (see Box 5.11).
  - To become acquainted with practical solution chemistry through an analytical scheme that has been used by many generations of chemists in qualitative analysis laboratories.
- **Literature**
  Use of the books by Burns *et al.* (suggested reading no. 6) is advised.
- **Tools needed**
  - The chemical tools typically used in Classical Qualitative Analysis.
  - Group, identification and masking reagents, with special emphasis on H$_2$S.
- **Briefly discuss and comment on the basic (chemical) and applied aspects of the scheme**
  - Preliminary tests.
  - Scheme layout. Group distribution. Place special emphasis on separations.
  - Ions involved.
  - Why are they here?
  - Identification reactions and their potential interferences.
- **Critically compare this analytical scheme with other available choices for inorganic cations**

Seminar 5.2

Global Assessment of Test Kits from a Specialized Manufacturer (Merck)

- **Learning objective**
  - To demonstrate the use of commercially available test kits for the rapid identification and quantitation of organic and inorganic species (non-clinical parameters).
  - To become acquainted with analytical tools of growing use.
- **Material**
  The detailed instructions supplied by the kit manufacturer.
- **Types of tests**

(A) Use of human senses	A.1 Reagent strips (immobilized reagents)
	A.2 Sample and blank vials (added reagents)

(B) Use of straightforward instruments

B.1 Reflectance measurements (reagent strips)

B.2 Spectrophotometric measurements (reagent-containing vials)

B.3. Miscellaneous

■ **Description and discussion of each case**

For one or two tests of each type:

- Describe the procedure.
- Discuss the chemical foundation of the test.
- Establish the scope of application (target analytes).
- Justify use according to sample type and specific information needs.

Seminar 5.3

Use of Reagent Strips in Routine Clinical Analyses (Combur-10-Test M, from Boehringer)

■ **Learning objective**

- To demonstrate the multiparameter determination capacity of a reagent strip for the rapid monitoring of 10 routine clinical parameters in urine.
- To emphasize the expeditiousness with which the response of this screening system can be obtained.

■ **Material**

The detailed instructions supplied by the manufacturer to describe the procedure, which includes the use of human senses (comparison with a colour scale) and of a reflection photometer. The results are not always co-incident. The output is (*a*) a binary response or (*b*) a concentration range (a semi-quantitative response).

■ **Parameters (urine)**

Density, pH, leucocytes, nitrite, protein, glucose, ketone bodies, urobilinogens, bilirubin and blood.

■ **Description and discussion**

- Describe the procedure (CMP).
- Discuss the (bio)chemical foundation of each test.
- Discuss the two forms of measurement (visual and instrumental).
- Comment on the two ways of expressing the results.
- Discuss the need for blanks.
- Identify potential interferences with each test.
- Comment on the clinical significance of the result for each test.

Seminar 5.4

The Use of Standards in Instrumental Qualitative Analysis

- ■ **Learning objective**
  To demonstrate different ways of using standards for comparisons in Instrumental Qualitative Analysis with a view to the identification of one or more analytes.
- ■ **Literature**
  - Books on various instrumental techniques.
  - Compilations of standard or official methods.
- ■ **Static instrumental techniques** [1]
  - Identification by FTIR spectroscopy.
  - Identification by mass spectrometry.
  - Mixed identification (IR-MS).
- ■ **Dynamic instrumental techniques** [1]
  - Identification by GC.
  - Identification by LC.
  - Identification by GC-MS.
- ■ **Discussion**
  Once the specific CMPs have been chosen and described, discuss the use of standards for the comparisons involved. Distinguish between method and equipment calibration in each CMP. Comment on the differences between the use of static and dynamic instrumental techniques in this context.

Suggested Readings

(1) "Quality Control in Analytical Chemistry", 2nd edn, G. Katemans and L. Buydens, *Wiley*, New York, 1993.
Contains the chemometric foundations of the binary response and its errors.

About Classical Inorganic Qualitative Analysis

(2) "Química Analítica Cualitativa", F. Burriel, F. Lucena, S. Arribas and J. Hernández Méndez, *Paraninfo*, Madrid (Spain), 1988.
(3) "Análisis Inorgánico Cualitativo Sistemático", F. Buscarons, F. Capitán García and F. Capitán Vallvey, *Reverté*, Barcelona (Spain), 1986.
(4) "Análisis Cualitativo Inorgánico sin Empleo del Acido Sulfhídrico", S. Arribas, *Gráficas Summa*, Oviedo (Spain), 1974.
(5) "Técnicas Experimentales de Análisis Cualitativo", F. Pino Pérez, *Ediciones Urmo*, Bilbao (Spain), 1979.

---

[1] Rather than describing the analytical technique involved, discuss the identification process in each case, with emphasis on the significance of standards.

(6) "Systematic Chemical Separation", D.T. Burns, A. Townshend and A.H. Carter, 3 vols, *Ellis Horwood*, Chichester, 1981.
Reference books for an expanded description of the procedures presented in this chapter.

About Classical Organic Qualitative Analysis

(7) "The Systematic Identification of Organic Compounds. A Laboratory Manual", R. Shriner, R. Fulson and D. Curtin, *Wiley*, New York, 1980.
(8) "Semimicro Qualitative Organic Analysis", N. Cheronis, J. Entrinking and E. Hodnett, *Wiley*, New York, 1964.
(9) "Spot Test Analysis", E. Jungreis, *Wiley*, New York, 1997.
Reference books for an expanded description of the procedures presented in this chapter.

About Instrumental Analysis

(10) "Instrumental Analysis", 4th edn, D. Skoog and J. Leary, *Saunders College Pu.*, New York, 1992.
(11) "Instrumental Methods of Analysis", H. Willard, L. Herrit, J. Dean and F. Settle, *Wadsworth, Inc.*, New York, 1988.
(12) "Instrumental Analysis", G.D. Christian and J.E. O'Reilly, *Allyn & Bacon*, New York, 1986.
Reference books for an expanded description of the procedures presented in this chapter.

About Chromatography

(13) "Técnicas Analíticas de Separación", M. Valcárcel and A. Gómez Hens, *Reverté*, Barcelona (Spain), 1988.
(14) "Chromatography Today", 5th edn, C.F. Poole and S.K. Poole, *Elsevier Science*, Amsterdam, 1991.
Reference books for an expanded description of the procedures presented in this chapter.

6 Quantitative Aspects of Analytical Chemistry

Objectives

- To describe the essential aspects of quantitation processes in Analytical Chemistry.
- To characterize the quantitative response and its forms of expression.
- To discuss the different choices available for calibration in Quantitative Analysis and their associated standards.
- To describe classical and instrumental analytical techniques for primarily quantitative purposes.
- To define and exemplify methodological approaches to analytical quantitation.

Table of Contents

6.1 Fundamentals

6.1.1 Introduction

Quantitative analyses involve the extraction of numerical information of absolute (*e.g.* mass units) or relative nature (*e.g.* a concentration or percentage) about one or more analytes present in a sample.

Quantitative Analysis is the intermediate link in the analytical chain of analysis types according to purpose (see Fig. 1.20). On the one hand, one must previously have ascertained whether not only the target analyte, but also other substances that might interfere with the analytical signal used for its quantitation, are present or absent. On the other, Structural Analysis can only be undertaken properly once the quantitative composition of a sample or object is known. Figure 6.1 places quantitation in a sequence of increasing analytical information content at the macroscopic level – the one studied in this chapter – and the microscopic level. Note that the last information link involves characterizing a qualitatively, quantitatively and structurally analysed sample with a view to determining other properties of internal (stability, constituent interactions and mobility) and external nature (reactivity, fitness for purpose).

The output of a quantitative analysis is a numerical result, preferably accompanied by its uncertainty. However, quantitative data are not limited to this type of analysis. Thus, as shown in Sect. 5.2.2, which discusses the quantitative connotations of the binary response in Qualitative Analysis, one of the steps that lead to such a response (*e.g.* whether a given substance is present above its threshold concentration) actually involves quantifying the analyte once it has been identified. Establishing structures, distributions and morphologies, all of which are "outputs" of Structural Analysis, also involves quantifying.

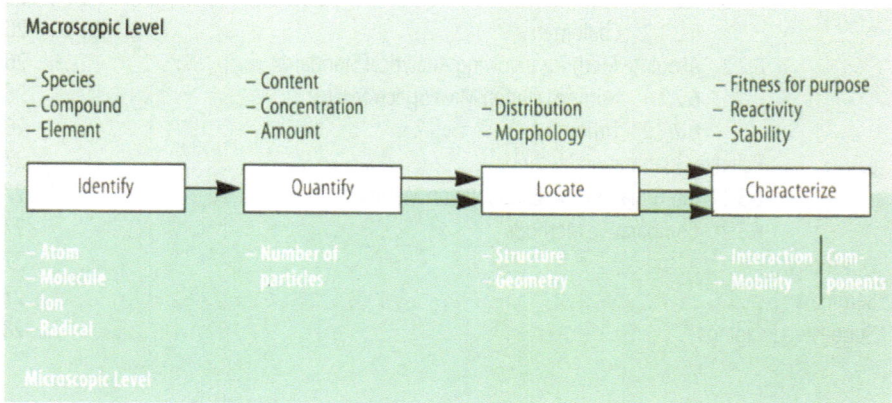

Fig. 6.1. Sequence of actions which, according to A. Zschunke, provide increasing analytical information about a sample or object at both the macroscopic and the microscopic level. Note that "to characterize" is used in a context other than that of Fig. 1.15

This chapter deals with the principles and fundamentals of Quantitative Analysis, with special emphasis on the equipment and methodology used.

6.1.2 Analytical Properties

Chapter 2, which discusses analytical properties in a systematic manner, focusses on Quantitative Analysis. Figure 2.2 is highly illustrative in this context; in fact, the chemical metrological ranking it shows is based on numerical data (individual results, means of results, the value held as true, *etc.*). Consequently, all the properties considered in Figs. 2.1 and 2.19 are relevant to the contents of this chapter.

A quantitative result can be thoroughly characterized in terms of the two capital analytical properties, *viz. accuracy* (closeness to the true value) and *representativeness* (the degree of consistency with the analytical problem addressed). These properties in turn rest on basic analytical properties (precision, sensitivity, selectivity and proper sampling); in fact, neither capital property can be realized with poor levels of the basic ones. One must also consider the accessory properties that characterize laboratory productivity (expeditiousness, cost-effectiveness, personnel safety). Even more important, one must bear in mind the complementary and contradictory relationships among these properties (see Sect. 2.7).

As shown in Sect. 3.6.1, a quantitative analytical result can also be characterized by its *traceability* (*viz.* its relationship to standards and the "history" of its production). In addition, it can be characterized by a specific *uncertainty* (*viz.* the numerical range where it can be expected to fall at a given probability level), which is a precision-related property. Finally, results are also appreciably affected in practice by *robustness* [*viz.* the ability to remain unaltered in response to slight changes in the experimental conditions of the chemical measurement process (CMP)].

6.1.3 Equipment and Method Calibration

Proper use of suitable measuring standards is crucial in Quantitative Analysis; without them, one would not be able to measure (or compare). All types of standards (basic, chemical and analytical) depicted in Fig. 3.4, as well as the different types of analytical chemical standards (Fig. 3.8), can be used for quantitative purposes in the first two steps of a CMP.

One important application of standards is calibration of equipment and methods (see Fig. 3.10).

The primary aim of *equipment calibration* is to ensure proper functioning of a quantifying instrument (*e.g.* a balance, burette, spectrophotometer, polarograph, mass spectrometer) by using standards not containing the analyte (*e.g.* transfer weights for a balance, a holmium filter for a UV-visible molecular absorption spectrophotometer). The example describing the calibration of a photometer in Box 3.3 is highly illustrative.

Method calibration in Quantitative Analysis is basically intended to establish a clear-cut, unequivocal relationship between the instrument response and the amount or concentration of analyte that produces it. Except in titrimetric standardization, such a relationship is derived by using standards containing the analyte (see Box 6.2).

Standards and calibration are two crucial concepts in Quantitative Analysis. Accordingly, the most important classification of CMPs for quantitative purposes is based on them (see Sect. 6.1.5).

6.1.4 Classical and Instrumental Quantitative Analysis

The distinction between these two types of analysis is not a categorical one as it rests on historical rather than on scientific or technical considerations. As can be seen from Fig. 1.21, both types of analysis can be used for quantitative purposes.

Classical quantitative analysis employs the burette and the balance – two instruments which have been in use for centuries– to implement the analytical gravimetric and titrimetric technique, respectively. Both can rely on human senses to make measurements. With former (two-pan) balances, human sight was used to directly read out the mass indication for the object being weighed; in modern balances, the mass is directly displayed as a digital readout. In classical titrimetries, human sight is used to

(*a*) zero the burette,
(*b*) stop the addition of titrant when the visual indicator changes colour, and
(*c*) read the titrant volume used off the millimetric scale on the burette.

In modern titrimetries, an instrumental system is used to identify the titration end-point and the titrant volume used is supplied in digital form by the auto-burette itself. As can be seen from Fig. 6.2, the interface between these two types of analysis is rather diffuse.

Instrumental quantitative analysis relies on qualitative information obtained in the second step of a CMP by using instruments other than the balance and the burette. An optical (*e. g.* molecular or atomic spectroscopy), electroanalytical (*e. g.* potentiometry, polarography, coulometry, anodic stripping voltammetry), thermal (*e. g.* differential thermal analysis, thermogravimetry), radiochemical (*e. g.* neutron activation, isotope dilution), magnetic (*e. g.* nuclear magnetic resonance) or mass technique (*e. g.* mass spectrometry) can be used for this purpose. Because they use on-line coupled detectors, gas and liquid chromatographs, and capillary electrophoretic systems, can also be considered instruments (see Boxes 5.16 and 5.17). Because the type of response obtained varies depending on whether or not the instrument is coupled on-line to a separation system, the signal will call for a specific quantitative treatment (see Fig. 6.2).

Not all instrumental techniques are equally applicable to the three basic types of analysis (qualitative, quantitative and structural). Figure 6.3 classifies the better known instrumental techniques into three categories according to whether they are preferentially used for qualitative or quantitative purposes. This is obviously a generalization and specific situations require using each technique in a way different from that depicted in Fig. 6.3.

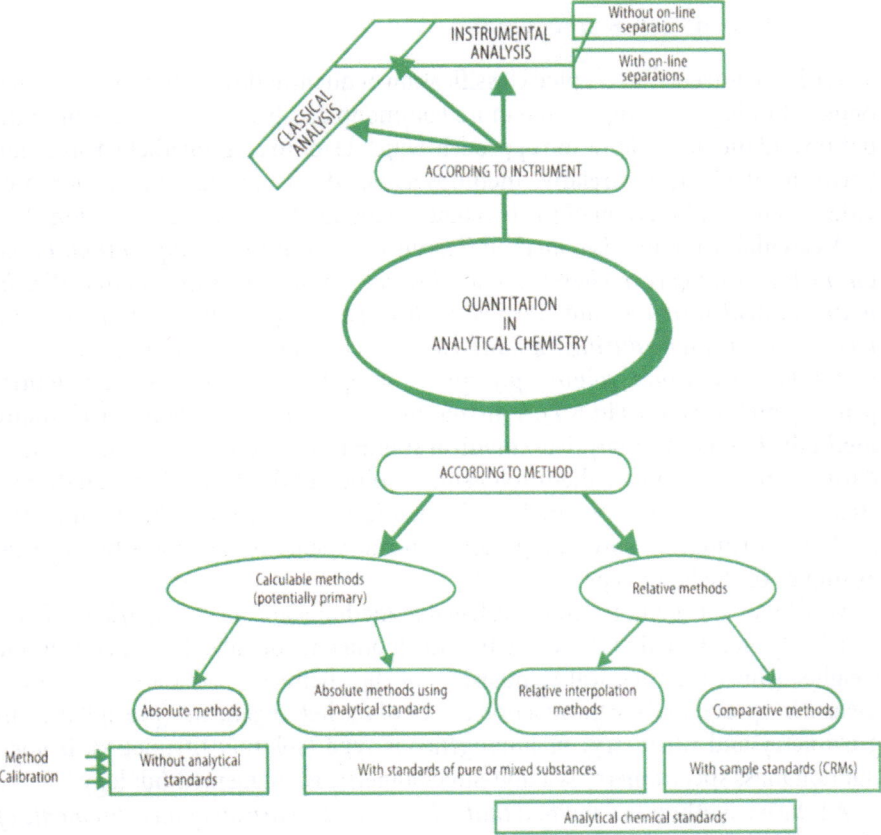

Fig. 6.2. Classification of quantitative analyses according to type of equipment and method used. For details, see text

QUALITATIVE ANALYSIS	BOTH	QUANTITATIVE ANALYSIS
Infrared (IR) molecular absorption spectroscopy	Atomic emission spectroscopy (ICP)	UV-visible molecular absorption spectroscopy
Nuclear magnetic resonance	Mass spectrometry	Molecular emission spectroscopy (fluorimetry)
Thermal techniques	Stripping voltammetry	Atomic absorption spectroscopy
X-ray diffraction	(Gas, liquid, supercritical fluid) chromatography	Coulometry
	Capillary electrophoresis	Potentiometry

Fig. 6.3. Classification of instrumental techniques into three groups according to applicability to CMPs for qualitative or quantitative purposes. The classification does not exclude the use of a given technique for a different analytical purpose in some cases

6.1.5 Quantitation Methodologies

There is no universally accepted classification of quantitation methodologies. That depicted in Fig. 6.2 compiles the more commonplace, basically in terms of foundation and method calibration approach, which leads to an immediate distinction between calculable and relative methods as per ISO Guide 32. This section provides a general discussion of the two that is expanded in subsequent sections.

A **calculable method** is *a method that yields a result from computations based on the laws that govern chemical and physical parameters, which materialize in mathematical formulae that involve both constants (e.g. atomic weights) and measurements made during the CMP such as the weight of the aliquot subjected to the process, titrant volume, precipitate weight, etc.* These are all potential primary methods (see Fig. 8.7). A distinction can be made here between **absolute methods that use no analytical chemical standard** (*e.g.* a pure substance, a standard sample) to produce the results and **absolute methods involving analytical standards**, which use an analytical chemical standard not containing the analyte. Titrimetries and isotope dilution mass spectrometry are two typical examples of the latter type.

Worthy of special note among calculable methods are *stoichiometric methods*, which rely on calculations based on the stoichiometric coefficients – and on atomic weights, which are chemical standards – of the chemical reaction on which the measuring process relies; the reaction can be of the heterogeneous (precipitation in gravimetry and titrimetry) or homogeneous type (solution titrimetry). Isotope dilution mass spectrometry is a non-stoichiometric absolute methodology.

A **relative method** is *a method that relies on a comparison of measurements of the (treated or untreated) sample with those provided by a set of analytical chemical standards from which a result can be derived without the need for calculations based on physical or chemical theories.*

Relative methods can be classified into two groups. *Interpolation methods* use instruments where the signal is unequivocally related to the analyte concentration and the relationship materializes in a "calibration curve", the ideal form of which is a straight line (one resulting from a linear relationship). Such a relationship is established by having the instrument process analytical standards (either pure or mimicking the composition of the sample in method calibration). There should thus be no differences in matrix between the sample and the set of standards; otherwise, such differences should affect the signal. The analyte concentration is determined by interpolating the sample signal into the calibration curve. Typical examples include photometric, fluorimetric, atomic absorption spectrophotometric and polarographic methods.

The other type of relative method, the *comparative method*, compares the sample signal with those produced by sample standards of similar composition (usually CRMs) containing the analyte (preferably at a variable concentration). As a rule, they are used when the instrument response depends not only on the analyte concentration but also on the sample matrix. These methods are usually employed in CMPs based on the direct insertion of (solid) samples (X-ray fluorescence spectroscopy included). While previous equipment calibration is not essential here, it invariably leads to improved results.

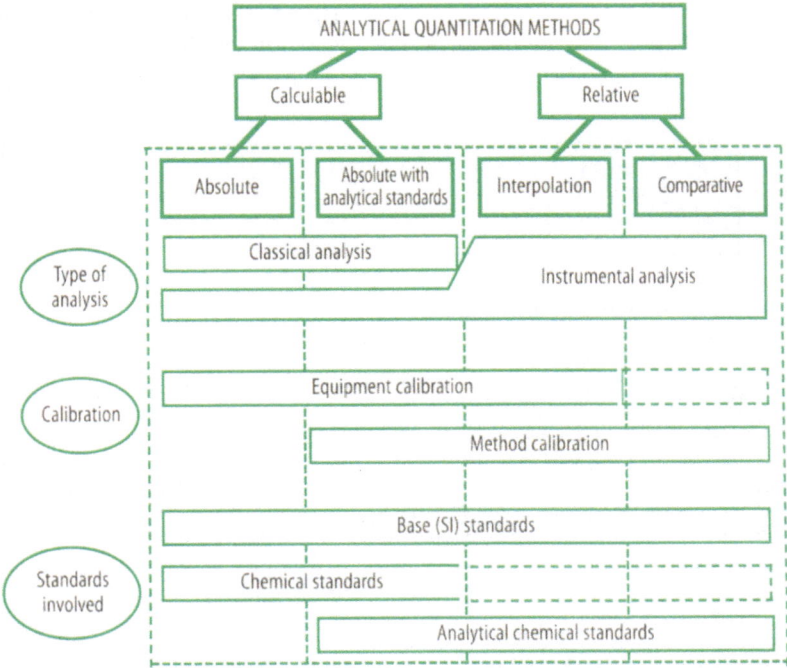

Fig. 6.4. Characterization of the four types of Quantitative Analysis methodologies in terms of analysis, calibration and standard types

There are thus four different approaches to Quantitative Analysis that depend essentially on the way the result is obtained (see Fig. 6.2) and stem from the type of method calibration and analytical chemical standards used. Figure 6.4 defines each approach in terms of three different criteria (*viz.* the type of analysis, calibration and standard involved).

6.1.6 Expressing Quantitative Results

The results of quantitative analyses are numbers (with their uncertainties) in specific units that are essential for their characterization.

Figure 6.5. shows the two more common ways of expressing quantitative analytical results. Thus, results can be expressed in an *absolute form* or referred to the amount (mass) of analyte present in the sample and appropriate units as per the scale of Box 6.1. Very often, results are expressed in a *relative form, i. e.* referred to the mass or volume of the aliquot subjected to the CMP. In this case, results can be expressed as proportions (analyte mass/sample mass or analyte mass/sample volume ratios), in the following forms (not endorsed by IUPAC but widely used in practice):

(*a*) Percentage (%) or parts per one hundred.
(*b*) Parts per million (ppm) or parts in one million.
(*c*) Parts per billion (ppb) or parts in 1000 million.
(*d*) Parts per trillion (ppt) or parts in 1000 billion.

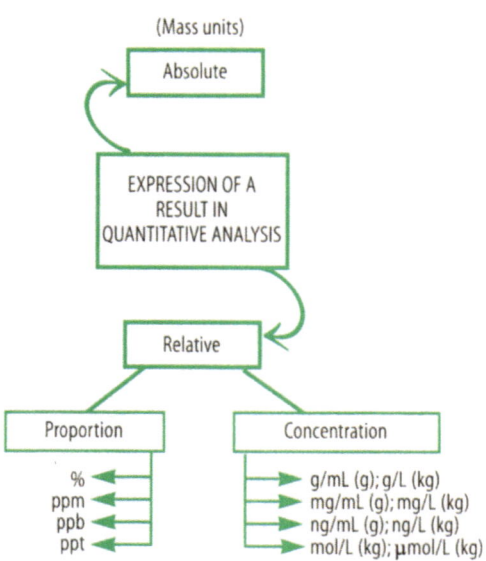

Fig. 6.5. Common ways of expressing results in Quantitative Analysis

_____ Box 6.1

Mass and Volume Unit Scales

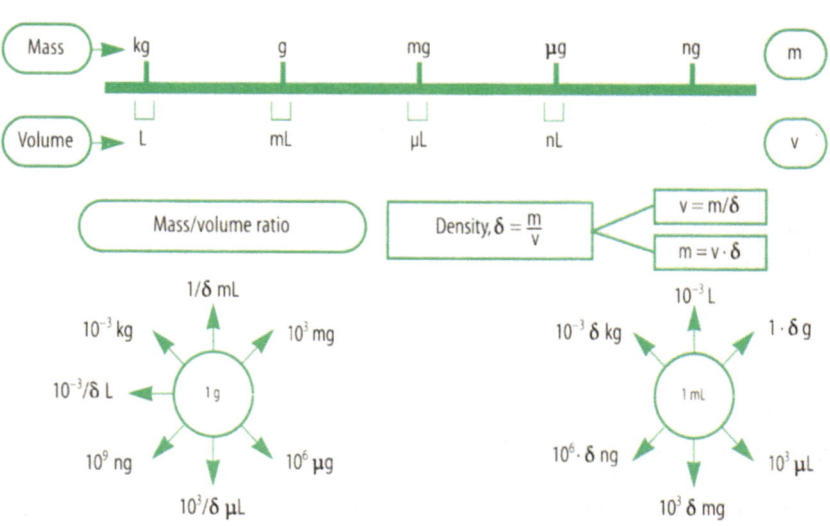

Example: The result of a CMP for determining traces of a pesticide in water is 3.10 ng/L. If the density of the water at a given temperature is known ($\delta = 1.024$ kg/L), the proper way of expressing the result will be in ng/kg (mass /mass):

$$\text{Result} = 3.10\,\frac{ng}{L}\,\frac{L}{1.024\,kg} = 3.03\,\text{ng/kg}$$

In addition, a relative result can be delivered as a concentration (the mass or moles of analyte per unit sample volume or mass), which can be expressed in a variety of ways including g (grams)/mL (g), mg/mL (g), ng/L (kg), mol/L (kg), *etc*. The different relative forms of expressing results are interrelated; thus, ppm is the same as mg/L or mg/kg, ppb as µg/L or µg/kg, and ppt as ng/L or ng/kg.

With liquid samples, it is much more rigorous to express concentrations relative to the sample mass than to the sample volume as the latter depends on the density of the liquid. In this way, results are expressed in SI base units (see example in Box 6.1).

Box 6.2

Equivalence Table for the Ways of Expressing Analyte Concentrations

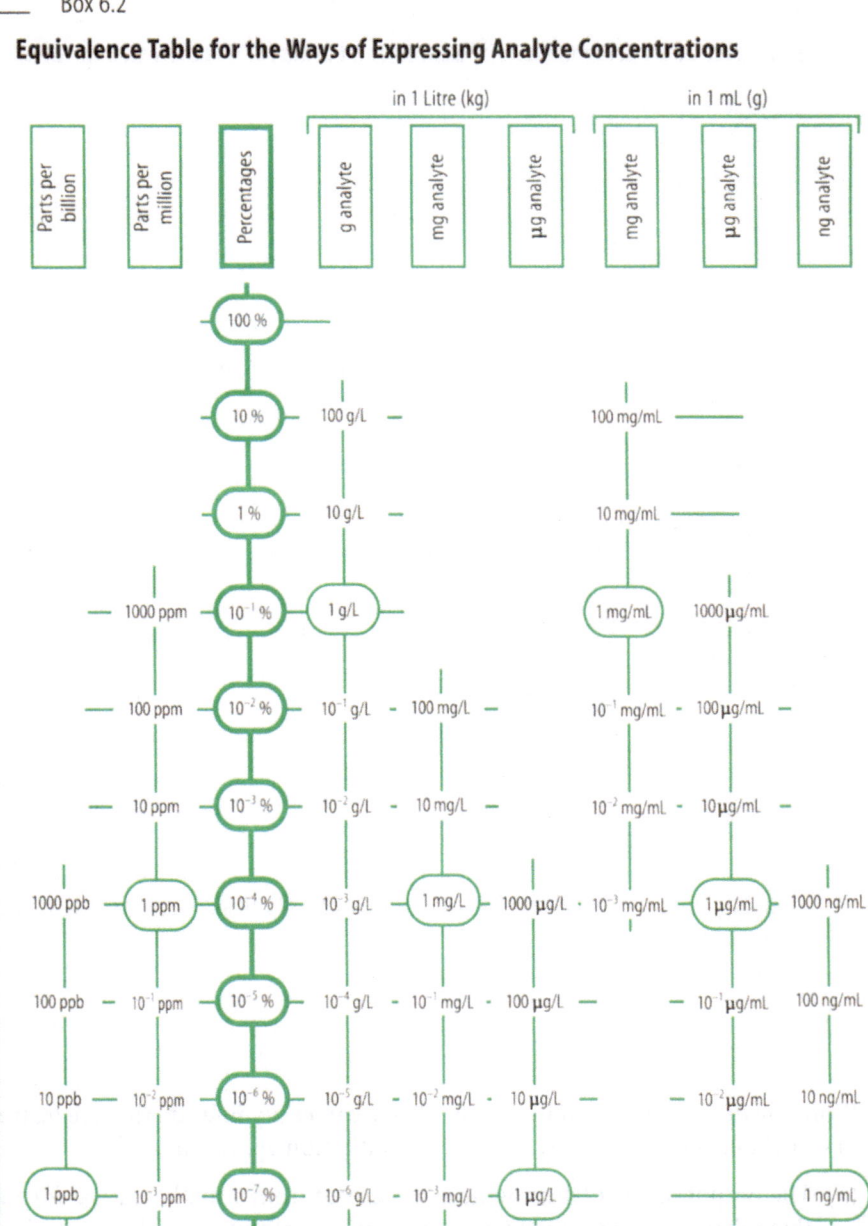

Box 6.2 contains an equivalence table for the different ways of expressing quantitative analytical results in a relative manner. The percentage (parts per one hundred) scale is used as reference; on the left are the ppm and ppb scales, and on the right various concentration scales referred to 1 litre (kg) or 1 millilitre (g) of solution. The unit value in each scale (1 ppb, 1 ppm, 1 g/L, 1 ng/L, 1 µg/L, 1 mg/mL, 1 µg/mL and 1 ng/mL) is highlighted. Box 6.3 shows typical examples of conversions among the different forms of expression.

------   Box 6.3

Examples of Transformations of Expression Forms for Quantitative Results

Example 1: Express the concentration 2.5 ppb in other units. Based on Box 6.2, this concentration corresponds to

$$2.5 \text{ ppb} \rightarrow 2.5 \text{ µg/L} \rightarrow 2.5 \text{ ng/mL} \rightarrow 2.5 \cdot 10^{-3} \text{ ppm}$$

a) Expression in µg/mL

$$2.5 \frac{\text{µg}}{\text{L}} \cdot \frac{\text{L}}{10^3 \text{ mL}} = 2.5 \cdot 10^{-3} \text{ µg/L}$$

b) Expression in g/mL

$$2.5 \frac{\text{µg}}{\text{L}} \cdot \frac{\text{L}}{10^3 \text{ mL}} \cdot \frac{\text{g}}{10^6 \text{ µg}} = 2.5 \cdot 10^{-3} \text{ µg/L}$$

c) Expression as a percentage

$$\left. \begin{array}{l} 2.5 \text{ parts} \text{ --- } 10^9 \text{ parts} \\ x \text{ --- } 100 \end{array} \right\} \ x = 2.5 \cdot 10^{-7}\%$$

d) Expression in moles/litre (M)
(molecular weight of analyte: 100)

$$2.5 \frac{\text{µg}}{\text{L}} \cdot \frac{1 \text{ mol}}{100 \text{ g}} \cdot \frac{1 \text{ m}}{10^6 \text{ µg}} = 2.5 \cdot 10^{-8} \text{ M}$$

Example 2: Express the concentration 3.1 µM in ppm and g/kg (analyte molecular weight = 200). Solution density = 0.921 kg/L.

a) $3.10 \dfrac{\text{µmol}}{\text{L}} \cdot \dfrac{10^{-6} \text{ mol}}{\text{µmol}} \cdot \dfrac{200 \text{ g}}{1 \text{ mol}} = 6.2 \cdot 10^{-4} \dfrac{\text{g}}{\text{L}} \cdot \dfrac{10^3 \text{ mg}}{1 \text{ g}} = 0.62 \text{ ppm (mass/volume)}$

b) $3.10 \dfrac{\text{µmol}}{\text{L}} \cdot \dfrac{10^{-6} \text{ mol}}{\text{µmol}} \cdot \dfrac{200 \text{ g}}{1 \text{ mol}} = 6.2 \cdot 10^{-4} \dfrac{\text{g}}{\text{L}} \cdot \dfrac{1 \text{ L}}{0.921 \text{ kg}} = 6.7 \cdot 10^{-4} \text{ g/kg}$

c) If the parts-per-million are expressed relative to density, then the result will be 0.67 ppm (mass/mass).

6.2 Calculable Quantitation Methods

From the definitions given in Sect. 6.1.5 one can easily infer distinctive features of calculable methods relative to other quantitation approaches. Thus:

(a) The output of calculable methods is derived from a mathematical formula that relates non-experimental parameters (*e. g.* atomic or molecular weights, the Faraday constant, the number of electrons exchanged in an electro-chemical reaction, stoichiometric coefficients for a chemical reaction) to primary data provided by the instruments used in the CMP (*e.g.* the mass of a precipitate in grams, the amount of electricity used in coulombs, the volume of titrant employed in mL).

(*b*) The equipment used should be calibrated using physical standards such as transfer weights, timer and ammeter checking systems, measurements of solution masses, *etc.*

(*c*) No tangible reference materials containing the analyte are employed.

(*d*) Chemical standards (*e.g.* atomic weights, the Faraday) and SI base units play a prominent role in the process by which the result is obtained.

(*e*) Calculable methods can be implemented both through classical (*e.g.* grav-imetry, titrimetry) and instrumental techniques (*e.g.* isotope dilution mass spectrometry, coulometry).

(*f*) They are potential primary methods (see Box 8.7).

As stated in Sect. 6.1.5, there are two different types of calculable methods, namely: *absolute methods involving no analytical standard*, which require no analytical chemical reference materials (*e.g.* pure substances or sample standards) to express the result; and *absolute methods involving analytical standards*, which use such standards in addition to others in order to express the result. This is the criterion adopted in this book to deal with these two types of methods separately.

6.2.1 Absolute Methods Involving no Analytical Standard

These quantitation methods provide results without the need to use analytical chemical reference materials such as pure substances or sample standards. Consequently, they require equipment calibration but not method calibration. This type of method is similar to quantitation methods (PMPs) for physical parameters (*e.g.* temperature, length, time, current intensity, pressure), which require calibration of the measuring instrument alone. As a result, this type of CMP is that with the shortest traceability chain; also, the last link in the chain can be an SI unit (a base standard). Both PMPs and CMPs use physical standards (*e.g.* transfer weights) for equipment calibration. However, CMPs are usually longer (multistep) as they entail conditioning the sample (see Sect. 4.3) prior to measurement (second step of the analytical process); by contrast, PMPs involve direct application of the instrument to the object or sample to be measured (see Box 6.4).

_____ Box 6.4

Absolute quantitation methods (absolute CMPs) using no tangible analytical chemical standards (*e.g.* gravimetry and coulometry) resemble physical metrological methodologies in that they only require calibration of the measuring instrument (typically, using physical standards). This box highlights the practical differences between absolute CMPs and PMPs.

In order to measure temperature (PMP) one can use, for example, a precalibrated thermocouple. The "sample" has little influence here; in fact, the same instrument can be used (over a certain temperature range) to measure this parameter in water, air and solids, with minimal changes. This is also the case with measurements of distances between trees, walls, locations, *etc.* In PMPs, the instrument may have to be adapted to different situations but will always rely on direct measurements.

Direct measurements in absolute CMPs are scant. The instrument (a balance) is calibrated with weight transfers (see Box 3.5). The gravimetric determination of calcium in a rock involves time-consuming operations such as grinding, dissolution, interference removal, addition of the precipitant ($C_2O_4^{2-}$), and filtration, washing, and calcination of the precipitate at $500\,°C$ prior to its gravimetric weighing (measurement). In addition, one must previously weigh the rock aliquot to be subjected to the CMP if the result is to be properly expressed (*e.g.* % Ca, mg Ca/kg).

Gravimetry and coulometry are the ultimate absolute calculable methods. Their most salient metrological features are summarized in Box 6.5 and commented on below.

_____ Box 6.5

General features of absolute calculable methods involving no analytical chemical standard

	Gravimetry		Coulometry
Instrument(s)	Balance		• Electrochemical assembly • Precision timer • Precision ammeter
SI base standards (traceability)	Kilogram (mass)		• Ampere (current intensity) • Second (time)
Standards used • Physical	Transfer weights		• Standard timer • Standard current intensity meter
• Chemical	Atomic weight	Gravimetric factor	Faraday (F) and atomic weights
• Analytical chemical	NONE		NONE
Formula	$W_a = F_g \cdot W_g$		$W_a = \dfrac{Q \cdot MW}{n \cdot F}$

6.2.1.1 Gravimetry

This is a classical quantitation technique that provides the mass (or concentration) of an analyte from weighings of the analyte itself, a stoichiometric reaction product or a derivative. The instrument used for this purpose is an analytical balance, which is calibrated with transfer weights (see Box 3.5). The CMP uses no analytical chemical reference materials and involves no method calibration, but only equipment (balance) calibration, by means of non-analytical standards. The short traceability chain, which leads to the kilogram prototype, usually results in high accuracy and precision that make gravimetry a potential primary method. On the other hand, accessory analytical properties are not so good: the CMP is labour-intensive, time-consuming and difficult to automate as a whole.

_____ Box 6.6

An Analytical Process Based on the Gravimetric Technique and Chemical Precipitation

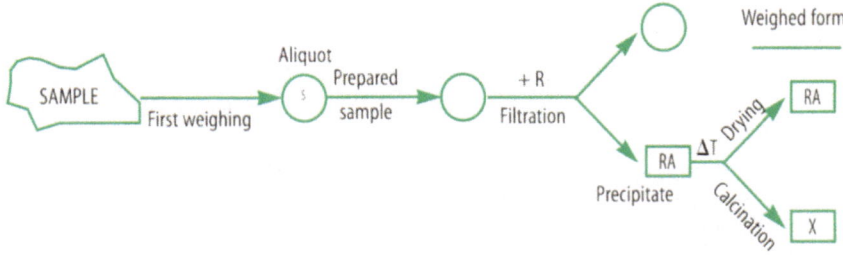

where S denotes the original sample, A the analyte, R the precipitant, RA the precipitate and X the product of the thermal treatment of RA.

It should be noted that the weighed form may coincide with the precipitated form or require drastic thermal treatment for stoichiometric traceability requirements to be met. These two types of gravimetry are rather different in practice. When the weighed form must previously be converted, the resulting CMP is slower, more labour-intensive and prone to error.

The precipitated form (RA) must meet the following requirements:

- It should be scarcely soluble ($s < 10^{-6}$ M).
- It should be free of impurities or easily freed from them (e.g. by washing).
- It should possess good mechanical properties (e.g. a low specific surface area) in order to facilitate filtering and reduce contamination.

The weighed form (RA or X) should have a well-defined, constant, accurately known composition as its stoichiometry will be the basis for calculations (traceability). Also, it should be stable (e.g. it should not react with atmospheric agents such as H_2O, CO_2 or O_2).

A gravimetric process involves three measurements, namely:

(a) the mass or volume of the aliquot to be subjected to the CMP;
(b) the mass of the vessel [a filtering plate for RA or a crucible for X] that is to hold the weighed form (this operation is known as "taring"); and

(c) the mass of the vessel containing the weighed form.
The difference between (b) and (c) provides the mass of the weighed form. Through the gravimetric factor, the mass of analyte – the proportion or concentration of which in the sample is calculated from the sample mass or volume, (a) – is determined.

Gravimetric methods can rely on three different types of preliminary operations, namely:

(a) *Chemical precipitation* by addition of an organic or inorganic reagent to the sample containing the analyte (see Box 6.6). This is the most common choice as it allows the determination of both cations (by precipitation as oxide hydrates, salts or metal chelates) and anions. One important practical consideration here is whether the precipitated form coincides with the weighed form (see Box 6.7).

_____   Box 6.7

Comparison of two Gravimetric Processes where the Weighed Form is the Same as or Different from the Precipitated Form

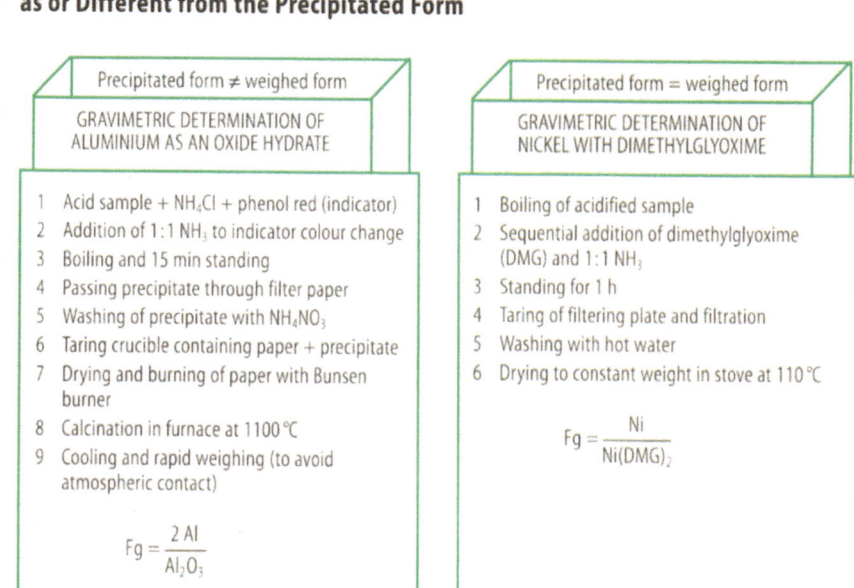

Note that the procedure in which the precipitated form coincides with the weighed form involves substantially fewer steps and that such steps are more simple.

(b) *Electrodeposition* by use of an electrochemical assembly consisting of a constant electrical power supply establishing a potential difference between an electrode (anode), where an oxidation takes place, and another (cathode), where a reduction occurs, both electrodes being immersed in a stirred sample solution. The cathode possesses a large surface area (typically that provided by platinum gauze) that increases the efficiency of the reduction reaction (*e.g.* $M^{2+} + 2e^- \leftrightarrow M$). The analyte gradually deposits onto the electrode until it virtually disappears from the solution. The cathode is weighed prior to and after electrolysis, the weight difference corresponding to the mass of analyte initially present in the sample.

(c) *Volatilization* of the analyte by heating of the sample. There are two different approaches to this gravimetric mode. One is based on the difference in sample mass before and after heating; in this way, sample moisture can be determined by using an automatic device which acts as the sample holder and is fitted to a commercially available balance. The other approach relies on the selective sorption (retention) of the analyte following volatilization. Such is the case with the gravimetric determination of carbonate by sorption of carbon dioxide released upon addition of a strong acid to the sample. The CO_2, held in a closed vessel, is retained by a trap containing sodium hydroxide and a desiccator:

$$CO_2\uparrow + 2\,NaOH \rightarrow Na_2CO_3 + H_2O$$

The weight difference of the trap prior to and after volatilization corresponds to the mass of CO_2 released from the sample.

The so-called *gravimetric factor* is the materialization of the stoichiometric relationship between the analyte and the weighed form, and is the traceability link established by chemical standards (atomic weights). The factor can be defined in two complementary ways, namely:

(a) the ratio of the formula weight of the analyte – depending on the way the results are to be expressed – to the molecular weight of the species being weighed; and

(b) a dimensionless number F_g by which the gravimetric weighing (W_g) must be multiplied in order to obtain the analyte weight (W_a):

$$W_a = F_g \cdot W_g$$

Box 6.8 illustrates the gravimetric factor concept with typical examples.

_____ Box 6.8

Gravimetric Factor

Let us complete the definition of gravimetric factor given in the text with selected examples.

The gravimetric factor is the result of a rule of three that establishes stoichiometric (traceability) and experimental relations. Thus, in the determination of dissolved Ag^+ by precipitation with Cl^- (the reagent) to obtain an AgCl precipitate (which is also the weighed form), one has:

TRACEABILITY $\rightarrow$ *Atomic weight of Ag (1:1)* $\rightarrow$ *Molecular weight of AgCl*

EXPERIMENTAL $\rightarrow$ *Analyte weight (W_a)* $\rightarrow$ *Gravimetric weighing (W_g)*

$$W_a = \frac{\text{Atomic weight of Ag}}{\text{Molecular weight of AgCl}} \cdot W_g = F_g \cdot W_g$$

A given precipitate (*e.g.* $BaSO_4$) can have different gravimetric factors depending on whether or not it is the analyte and on the way the results are expressed:

DETERMINATION OF →	Ba^{2+}	$BaCl_2$	S	Na_2SO_4	C_6H_5SH
→	Ba	$BaCl_2$	S	Na_2SO_4	C_6H_5SH
FACTOR	$BaSO_4$	$BaSO_4$	$BaSO_4$	$BaSO_4$	$BaSO_4$

The gravimetric factor must consider the differential stoichiometric composition of the form used to express the analyte and the weighed form. Thus, in the gravimetric determination of iron, the metal is precipitated as a basic salt and weighed as ferric oxide, so the gravimetric factor will be

$$F_g = \frac{2\,Fe}{Fe_2O_3}$$

The gravimetric factor dictates the sensitivity and precision of a CMP. The sensitivity is defined here through the limit of quantitation, *viz.* the smallest amount of analyte (W_a) that can be determined with a given, acceptable error. The lower is F_g (*i.e.* the larger is the molecular weight of the weighed form), the more sensitive will be the gravimetric method concerned. Thus, the gravimetric determination of aluminium can be addressed in two different ways, namely: by precipitation as an oxide hydrate (and weighing as Al_2O_3) or as an oxinate (and weighing as such). The respective gravimetric factors will be

$$F_g = \frac{2\,Al}{Al_2O_3} = 0.5292 \qquad F_g = \frac{Al}{Al(Ox)_3} = 0.0587$$

The oxinate procedure is nearly ten times more sensitive than the oxide procedure. On identical gravimetric weighings (W_g), W_a will decrease with decreasing gravimetric factor (F_g).

6.2.1.2 Coulometry

This is an electrochemical technique that quantifies the amount (or concentration) of analyte by measuring the amount of electricity Q (in coulombs, $Q = I \cdot t =$ amperes × second) required to transform the analyte at an electrode. It is based on Faraday's law, according to which transforming one electrochemical equivalent (MW/n) of a substance requires using a fixed amount of electricity called "the Faraday constant", $F = 96.485 \pm 0.029$ coulombs. The ensuing methodology is of the absolute calculable type as the amount of analyte is derived from the expression

$$W_a = \frac{Q \cdot MW}{n \cdot F} = \frac{I \cdot t \cdot MW}{n \cdot F}$$

where W_a is the weight of analyte present in the sample, MW the molecular (or atomic) weight of the analyte, n the number of electrons exchanged, F the Faraday constant, Q the amount of electricity used (in coulombs), I the current intensity (in amperes) and t time (in seconds). The experimental assembly used to implement the coulometric technique is of the potentiostatic type; the potential of the electrode where the analyte is oxidized or reduced is kept constant during the electrolysis. The amount of electricity used is usually measured by means of an electronic integrator as the current intensity decreases gradually with time:

$$Q = \int_0^t I \cdot dt$$

This methodology therefore uses no analytical chemical standards. Only those instruments required to measure the time and current intensity need be calibrated. Consequently, the coulometric technique is directly linked to SI units. Because of its low selectivity, however, its sole practical use is in establishing the traceability of pure substances employed as analytical chemical standards for relative methods.

One alternative instrumental approach involves using a constant current intensity for the electrochemical generation of titrants interacting in a stoichiometric manner with the dissolved analyte via an acid-base, complex-formation, precipitation or redox reaction. The end-point is detected by means of a visual or physico-chemical indicator (*e.g.* a potentiostat). This approach shares some features of absolute methods (*e.g.* it uses no analytical chemical standards) but can also be placed among titrimetric methods, which are dealt with in the following section.

6.2.2 Absolute Methods Involving Analytical Standards

These methods quantify the amount (or concentration) of analyte by using an analytical chemical standard (usually a pure substance) to obtain the result. Consequently, they require both equipment and method calibration, the latter of which, however, is approached differently from relative methods. The two most representative methods of this type (isotope dilution mass spectrometry and titrimetry) are discussed below.

6.2.2.1 Isotope Dilution Mass Spectrometry

This methodology uses a standard in the form of an enriched analyte isotope that is added to the sample. It measures the isotope ratios for the sample (S), standard (St) and spiked sample (S + St) by mass spectrometry. This is thus a form of internal standardization such as those of Box 3.14. The amount of ana-

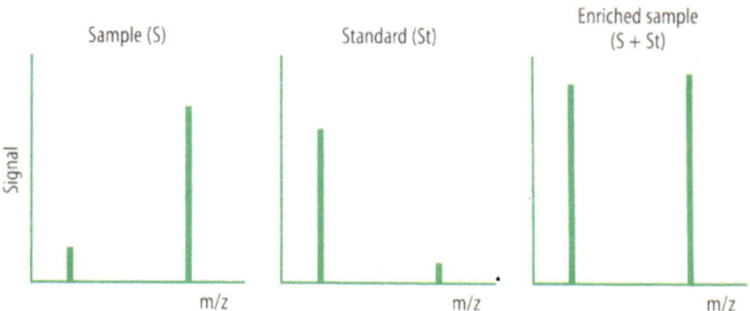

Fig. 6.6. Mass spectra obtained from aliquots of the sample (S), a standard of an isotope of the analyte (St) and a spiked sample (S+St) with a view to determining the analyte concentration using the absolute methodology isotope dilution mass spectrometry. For details, see text

lyte present in the sample is derived from the expression

$$W_a = \frac{R_{St} - R_{S+St}}{R_{S+St} - R_S} \cdot \frac{1 + R_S}{1 + R_{St}} \cdot W_{St}$$

where W_a and W_{St} are the masses of analyte (in the sample, S) and standard added, respectively, and can be replaced with the respective concentrations. The ratios in the above expression, R, are established from discriminate measurements of each mass or charge upon processing of the three above-mentioned aliquots with the instrument to obtain mass spectra such as those of Fig. 6.6.

The IDMS technique can be considered an indirect type of gravimetry: measured masses are directly traceable to a base (SI) standard and relative measurements of the isotopes are related to atomic weights (chemical standards).

6.2.2.2 Titrimetry

This is a classical technique which has given rise to an absolute methodology based on the use of analytical chemical standards that react in an unequivocal, stoichiometric manner, with the analyte. Calculations are based on the stoichiometric relation involved in the main titrimetric reaction.

Technically, titrimetry is based on the accurate (precise) measurement of the volume of a standard solution (called "titrant solution") of the analytical reagent needed to have the analyte present in the sample solution react quantitatively with it. The *equivalence point* of the titration is the theoretical titrant volume needed to react stoichiometrically with the analyte. The *end-point* of the titration is reached when the titrant volume used in practice (*i.e.* until an indicator system provides the visual signal to stop the addition process) has been added over the analyte solution.

The instrument is a burette. Calibration is both of the equipment (burette) and method type: the titrant concentration must be accurately known, which can be ensured by using a solution of a (directly prepared) primary standard or a secondary standard linked to a primary one via a preliminary titration. Box 6.9 describes two examples of titrimetric procedures that represent two different situations as regards traceability. Figure 6.7 provides a schematic depiction of the titrimetric process.

_____ Box 6.9

The following two examples illustrate the more usual ways of establishing the traceability chain in methods based on the *titrimetric technique*.

Direct use of a primary standard

In the titrimetric determination of water hardness (*viz.* the Ca^{2+} and Mg^{2+} concentrations) by formation of metal chelates with EDTA (H_4Y), a commercially available substance ($Na_2H_2Y \cdot 2H_2O$) exists that can be used to prepare a titrant (R) solution of accurately known concentration. An aliquot of sample is supplied with NH_4Cl/NH_3 buffer at pH 10 and a metallochromic indicator (neT). The burette is zeroed with the R solution and added dropwise over the sample until the indicator changes colour.

Use of a secondary standard

When no primary standard for the intended purpose exists or those available are very expensive, one can use a secondary standard, of poorer purity, stability, *etc.*, than a typical primary standard (see Sect. 3.5.1). A solution of the secondary standard is prepared and an aliquot titrated with a primary standard. Such is the case with the sodium hydroxide (NaOH) solutions widely used in acid-base titrations. Solid NaOH cannot be a primary standard because it is easily carbonated. An NaOH solution is made after washing the solid several times that will contain an approximate reagent concentration $[R]_{appr}$. The solution is used to zero the burette and is standardized with a solution of a primary standard (sodium hydrogen phthalate) prepared by weighing to the tenth of a milligram. A volume (practical mL) is thus measured. The number of theoretical mL corresponding to the amount of acid standard is calculated and the factor

$$f = \frac{\text{theoretical mL}}{\text{practical mL}}$$

is obtained and multiplied by the approximate concentration of the secondary standard solution in order to determine its actual concentration:

$$[NaOH] = [NaOH]_{appr} \cdot f$$

After this preliminary standardization, the sodium hydroxide solution can be used to titrate acid analytes (*e. g.* to determine the acidity of commercially available vinegar).

The *main titrimetric reaction* ($A + R \rightarrow B$), which can be an acid-base, complex-formation, precipitation, redox, addition, condensation or substitution one, must meet five essential requirements, namely:

(*a*) it should have a well-defined stoichiometry as this will be the basis for calculations;

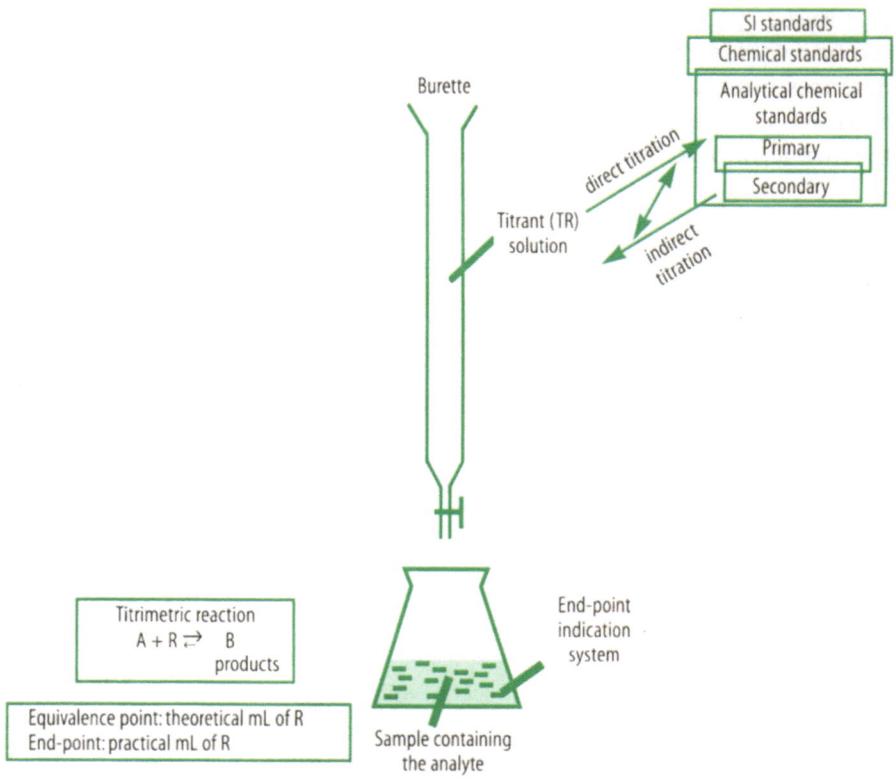

Fig. 6.7. Schematic depiction of a typical titration. For details, see text

(*b*) it should be as selective for the analyte as possible;
(*c*) it should be highly prone to completion (*i.e.* it should possess a large equilibrium constant);
(*d*) it should be fast (a slow kinetics is unsuitable for titrimetric use); and
(*e*) a suitable indicator for its end-point should exist.

Regarding analytical properties, titrimetries provide three substantial advantages (*viz.* simplicity, expeditiousness and a wide scope) that make them highly useful tools for routine analytical use. They share the scant selectivity of gravimetries but lack their accuracy and precision.

End-point indication systems should exhibit an abrupt, well-defined change in order that addition of the titrant solution may be timely stopped, or continuously monitor one or more ingredients of the titrimetric reaction. They should respond to or expose an abrupt change in the concentration of A, R or B (*i.e.* a change in $[H^+]$ in acid–base titrations, in $[M^{n+}]$ in precipitation and complex-formation titrations, and in the potential in redox titrations). One can thus use two different types of indicators in this context, *viz.* visual and physico-chemical (or instrumental) indicators. Visual indicators, typical examples of which are described in Box 6.10, allow a human operator to detect a change (the appearance of a colour, fluorescence, a precipitate). A wide variety of instrumental (optical, electroanalytical and radiometric) indicators also exist that monitor changes in the ingredients of the titrimetric reaction via continual measurements of a physico-chemical property of A, R or B (see Box 6.11).

_____ Box 6.10

Visual Indicators in Titrimetries

Visual indicators allow one to identify the end-point of a titration and hence to stop the addition of the titrant solution from the burette. As a rule, visual indicators are chemical substances (Ind) that are added at low concentrations to solutions containing the analyte. No self-indicating titrimetric system need be added to the titrand. Several typical examples follow.

TYPE 1. **Self-indicating systems.** These require no addition of Ind. One case in point is permanganimetry for the determination of reducing analytes (*e.g.* Fe^{2+}, H_2O_2, I^-) in an acid medium. As the titrant is added, its deep violet colour gradually disappears; when the endpoint is reached, the addition of a further drop of MnO_4^- gives a persistent pink colour.

TYPE 2. **Addition of chemical substances interacting with the analyte.** Before the titration proper, a small proportion of the analyte ($<0.1\%$) interacts with the indicator to form a chemical species IndA (colour 2). During titration, the remaining analyte interacts with R. At the end-point, R displaces the amount of analyte associated to Ind, which is released and changes the colour from 2 to 1 (Ind):

$IndA + R \rightarrow Ind$
colour 2   colour 1

Examples:

- Titration of Mg^{2+} with EDTA. The indicator (neT) forms a chelate with Mg^{2+}. At the endpoint, neT, which has a different colour, is released.
- Titration of HCl with NaOH in the presence of phenolphthalein (Phen) as indicator. At the beginning of the titration, Phen is associated to protons (PhenH) and colourless. A slight excess of NaOH at the end-point releases Phen, which gives its red-violet colour to the solution.

TYPE 3. **Addition of chemical substances interacting with the reagent.** The indicator, Ind (colour 1), is added at the start of the titration. Because it does not interact with the analyte, the colour is preserved throughout the titration. At the end-point, a small excess of R causes an abrupt colour change:

$Ind + R \rightarrow IndR$
colour 2   colour 1

Examples:

- Titration of Cl^- with Hg (II). The indicator, diphenylcarbazide, forms an orange-coloured chelate with the metal cation (R).

———   Box 6.11

Instrumental Indicators in Titrimetries

Instead of the human senses, one can use instruments to monitor changes in the ingredients (A, R or P) of the main titrimetric reaction during the process. The instrument response is a titration curve, *viz.* a plot of the variation of the measured signal (instrumental parameter, P) with the volume (mL) of titrant added. Depending on the relationship between the instrumental parameter, P, and the ingredient concentrations, two types of curve (logarithmic or linear) can be obtained:

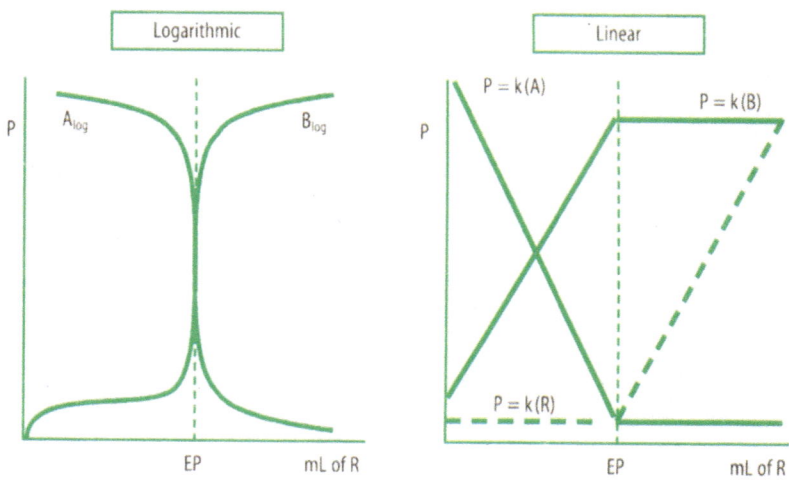

Non-linear or logarithmic curves arise when the monitored instrumental parameter and the logarithmic concentration of the ingredients (A, R, B) are proportional. This is the way pH $(= -\log [H^+])$ is monitored in acid–base titrations, pM $(= -\log [M^{n+}])$ in complexometric titrations and the potential, $E\{= E_0 + 0.059/n \cdot \log ([Ox]/[Red])\}$, in redox titrations. As can be seen from the figure above, the curve is S-shaped and increases or decreases depending on the way the concentration of the monitored species varies during the process. Thus, in the titration of an acid with a base, the pH is initially acid (low) and rises during the titration (B_{log} type); if the analyte is a base, then the pH decreases as the acid is added (A_{log} type).

When P is directly proportional to the concentration of any reaction ingredient (A, R or B), continuous monitoring of the former during the titration provides two straight lines that intersect at the end-point. These are linear curves. The intersection is not so abrupt as shown in the figure above; in fact, a curved segment arises near the end-point (EP) instead. The shape of the curve depends on the particular reaction ingredient being monitored – the figure shows the most usual. P can vary with the concentration of A, R and B, giving rise to two straight lines intersecting with an opposite trend (increasing or decreasing P). These linear profiles are also obtained when the monitored instrumental parameter (*e. g.* colour, fluorescence) corresponds to the indicator.

Depending on the procedure used, titrimetries are implemented as direct titrations or as back-titrations. In direct titrations (Fig. 6.7), the titrant, R, is added from the burette over the analyte solution, A, to obtain the product, B.

If the reaction fails to meet any of the above-mentioned requirements, one can still use an indirect procedure to determine A (a back-titration). To this end, a conventional analytical reaction is previously conducted by adding an accurately known excess of a standard solution of reagent R to the sample solution:

$$A + R \leftrightarrow B \quad \text{(excess R remains)}$$
controlled
excess

Then, excess R not reacted with A is titrated by adding a new reagent R' from the burette until a new end-point is reached:

$$R + R' \to H$$
excess

The analyte concentration is obtained from the difference between the millimoles (volume × concentration) of the two analytical chemical standards used (R and R'), which are indispensable with a view to establishing the traceability chain. Box 4.12 describes two typical examples of back-titration.

  Box 6.12

Back-Titrations

A large number of analytes (A) cannot be determined by direct titration with a standard reagent (B) for a variety of reasons the most common of which are (*a*) the lack of stable titrants in solution, (*b*) slowness in the A + R reaction and (*c*) the lack (or high cost) of a suitable end-point indicator. Under these circumstances, one must resort to an indirect titrimetric procedure based on a preliminary analytical reaction. Below are discussed two typical examples.

Example 1

When no metallochromic indicator is available to titrate a metal ion M^{2+} with EDTA or the chelate formation reaction is too slow, two standard solutions (one of EDTA and the other of a different metal ion, M'^{2+}) are prepared and used as follows: (1) an accurately measured volume of the EDTA standard solution (a controlled excess) is added to the analyte (M^{2+}) solution, which is then pH-adjusted with an appropriate buffer – if required, the mixture is heated –; (2) excess EDTA remaining after reaction with the analyte is titrated by addition of M'^{2+} standard from the autoburette, in the presence of a suitable indicator for the EDTA-M'^{2+} titrimetric reaction. The analyte concentration is obtained by difference as explained in the text.

Example 2

The $I_2/2I^-$ redox system is widely used to determine both oxidants and reductants; however, standard solutions of both species cannot be prepared in practice. Instead, an indirect titrimetric procedure is normally employed. One of the most commonplace is that used in the determination of oxidizing analytes such as dissolved chlorine. Pool water is supplied with an unmeasured excess of KI to form iodine according to

$$Cl_2 + 2I^- \leftrightarrow I_2 + 2Cl^-$$

If any hypochlorite is formed by hydrolysis of the chlorine ($Cl_2 + H_2O \rightarrow ClO^- + Cl^- + 2H^+$), the original stoichiometry will still be preserved since the reaction between ClO^- and I^- also yields one molecule of iodine:

$$ClO^- + 2I^- + 2H^+ \leftrightarrow I_2 + Cl^- + H_2O$$

The iodine thus formed is titrated with $S_2O_3^{2-}$, using starch as indicator:

$$2S_2O_3^{2-} + I_2 \leftrightarrow S_4O_6^{2-} + 2I^-$$

At the end-point, the blue colour of the I_2-starch system disappears.

The $S_2O_3^{2-}$ solution is the secondary standard, which must previously be standardized with a primary standard. This is done by using KIO_3, which, in the presence of an unmeasured excess of I^-, produces I_2 in stoichiometric amounts:

$$IO_3^- + 5I^- + 6H^+ \leftrightarrow 3I_2 + 3H_2O$$

Figure 6.8 illustrates various technical approaches to titrimetry that differ in the degree of human involvement and the type of indication system used. In the *manual procedure* (Fig. 6.7), a visual indicator is usually employed to detect the end-point – a physico-chemical indicator such a pH electrode in acid-base titrimetries is also a common choice –; the number of millilitres used is read off the burette's multimetric scale. The *semi-automatic procedure* uses a piston burette that is usually controlled by the operator and has the advantage that it is self-zeroing and provides digital readings – which avoids human errors –; the proce-

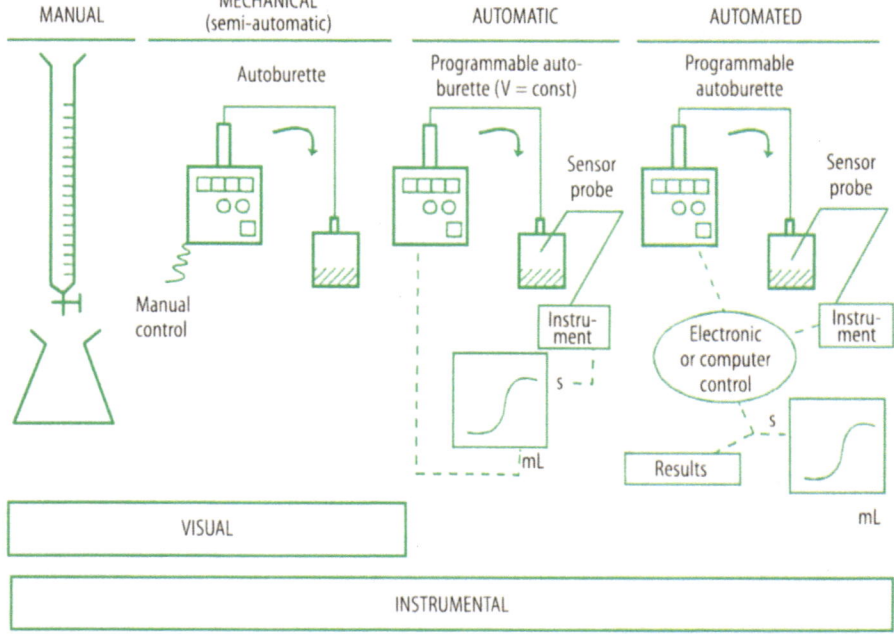

Fig. 6.8. Titrimetric assemblies involving human participation to different degrees and also different indication systems. For details, see text

dure is compatible with both visual and physico-chemical indicators. In the *automatic procedure*, an autoburette is programmed to add the titrant, at a constant flow-rate (*e.g.* 0.5–1 mL/min), over the sample solution, into which an instrumental indicator (*e.g.* a glass-calomel combined electrode, a fibre-optic photometric probe) is immersed. The output is a signal-volume (time) recording that is used as the titration curve. In this procedure, the titrant addition rate need not be decreased near the end-point as the kinetics of the titrimetric reaction is quite slow under these conditions. The *automated procedure* uses a similar assembly but includes an electronic or computer-controlled device to alter the titrant addition rate without human intervention. The titration is started at a constant rate that is gradually reduced – on the basis of the monitored signal increments – as the end-point is approached; thus, when signal increments exceed the preset value, the electronic or computer-controlled device "commands" that the addition rate be lowered.

6.3 Relative Quantitation Methods

Relative quantitation methods compare the signals yielded by one or more standards containing the analyte with those resulting from samples also containing it. The result (an amount or concentration of analyte) is directly obtained from the comparison, without the need to use formulae from any physico-chemical laws.

Both equipment and method calibration are mandatory here. In fact, relative methods are characterized by the use of method calibration with analytical chemical standards.

As a rule, relative methods are based on instrumental (optical, electroanalytical, thermal, mass-based) techniques; by contrast, calculable methods can also use classical techniques. In fact, relative methods constitute the most widely used quantitation approach in Analytical Chemistry.

Because they involve a longer traceability chain, relative methods have little potential to be primary methods (see Box 8.7).

As stated in Sect. 6.1.5, there are two different types of relative methods (Fig. 6.9), both of which are discussed below.

6.3.1 Interpolation and Extrapolation Methods

These are the most widely used methods for quantitation in Analytical Chemistry as they are the best suited to most affordable instrumental techniques. There are two possible approaches depending on the way the results are obtained, *viz.* by interpolation or by extrapolation.

Direct or *interpolation methods* compare sample signals with signals provided by "calibration samples" that are in fact standards containing variable concentrations of the analyte prepared in such a way as to mimic the matrix of

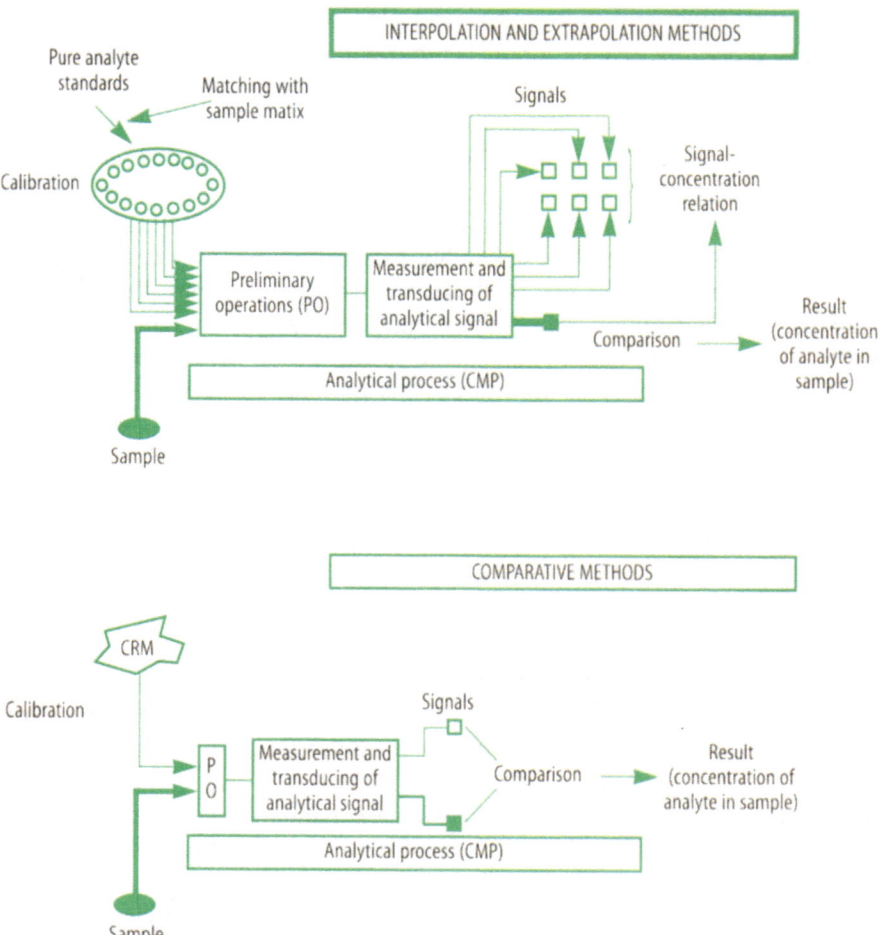

Fig. 6.9. Schematic depiction of the two main types of relative quantitation methods: interpolation, extrapolation and comparative methods. For details, see text

the real sample – using pure standards for this purpose can give rise to substantial systematic errors. As can be seen in Fig. 6.9, the calibration samples and the real sample are subjected to the analytical process separately. The calibration signals thus obtained allow one to unequivocally relate the signal to the analyte concentration through the so-called "calibration curve", the ideal form of which is a linear signal-concentration relation over a given concentration range. A detailed description of such a relation is provided in Sect. 2.5.2 and Fig. 2.11. The analyte concentration in the sample is determined by interpolation. This can be done graphically, once the calibration curve has been constructed, or chemometrically, after establishing the equation for the curve, which, for a linear relation, will be of the form signal (x) = intercept (y) + S · analyte concentration (S being sensitivity as defined by IUPAC, *i.e.* the slope of the calibration curve). In simple terms, the aim is to find the concentration corresponding to the sample

signal by using a set of known signal-concentration pairs. Figure 6.10A illustrates the graphical procedure. Six calibration samples containing increasing
concentrations of analyte (0–6 arbitrary units) were prepared and subjected to
the CMP to obtain six signals (0.2, 0.4, 0.6, 0.8, 1.0 and 1.2 arbitrary units). The
sample was also subjected to the CMP and the signal it yielded, 0.5, was interpolated to obtain the analyte concentration: 2.5. Box 6.13 describes a typical
example.

Although the matrix of "calibration samples" mimics that of the real sample,
there can still be differences potentially leading to major errors if the resulting

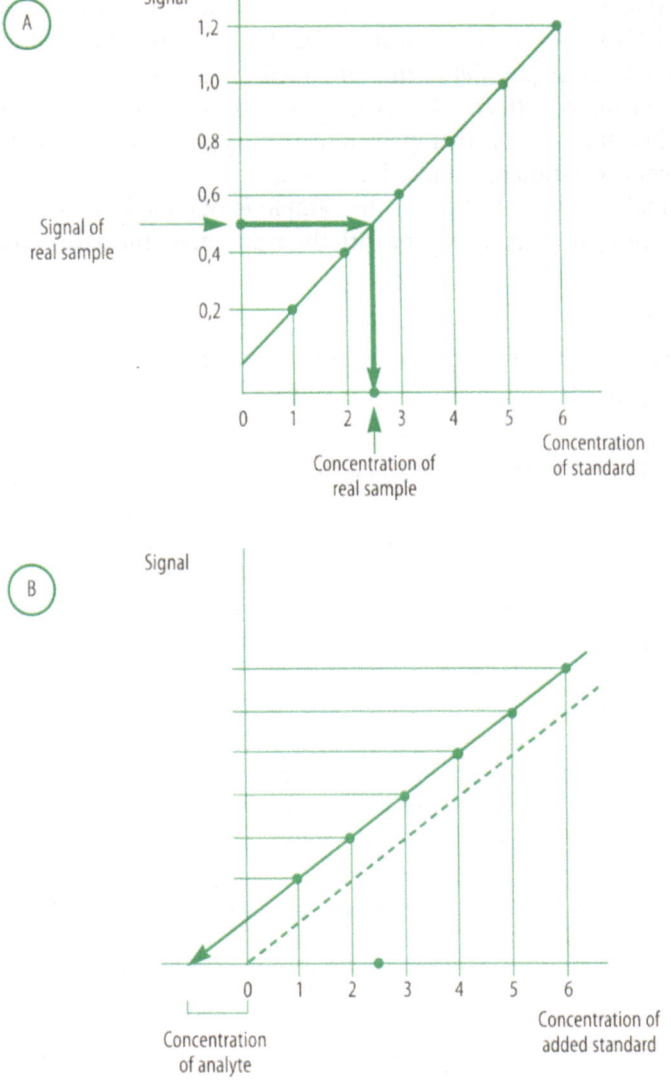

Fig. 6.10. Graphical determination of the analyte concentration in relative quantitation
methods. (A) Interpolation method. (B) Multiple standard addition method based on
extrapolation of the results. For details, see text

interferences are of the multiplicative type, *i.e.* if they primarily affect the sensitivity (see Sect. 2.5.3). These errors can be minimized by using the *extrapolation* or *multiple standard addition method*, which involves the following steps:

(a) A number n of sample aliquots are supplied with increasing concentrations (*e.g.* 0–6 arbitrary units) of a pure standard of the analyte, all containing an analyte concentration equal to the combined original (sought) and added concentrations.

(b) The n aliquots of spiked real samples are subjected to the analytical process in order to obtain one signal per aliquot as shown in Fig. 6.9.

(c) The signals thus obtained are plotted against the concentrations of standard added (0–6). As can be seen from Fig. 6.10B, a straight calibration line is obtained that is parallel to that provided by the pure standards in the absence of matrix effects. Otherwise, the two lines will not be parallel. The signal on the y-axis is that yielded by the unspiked sample (that to which a zero concentration of standard is added).

(d) The analyte concentration in the sample is obtained by extrapolating the calibration curve; its intercept with the x-axis gives the datum sought.

_____   Box 6.13

Relative Interpolation and Extrapolation Methods

The practical implementation of these two types of method is illustrated here with the same analytical problem (*viz.* the determination of iron in wine). The choice in each case will be dictated by the characteristics of the instrumental technique to be used.

With atomic absorption spectroscopy, one can use the direct interpolation mode, which is highly selective and provides signals that are not significantly affected by the sample matrix (usually not very complex in any case). Six iron standard solutions of variable concentration (0.5–5 ppm) are prepared and delivered to the flame by nebulization, their absorbances then being recorded (see Fig. 6.10). Next, the sample is aspirated to obtain a signal that is interpolated into the curve to determine the iron concentration in the wine.

The CMP can also be based on UV-Visible absorption (photometry), for example. This entails supplying the sample with a ligand L (*o*-phenanthroline) to form the chelate FeL_3^{2+}, which is deeply coloured ($\lambda_{max} = 510$ nm); a reductant (hydroxylamine or ascorbic acid) is previously added to convert all iron present into Fe^{2+} ion. The selectivity is lower here as the procedure can be interfered by a number of factors including the wine's base colour, the presence of metal ions forming chelates with the ligand, the addition of an inadequate amount of reductant, the presence of iron-masking ligands, *etc.* It is therefore advisable to use the extrapolation approach based on multiple standard additions. Thus, six aliquots of the wine are supplied with increasing concentrations (0–4 ppm) of Fe^{2+} and subjected to the CMP. Six signals are thus obtained such as those of Fig. 6.10B. The amount of analyte present in the unknown is determined by extrapolation.

6.3.2 Comparative Methods

These quantitation methods are essentially similar to relative methods but have some special connotations that have led ISO to deal with them separately. They are also based on comparisons of the signals yielded by the sample and a standard. However, they differ from relative methods in the following respects:

(a) They are used when the response of the instrumental technique of choice is highly sensitive to changes in sample matrix.
(b) As a rule, they are employed when available equipment is amenable to direct insertion of untreated solid samples.
(c) The standards used are certified reference materials (CRMs) the matrices of which are as similar as possible to that of the sample.

In the ideal situation, CRMs spanning a range of analyte concentrations would be available. In such a case, the above-described interpolation method is used. In daily practice, however, only a single CRM is usually available, so the analyte concentration has to be determined by comparing the sample signal with that provided by the CRM, which will contain a certified concentration (or amount) of analyte.

This approach is frequently associated to such techniques as arc or spark atomic emission spectroscopy and X-ray fluorescence.

Questions

(1) Why does quantitation in Analytical Chemistry affect all three types of analysis (Qualitative, Quantitative and Instrumental)?

(2) What is the output of a quantitative analysis? Give meaningful examples.

(3) What analytical properties characterize, in a direct manner, the result of a quantitative analysis?

(4) Distinguish equipment and method calibration in Quantitative Analysis.

(5) Comment on the metrological properties of a quantitative result in Analytical Chemistry.

(6) Place Quantitative Analysis in the panoramic view of Fig. 6.1.

(7) Discuss the relationships between the metrological quality of a quantitative result and accessory analytical properties.

(8) What are the differences between Classical and Instrumental Quantitative Analysis?

(9) Are base (SI) standards relevant to Quantitative Analysis? Are chemical and analytical chemical standards?

(10) Why a given instrumental technique is normally better suited to one of the analysis types shown in Fig. 6.3 than to the others? Give some examples.

(11) Can quantitative analyses be carried out via classical techniques involving no human participation? Give an example.

(12) Explain the differences between a calculable and a relative quantitation method.

(13) Why must a concentration expressed as a mass/volume ratio be converted into a mass/mass one before an analytical result is delivered?

(14) What are the differences between relative interpolation methods and comparative methods?

(15) In what way are absolute methods involving some tangible analytical chemical standard different from those that use none?

(16) What are stoichiometric quantitation methods?

(17) Can the temperature influence the expression of the analyte concentration in mass/volume units? Why?

(18) Relate a 0.1% concentration to the ppm scale and express it in μM (analyte molecular weight = 124).

(19) Which concentration is correct, 1.30 g/kg or 1.25 g/L? Why?

(20) The result of a calculable method, obtained by solving the pertinent formula with the aid of a pocket calculator, is 6.0324531242 g/kg. Is it correctly expressed?

(21) Express the concentration 1.23 millimoles/litre in ppb and ppm.

(22) Convert a quantitative analytical result of 0.3 ppm into ppb, g/kg and a percentage.

(23) Name six features of a calculable quantitation method used in Analytical Chemistry.

(24) What are the most salient differences between the two types of calculable methods? Give some examples.

(25) Briefly comment on gravimetry in relation to the use of standards, calibration and the ways of expressing results.

(26) What are the generic elements of the mathematical expression that allows one to obtain the result of a quantitative analysis by use of a calculable method? Give meaningful examples.

(27) Briefly discuss procedural approaches to determining the analyte concentration from mass measurements.

(28) What are the most salient differences between the two types of relative methods? Give several examples.

(29) Why does the gravimetric technique involve measuring at least two masses? Use meaningful examples to explain it.

(30) Establish similarities and differences between physical metrological methods and absolute chemical methods.

(31) Why is whether the precipitated form coincides with the weighed form so important in gravimetry?

(32) What requirements must the precipitated form and the weighed form meet in gravimetry?

(33) What are the generic differences between gravimetry and coulometry?

(34) Compare the gravimetric factor and the Faraday in the context of calculable methods.

(35) What is titrimetric standardization?

(36) Why is isotope dilution mass spectrometry not a purely absolute method? Why is it not stoichiometric?

(37) Why is the gravimetric factor related to sensitivity in gravimetries?

(38) Distinguish the two methodological approaches to coulometry.

(39) Compare titrimetries and gravimetries in metrological terms. Comment on any practical differences between the two.

(40) Why can classical techniques not be considered relative quantitation methods?

(41) What is the difference between automatic and automated titrimetries?

(42) Compare direct titrations and back-titrations in metrological terms.

(43) Why are there two general types of titration curve?

(44) In what way is the titration curve related to end-point indication?

(45) What are the differences between interpolation and extrapolation methods?

(46) What types of standards are used to construct the calibration curve?

(47) What are the differences between interpolation and comparative methods?

(48) What is the most salient advantage of relative methods involving multiple additions of an analyte standard to a number of sample aliquots?

(49) What type of instrumental technique uses comparative methods most frequently?

(50) Compare quantitation methods using analytical chemical standards containing the analyte and the types of calibration involved.

Seminar 6.1

Calculating the Results in Quantitative Analysis

- **Learning objective**
  To describe the third step of CMPs intended to quantify the amount or concentration of analyte in a sample the mass (or volume) of which will have to
  be considered in expressing the quantitative result.
- **Literature**
  Analytical Chemistry handbooks and textbooks describing CMPs for quantitative purposes. Select CMPs based on calculable (*e.g.* gravimetries, titrimetries) and relative methods.
- **General procedure**
  (1) Compile and justify the formulae used by the different calculable
      methods.
  (2) Compile and justify the procedures used to process the method calibra-
      tion data provided by the different relative methods.
  (3) Select quantitation methods, preferably typical of each category. For each
      method:
      (*a*) identify standards,
      (*b*) describe the calibration processes,
      (*c*) describe the procedure used to calculate the quantitative result, and
      (*d*) solve numerical problems from simulated data for the analytical
           chemical standards and measurements.
  (4) Describe three or four quantitation procedures of different types and
      discuss their similarities and differences in terms of the way the output is
      obtained.

Seminar 6.2

Forms of Expressing Quantitative Results

- **Learning objective**
  To become acquainted with the different ways of expressing quantitative ana-
  lytical results and their mutual relationships.
- **Materials**
  Use the information given in this chapter (especially in Boxes 6.1–6.4).
- **Procedure**
  (*a*) Systematically describe the different ways of expressing a quantitative
       result.
  (*b*) Establish the principal links between the different forms of expression:
       (1) Absolute ↔ relative.
       (2) Mass/volume ↔ mass/mass.

(3) Percentage ↔ ppm–ppb.

(4) Mass/mass ↔ moles (milli, micro)/litre.

(c) Perform 20 numerical simulations of conversions between different ways of expressing a result.

■ **Discussion**

Establish generic conclusions from the previous conversions. Comment on the significance of the mass (volume) of the sample aliquot that is subjected to the CMP. Bear in mind Figs 1.23 and 1.24, which classify chemical analyses according to initial sample size and relative proportion of analyte mass in the sample.

Seminar 6.3

Use of Standards in Quantitative Analysis

■ **Learning objective**

To explain the different approaches to the use of references with a view to quantifying in Analytical Chemistry, approaches with dictate the type of method to be employed.

■ **Literature**
- Monographs and textbooks on classical and instrumental techniques.
- Compilations of standard and officially endorsed methods.

■ **Calculable methods (examples)**
- Gravimetric determination of moisture.
- Titrimetric determination of moisture (Karl Fischer method).
- Titrimetric determination of the silver content (fineness) in a jewel.
- Gravimetric determination of calcium in a rock.
- Titrimetric determination of water hardness.

■ **Relative methods (examples)**
- Determination of metal traces in sea water by atomic absorption spectroscopy.
- Determination of the iron content in an alloy by X-ray fluorescence.
- Photometric (UV-visible) determination of the aldehyde and ketone contents in mixtures by reaction with dinitrophenylhydrazone.

■ **Discussion and description**

Once the method is selected and the procedure involved in the CMP described, specify the standards (classified by type) and calibration (equipment or method) used. From this information, place each method studied in the general classification of quantitation approaches.

Seminar 6.4

Detailed Comparison of CMPs for the Determination of the Same Analyte

(I) Determination of lead in environmental samples

■ **Learning objective**

As part of an integral proposal of student practices based on the use of various instrumental techniques for the same type of sample (environmental) and analyte (lead), compare the different analytical procedures for quantifying traces of the pollutant.

■ **Literature**
  ● "Lab Lead: Teaching Instrumentation with One Analyte", A. Fitch, Y. Wang, S. Mellican and S. Madra, *Anal. Chem.*, 1996, 727A.
  ● Handbooks describing methods based on the use of various instruments dealt with in the previous paper.

■ **Procedure**
  (1) Carefully read the recommended paper and identify CMPs.
  (2) Find the descriptions of the CMPs in handbooks.
  (3) Assess the results. Examine the influence of the "client" and other factors.

■ **Discussion**
  ● Discuss the environmental impact of lead (analytical problem).
  ● Critically compare the quantitation techniques used in Analytical Chemistry.
  ● The role of the "client" in the laboratory: discussing the results.
  ● Comment on the proposed integral learning approach.

Seminar 6.5

Detailed Comparison of CMPs for the Determination of the Same Analyte

(II) Determination of organochlorine pesticides in various matrices

■ **Learning objective**

To know the wide variety of available analytical approaches to the determination of each analyte, as well as their varying suitability to sample types, analyte concentrations, instrumental techniques, standards and quality compromises.

■ **Literature**

Handbooks and monographs describing determinations of organochlorine pesticides. The CMPs selected should involve classical techniques, photometries, fluorimetries, chromatographies, electroanalytical techniques, *etc.*

■ **Procedure**
Provide a detailed description of the selected methodologies, including, at least, the following:
- Analytical problem. How will the quantitative output obtained in the determination of the analyte be used?
- Characteristics of the sample, including the estimated analyte concentration levels. Sample treatments to be applied.
- Describe the implementation of each CMP in detail, specifying the tools and processes involved.
- Describe the calibration operations involved in each CMP and the standards used.
- Calculate the results, if applicable.

■ **Discussion**
Critically compare the selected approaches in terms of analytical problem, analytical properties, complexity of the CMP, standards used, types of calibration, calculation of the results, *etc.*

Suggested Readings

General

(1) "Analytical Chemistry", G. D. Christian, 5th edn, *Wiley*, New York, 1994.
This book deals in reasonable detail with the theory and practice of Analytical Chemistry. Its is clear and straightforward, illustrated with many examples and completed with exercises. It can be used to obtain supplementary information on specific aspects of this chapter.

(2) "Exploring Chemical Analysis", D. H. Harris, *Freeman & Co.*, New York, 1996.
An innovative teaching approach to the general principles of Chemical Analysis.

(3) "Fundamentals of Analytical Chemistry", 6th edn, D. A. Skoog, D. M. West and F. J. Holler, *Saunders College Publishing*, New York, 1995.
In its two volumes, the authors provide a generic, practical view of Analytical Chemistry and deal with the topics covered in this chapter. The many examples given are a high pedagogical asset.

(4) "Analytical Chemistry", R. Kellner, J.M. Mermet, M. Otto and H. M. Widmer (Eds), *Wiley-VCH*, Weinheim, 1998.
In its nearly one thousand pages, this textbook materializes the so-called "Eurocurriculum" in Analytical Chemistry, which is the product of the collective endeavour of authors from many universities, coordinated by the Analytical Chemistry Division of the European Federation of Chemical Societies (DAC-FECS). It contains abundant information that completes the contents of this chapter.

About Classical Quantitative Analysis (Gravimetries and Titrimetries)

(5) "Classical Methods", J. Mendham, D. Copper and C. Doran, *Wiley*, New York, 1987.

   This is one of the books in the Analytical Chemistry Open Learning (ACOL) series, written from a definitely pedagogical point of view. Its many exercises make it especially useful.

About Instrumental Quantitative Analysis

See the literature cited in Chap. 5.

7 The Analytical Problem

Objectives

- To introduce the "problem" concept in Analytical Chemistry.
- To describe, in a systematic manner, the general steps of the analytical problem-solving process via a variety of examples.
- To compare the information required from and supplied by the analytical laboratory.
- To emphasize the applied side of Analytical Chemistry and its social-economic role.

Table of Contents

7.1 Introduction

As noted in Chap. 1, Analytical Chemistry aims to extract the (bio)chemical information latent in an object or system with a view to making well-founded, efficient, timely decisions (see Fig. 1.5). In summary, the primary goal of Analytical Chemistry is to reduce the generic uncertainty (see Figs 2.3 and 2.4) in the qualitative, quantitative and structural (bio)chemical information of matter in space and time.

Any analytical information obtained by subjecting a sample to a CMP that is not fit for the intended purpose will obviously be of poor quality, even if it possesses any metrological features of excellence (*e.g.* a high accuracy and a low specific uncertainty). If any of the attributes of the information required by the client (*e.g.* expeditiousness, low cost) is not achieved, the analytical chemist in charge of the laboratory supplying it will obviously have failed. As shown in Fig. 7.1, the essential ingredients of quality in the results (*viz.* the capital properties accuracy and representativeness, illustrated in Fig. 2.7) must be completed with a third cornerstone that is, in fact, the top representativeness level (see Fig. 2.8): suitability to the information needed by the client, which is the primary source of the analytical work. This aspect is rarely considered systematically in quality standards despite its strategic significance. Chemical metrology, on which accuracy relies, has devoted little effort to improving representativeness; also, it does not regard suitability in the information as an essential part of it.

This chapter defines the analytical problem as the third cornerstone of analytical quality. The analytical problem has superseded samples and analytes as the primary target of current Analytical Chemistry. In fact, a number of modern definitions of this discipline include solving the analytical problem as an essential part (see Sect. 1.1).

___ Box 7.1

The analytical problem is the realization of the applied side of Analytical Chemistry (see Fig. 1.8), *i.e.* the ability of this discipline to provide the (bio)chemical information required for a specific purpose such as the following:

- Accepting or rejecting a batch of bottled wine.
- Deciding whether water from a given source is fit for drinking.
- Confirming whether a given food deserves the attribute "light".
- Checking whether an athlete has taken forbidden drugs.

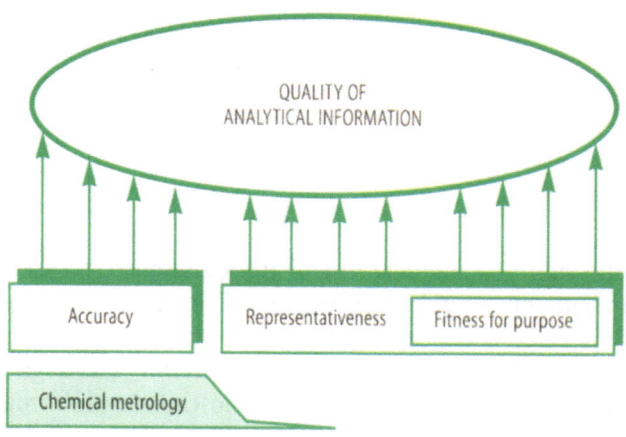

Fig. 7.1. Foundations of supplied analytical information

- Appraising a material.
- Determining whether a given location is polluted.
- Issuing a clinical diagnosis.
- Determining whether animals for human consumption have been given forbidden hormones.
- ...

The analytical data on which these decisions are to rest should be clear-cut and unequivocal. In addition, the information should be timely delivered, cost-effective, *etc.*, if it is to help solve the social-economic problem addressed. Analytical information that is delivered too late will be poor information as it will be of no use in making a decision subjected to a deadline; the results supplied, however metrologically excellent, will be absolutely useless for the intended purpose. Such is the case with the analysis of perishables, which decay with time: the longer obtaining the analytical results take, the poorer will be the quality of the product reaching the consumer.

Following a discussion of the meanings of "problem" in Analytical Chemistry, the analytical problem is defined as an interface between the *client* and the *analytical chemist*, and its essential elements and the generic steps involved in its resolution are described. Consistency and divergence between the information required and that supplied is then analysed, and the analytical problem is placed in the context of analytical quality.

_____ Box 7.2

Who is the "client" of an analytical laboratory? The easiest answer to this question is "the person who requests the (bio)chemical information about an object (system) or sample". However, this question necessitates a more specific answer since the client is a key element in defining the analytical problem and assessing the output (reports). The client's opinion is essential as quality, in its broadest sense, involves satisfying him or her.

First, one must consider whether the analytical laboratory is a part of the body requesting the information, *i.e.* whether it is administratively bound to the client. Obviously, the relationship (interface) between the two parties will be rather different depending on it. As a rule, the more independent of the client the laboratory is, the greater will be the freedom of the latter to act objectively and its scientific and technical consistency in approaching the analytical problem.

Second, one must consider the nature of the body requiring the analytical information to solve its problem. A wide range of situations between those of multinationals and public bodies exist that obviously call for different approaches to the analytical problem.

Third, one must consider the human factor, which is crucial to the client-analytical chemist relationship. The interested body requests information via its officials, who should be competent enough to accurately state their needs if the analytical problem is to be properly approached.

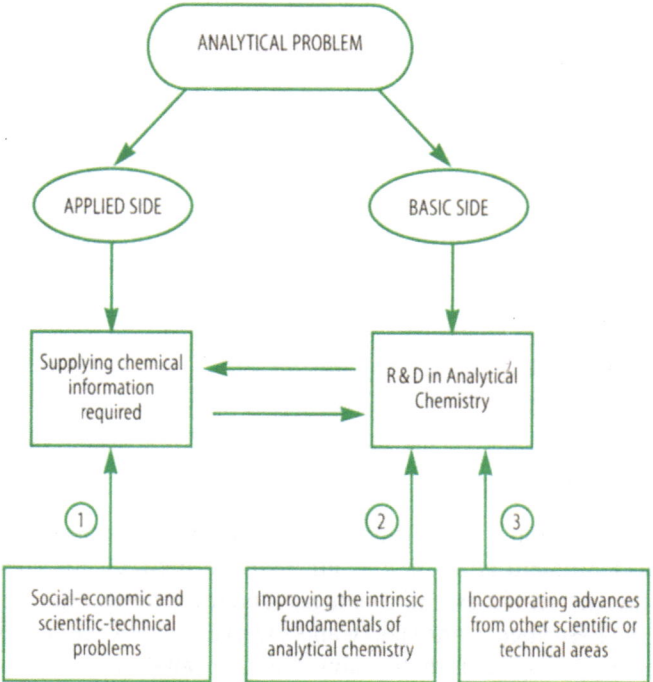

Fig. 7.2. The two sides of the "problem" in Analytical Chemistry. The traditional boundaries of the analytical laboratory are broken at 1 and 2

7.2 Meanings of "Problem" in Analytical Chemistry

As a scientific discipline, Analytical Chemistry possesses a specific foundation and developments that distinguish it from other scientific and technical areas of knowledge (see Figs. 1.9 and 1.11) and relate two of its essential elements – in addition to education –, *viz.* research and development (R&D) and the body of reported CMPs and tools (see Fig. 1.8).

In this context, the word "problem" has the two meanings illustrated in Fig. 7.2, which consider the basic and applied sides on Analytical Chemistry, namely: (*a*) a request for (bio)chemical information arising from a social-economic problem (see Box 7.1), which is the side dealt with in this chapter (Fig. 7.2.1); and (*b*) an innovative approach related to the intrinsic foundation of Analytical Chemistry and to the incorporation (adaptation) of developments from other scientific and technical areas.

7.3 Integral Definition of Analytical Problem

The term "analytical problem" was implicitly defined in Chap. 1. This section describes and integral approach to this concept that takes account of (*a*) its interfacial nature, (*b*) its relationships in the context of the rankings inherent in Analytical Chemistry, (*c*) its relationship to the capital property traceability and (*d*) its involvement in the collapse of traditional laboratory barriers.

The analytical problem can be viewed as an active two-way *interface* between the "client" requiring the information and the analytical chemist supplying it. Quality in this context can be defined in a practical, categorical manner, as satisfying the client; the analytical problem therefore possesses a high strategic significance. Figure 7.3 depicts the triple connection between the client's and the analytical chemist's domain. The social-economic problem faced requires meeting specific information needs in order to ensure quality from a body outside the laboratory. The analytical problem can be viewed as the connection (consistency) of these three aspects with the analytical process, the three types of analytical properties and analytical quality, respectively. The CMP of choice will have to be designed, applied and validated with the information needed in mind. Analytical properties and their mutual relationships allow one to characterize the information required and produced. The two components of analytical quality (*viz*. quality in the results and quality in the CMP) are related to external quality (*viz*. quality from outside the laboratory) via the analytical problem.

The conceptual an technical rankings of Sect. 1.6 can be considered in an integral, orderly manner (see Fig. 7.4). The analytical problem is at the top of the ranking in Fig. 2.8, which is discussed in detail in the following section. Also, it is linked to three other top levels, namely: external quality, reports (see Fig.

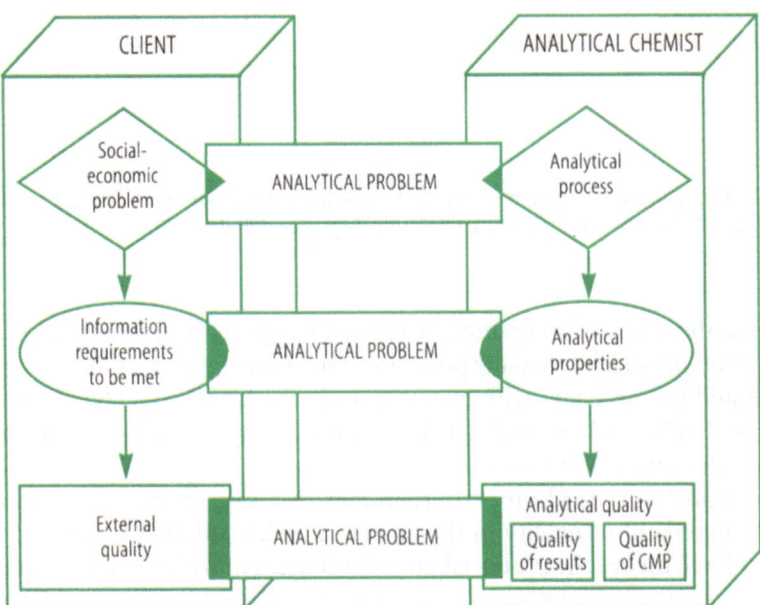

Fig. 7.3. The analytical problem as an interface between the social-economic problem of the client, who requests the analytical information, and the analytical chemist, who produces it. For details, see text

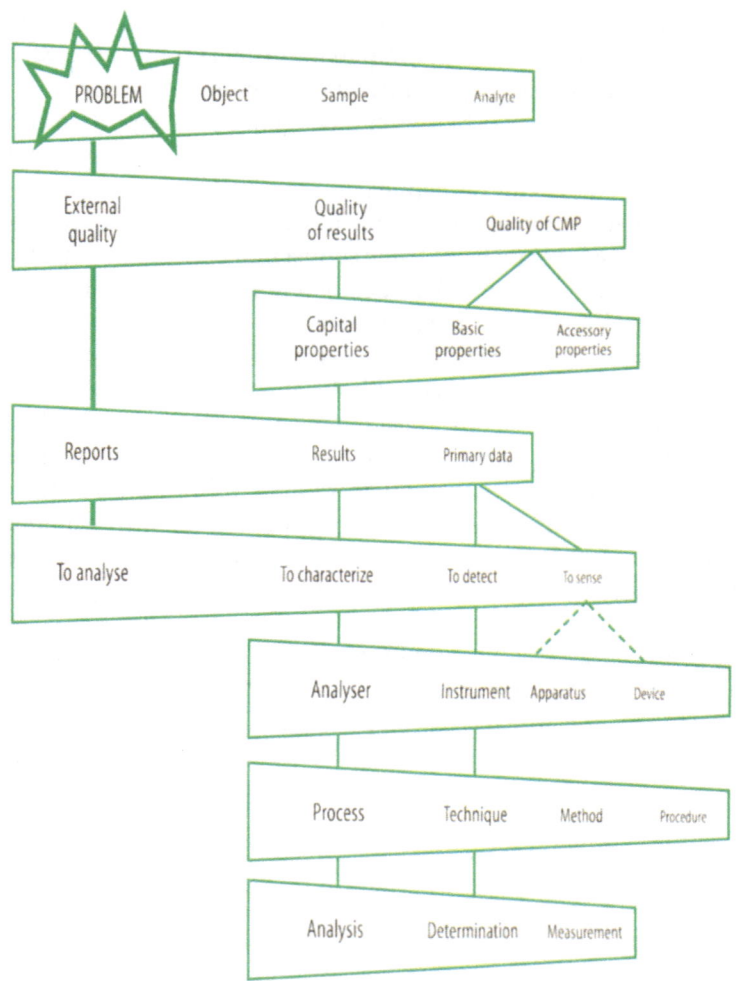

Fig. 7.4. The analytical problem in the context of the conceptual and technical rankings of Analytical Chemistry and their mutual relationships. For details, see text

1.15) and the action of analysing (see Fig. 1.15). Reports (*viz.* results discussed in the light of the social-economic problem addressed) are the outputs of analysing and should be consistent with quality outside the laboratory. Figure 7.4 shows additional relationships that can be used to complete, in a global, integral manner, the contents of Sect. 1.6.

The capital analytical property representativeness (see Sect. 2.4.2) is an essential ingredient of quality in the analytical information (see Fig. 7.1). At its highest level, the results produced are fully consistent with the social-economic problem, so the ensuing analytical problem will have been correctly solved (third level in Fig. 2.8). Thus, properly solving the analytical problem entails ensuring that the results are fully representative. The analytical problem is also related to traceability of a sample aliquot (see Box 7.3).

Fig. 7.5. Integral definition of "analytical problem". For details, see text

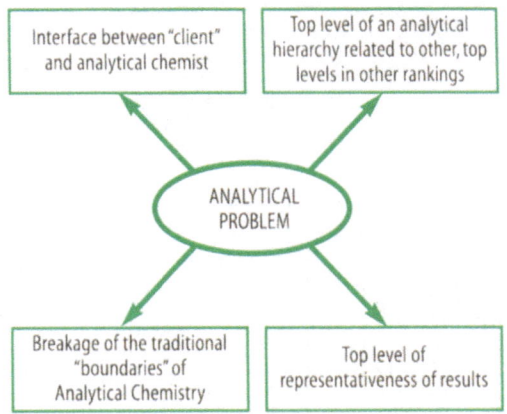

Interface between "client" and analytical chemist

Top level of an analytical hierarchy related to other, top levels in other rankings

ANALYTICAL PROBLEM

Breakage of the traditional "boundaries" of Analytical Chemistry

Top level of representativeness of results

Box 7.3

Relationship Between Analytical Problem and Traceability

Ensuring cyclic traceability in a sample aliquot subjected to a CMP entails relating it unequivocally to the output (sample custody chain) and, in a clear-cut, also unequivocal manner, to the specific information needed (representativeness). Through these two relationships, the results are made consistent with the social-economic problem addressed (see Sect. 3.6.2).

There is thus a direct relationship between traceability, as the property of a sample aliquot, and the analytical problem. Provided the analytical problem is properly solved, the results will be traceable to the social-economic problem concerned.

The analytical problem has broken and expanded the traditional boundaries of Analytical Chemistry, which used to be the laboratory walls (see Sect. 1.4.6). This has entailed a change from a passive role (*e.g.* receiving samples and producing results) to an active one that rests on a two-way, symbiotic relationship to the social and economic bodies that raise the information needs prior to and after implementation of CMPs. Only in this way can Analytical Chemistry acquire its true dimension.

Figure 7.5 depicts the four aspects of the analytical problem described in this section. Obviously, all are mutually related. Thus, the interfaces in Fig. 7.3 are consistent with the top representativeness level, with the top of the ranking and with collapse of the traditional barriers of Analytical Chemistry.

7.4 Elements of the Analytical Problem

As schematically shown in Fig. 7.6, the analytical problem encompasses tangible and intangible elements, both of which are required for an accurate, integral definition.

The *tangible elements* are those in the ranking of significance and scope of Fig. 1.18, and materialize in examples such as those of Box 1.11. The object is the

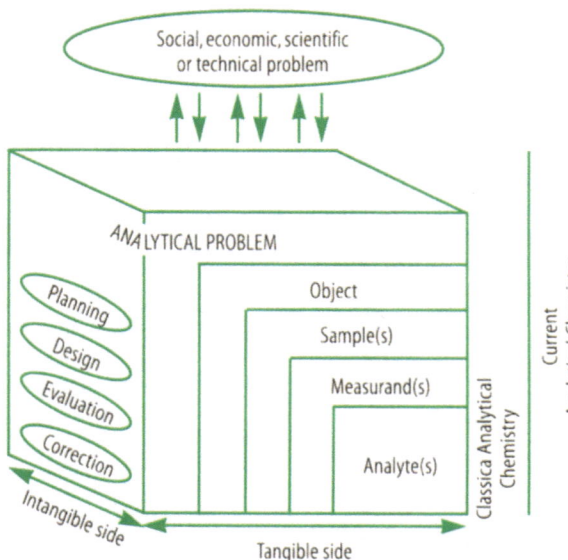

Fig. 7.6. Tangible and intangible elements of the analytical problem. For details, see text

entity that is to be chemically described via the analytical results and their interpretation; typical examples include bay water, river water, a mineral cargo or gas flowing along a pipe. As a rule, describing an object entails using four coordinates (three in space and one in time); frequently, however, additional parameters such as pressure, temperature, *etc.*, are also required. The object should be a faithful reflection of the social-economic problem that raises the information needs. The sample can be defined as a part of the object that possesses all the connotations described in Sect. 4.3.3. Finally, the measurand is the quantity to be measured, which often coincides with the amount or concentration of analyte.

The *intangible elements* are also an essential part of the analytical problem. Planning, design, evaluation and correction are four key activities without which the analytical problem could never be properly solved. These elements are discussed in detail in the following sections.

Figure 7.6 also shows the changes involved in the shift from the traditional perception of Analytical Chemistry to a modern approach consistent with its goals.

7.5 Steps of the Analytical Problem-solving Process

The integration of the tangible and intangible elements of an analytical problem and its relationship to the social-economic problem (Fig. 7.6) materialize in the general steps to be taken for its proper planning and resolution; this requires that the analytical information produced be effective with a view to solving the social-economic problem. The five steps involved, which are executed in a cyclic manner, are schematically depicted in Fig. 7.7 and commented on below.

Fig. 7.7. General steps involved in planning
and solving the analytical problem. For details,
see text

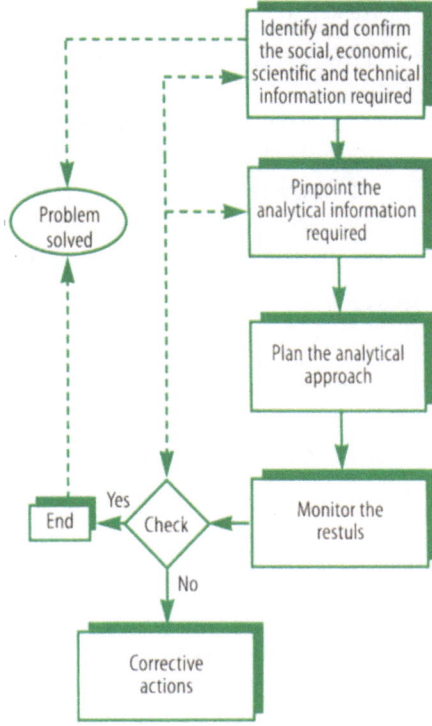

7.5.1 Identifying and Confirming the Information Required

The analytical problem cannot be approached from a single side; in fact, proper planning and resolution here entails the involvement of the two parties (the client and the analytical chemist), particularly in the first step (*i.e.* the interface of Fig. 7.3 is crucial). The analytical chemist should play an active role and collaborate with the client in pinpointing the information required to make a well-grounded, efficient, timely decision. This requires that the analytical chemist know the basic characteristics of the social, economic, scientific or technical problem, and have an answer for questions such as what?, why?, how?, when? and where? (see Box 7.4). Communication between the two parties should be fluent, interference-free, both at the beginning and at the end of the cycle implicit in Fig. 7.7.

_____ Box 7.4

First Step of the Analytical Problem-solving Process

Pinpointing the (bio)chemical information required by the client to solve his or her problem and achieving the level of quality (external to the laboratory) sought are crucial. They entail providing unequivocal answers about the social-economic problem as illustrated in the following examples.

Example 1

Problem: Appraising a drug whose active principle contains two enantiomers (optical isomers) one of which (D) is pharmacologically inactive (why?) while ensuring the presence of the highest possible proportion of the active one (L) (what for?).

Information required: The proportion of each enantiomer in the finished product (what?) and in the intermediate steps of the production process (how?). This requires the knowledge of the raw materials and intermediate products involved (where?, when?) if the proportion of the active isomer in the product is to be maximized (what for?).

Example 2

Problem: A river is polluted with organic waste from neighbouring factories (why?).

Information required: Type of industry or raw materials and products involved (what), production steps (how?), waste purifying systems used (how?), production temporality (when?) and location of the discharge along the river bank (where?).

Example 3

Problem: A forest enclave has been deteriorated by acid rain (why?). The aim is to reduce the levels of atmospheric pollutants (SO_2, NO_x) released by the neighbouring industries in order to preserve the trees (what for?, how?).

Information required: Changes in the contents of sulphur, phosphorus, nitrogen, *etc.*, in the tree leaves and pollutant levels in the atmosphere (what?, when?).

Only with a sound knowledge of the social-economic problem can the analytical problem be properly approached and the analytical information supplied useful with a view to solving it.

The degree to which the analytical chemist is to be involved in this step will depend on the technical expertise of his or her interlocutor, who may be an experienced analytical chemist or a professional with no technical qualifications. This step is undoubtedly crucial; solving an analytical problem that has not been correctly approached makes absolutely no sense. Frequently, the analytical chemist is minimally involved in this step – either by tradition or by imposition. One of the differences between classical and modern Analytical Chemistry is the shift from the traditionally passive role of the analytical chemist in this step to an active one.

7.5.2 Pinpointing the Analytical Information Required

In this second step, the analytical chemist must convert the generic chemical information identified in the first into the analytical information required to help solve the social-economic problem. By the end of this step, a series of specific details must be known (see Fig. 7.8). *First*, the object should be accurately defined and the sampling plan decided upon. *Second*, the measurands (analytes) whose presence or concentration is to be determined should be established. *Third*, the type(s) of analysis to be performed [*viz.*

(*a*) qualitative, quantitative or structural;
(*b*) static, temporal or spatial; and

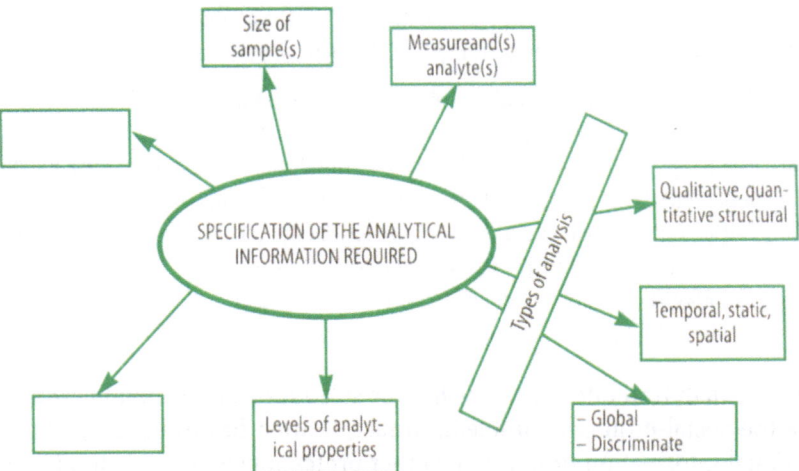

Fig. 7.8. Features of the analytical information required to solve a social–economic problem. Once these have been established, the second step in the planning and solving of the analytical problem is executed

(c) global (total content) or discriminate (speciation)] should be determined. *Finally*, the quality requirements for the analytical information to be obtained (in terms of the analytical properties and their mutual relationships) should be known.

This last aspect is crucial as an overall quality compromise between accuracy and representativeness (reliant on precision, sensitivity, selectivity and proper sampling) on the one hand, and productivity (expeditiousness, cost-effectiveness and personnel-related factors) on the other, must be adopted. Such a compromise should be consistent with the definition of the analytical problem. Both the properties that characterize quality and those that define productivity (or rather, their levels) should be set as the analytical goals to be reached in order to properly respond to the analytical problem. The quality policy of the laboratory itself (see Chap. 8) is also subject to consistency between the definition of the analytical problem and its materialization (under quality compromises) in the significance and property levels to be achieved. The suitability of the information supplied relies on such consistency, as does its actual usefulness for solving the social, commercial or economic problem that raised the analytical problem.

_____ Box 7.5

Second Step of the Analytical Problem-solving Process

Converting the generic (bio)chemical information identified and confirmed in the first step into specific analytical information (Fig. 7.8) is the goal of the second step in the analytical problem-solving process and also a prerequisite for tackling the third (planning the analytical approach and selecting the CMPs to be used).

A non-government organization (NGO) for the defense of consumers' rights has set to check the reliability of the fat contents listed by the manufacturers of skim yoghurt on their packages (light, < 0.1%; skim, 0.0% fat matter). This is the social-economic problem. Defining the ensuing analytical problem requires specifying various factors such as:

(a) Samples: light and skim yoghurt (plain, flavoured and fruit varieties).
(b) Analytes: a family (total fat, which will require a global analysis).
(c) Sampling: a crucial aspect of the analytical problem. The samples studied should be representative of the different brands and purchased at different outlets.
(d) Analysis type: qualitative (presence/absence of fat), quantitative (percentage of fat) or semi-quantitative (whether or not the fat content exceeds 0.1%).
(e) Analytical properties: acceptable accuracy and precision (reliability in the binary response), at the expense of laboratory productivity.

However, analytical information is not the sole type of information required to solve the social-economic problem. In fact, it often has to be integrated with additional information produced by other professionals (*e.g.* economists, geologists, pharmacologists, physicians).

7.5.3 Planning the Analytical Approach

After the precise output expected from it is established, the laboratory must conduct the analytical process(es) required to fulfill the objectives. To this end, the following four factors should be considered:

(*a*) the nature of the information required,
(*b*) the properties of the object (or sample) and measurands,
(*c*) the material and human resources available, and
(*d*) the price the client is willing to pay for the analytical information requested.

The analytical process designed to fulfill the above-mentioned objectives must encompass every step between the uncollected, unmeasured, untreated sample and the results, expressed as required (see Chap. 4). The choice of classical, instrumental, hybrid, automatic and screening systems, among others, depends critically on the above-mentioned factors, which should be considered in a systematic manner. The analytical approach may raise a need for specific or global changes with a view to fulfilling the information requirements (the basic side of Analytical Chemist).

_____ Box 7.6

Third Step of the Analytical Problem-solving Process

This step involves choosing the CMPs to be used in order to obtain the analytical information specified in the second step. The exact approach to be used will depend on the factors depicted in the figure below.

Thus, the CMPs of choice will depend on a variety of factors. In fact, the choice will be rather different whether, for example,

● The laboratory is a poorly-equipped "peripheral" unit or a self-contained facility (see Box 7.7).

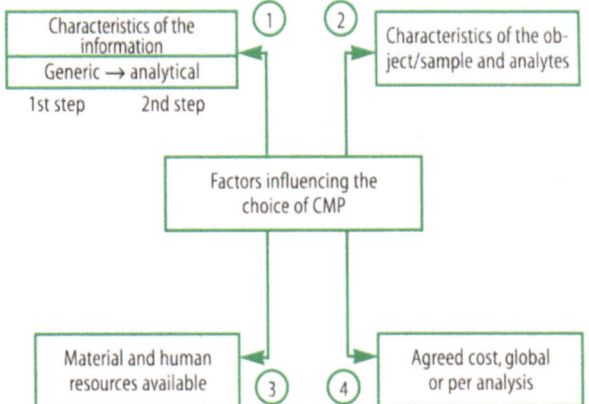

- Delivery of the information is subject to a tight deadline or a small uncertainty is acceptable at the expense of a longer turnaround time.
- Global or discriminate information is needed.
- The client has placed a cost restriction on the analyses.
- The analytes are few and similar in nature or many and widely different; macro components or traces.
- The samples are solid, liquid or gaseous; stable or unstable.
- The samples are readily (sea water) or hardly accessible (baby's blood, a painting by El Greco).

All these factors will dictate the CMP of choice. The experience of the analytical chemist is ultimately essential in this planning step.

This planning step is referred to by some authors as *choosing a suitable analytical methodology*. Some textbooks deal with it in their closing chapters.

Planning the analytical work to be conducted inside or outside the laboratory inevitably includes developing quality control and assessment systems within the framework of Quality Assurance (see Chap. 8). This will be of great assistance in fulfilling the objectives of the analytical problem.

_____ Box 7.7

The CMP to be chosen in the third step of the analytical problem-solving process depends critically on the available means.

Example 1

Problem: Metal contamination in brine.

If an atomic absorption or emission instrument is available, the CMP will be quite simple thanks to the high sensitivity and selectivity of atomic spectrophotometric techniques. It will suffice to the dilute the sample for direct insertion by aspiration or to inject a microvolume into an automatic continuous system coupled on-line to a multi-parameter instrument. If only a straightforward UV-visible molecular absorption instrument (a photometer) is available, a CMP for each trace element potentially present as a contaminant in the brine will have to be developed; this will be labour-intensive and time-consuming as it will call for preliminary operations such as interference removal and preconcentration in order to boost the selectivity and sensitivity, respectively (see Sect. 4.3.4).

Example 2

Problem: The presence of a preservative banned by European legislation in a soft drink.

If a liquid chromatograph (LC) is available (see Box 5.16), then the CMP will be very simple: it will suffice to degas the sample and insert an aliquot into the chromatograph (identification can be done as described in Box 5.18). If only a rudimentary photometer is available, however, interferences will preclude direct measurements of the preservative absorbance, so some preliminary operation (*e. g.* a liquid-liquid or solid-phase extraction) will be required to partly overcome this drawback; alternatively, a derivatizing reaction can be used to accomplish the effect depicted in Fig. 5.12.

7.5.4 Monitoring the Results

The fourth step of the analytical problem-solving process involves two distinct actions both of which are related to validation (see Sect. 4.6) and to the first two steps of the process. *First,* one must evaluate internal quality, which entails validating the results with respect to the levels sought in the analytical properties and their compromises; this is the responsibility of the analytical chemist, in the second step of the process. *Second,* one must validate the results in relation to the information requested by the client and to data supplied by other professionals; the purpose is to check for external quality, which, again, requires establishing the interface involved in the first step. If both evaluations (comparisons with references) are favourable, the analytical problem will have been solved (see Fig. 7.7).

_____   Box 7.8

Fourth Step of the Analytical Problem-solving Process

The evaluation of the results produced by the CMPs designed in the third step of the process should rely on comparisons with two types of reference, namely: the characteristics of the analytical information specified in the second step and the information needs confirmed in the first. This evaluation will reveal whether the problem is solved or a fifth step is required.

Example 1. Presence of cadmium in yellow toys

After the CMP is conducted, the final concentration in the treated sample is found to fall beyond or near the limit of quantitation for the linear range of the calibration curve. Obviously, the methodology cannot be validated for quantitation purposes as the uncertainty of the result in this concentration region will be very high. However, if the information requested is of the purely qualitative type and such a concentration exceeds the limit of detection, the binary response provided will be quite valid.

Example 2. Contamination of fruit with pesticides

After the CMP is carried out, real samples are found to provide no signals (peaks) in the chromatogram. The easiest interpretation is that the fruit is "clean". However, as in the previous case, the evaluation of this fourth step should include the systematic consideration of the limits of detection and quantitation for the CMP concerned, as well as the highest legally accepted level (*e.g.* 0.1 ng/kg). After this stricter evaluation, the immediate answer may not be so clear and the CMP may have to be modified.

7.5.5 Implementing Corrective Actions

When the results do not meet the client's requirements or fail to reach the sought levels in the analytical properties, the analytical approach used should be revised following careful examination of the data provided by the quality systems employed, which will be of great assistance in pinpointing the sources of divergence. This may call for minimal (*e.g.* standards, instrumental settings), partial (*e.g.* calibration system, instrument type) or global changes (*e.g.* methodology) in the analytical processes(es). The output of the new analytical approach should also be subjected to the dual validation involved in the fourth step of the analytical problem-solving process. These corrective actions are frequently implemented with no prior planning and result in a considerable waste of time and work.

_____ Box 7.9

Fifth Step of the Analytical Problem-solving Process

A negative result in the evaluation of the fourth step (unvalidated results) entails the implementation of corrective actions on the analytical approach developed in the third step.

Thus, in the examples of Box 7.8, inadequate sensitivity results in uncertainty. The corrective actions to be taken involve including an analyte concentration sub-step among preliminary operations. With Cd^{2+}, a chelating ion-exchange resin (*e.g.* Chelex 100) can be used for this purpose; however, a larger amount of sample (*e.g.* scrapings of the toy's yellow coating) will have to be taken. Pesticides in the fruits will also call for an additional step (*viz.* the solid-phase extraction of the solvent used to leach the analyte from the surface of the pears, apples, oranges, *etc.*).

7.6 Consistency Between Required Information and Supplied Information

Frequently, a divorce exists between the information sought and that provided by the laboratory as a result of the analytical problem being incorrectly planned and solved. Figure 7.9 compares these two types of information, which can be related in various ways as shown below.

In the ideal situation (Fig. 7.9, a), the two types of information coincide in both qualitative (measurands, analysis types) and quantitative terms (global or discriminate determination, uncertainty level).

Divergences between the information required and that supplied can be of various types. Thus, the may not coincide fully (Fig. 7.9, b) owing primarily to a

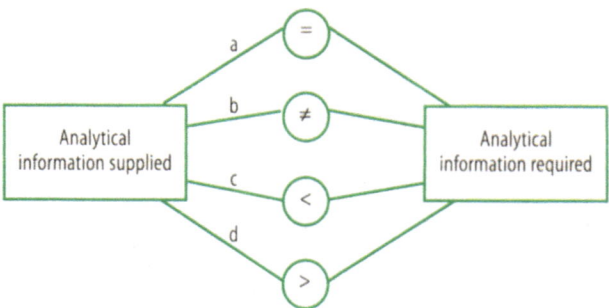

Fig. 7.9. Different possible situations arising from a comparison of the analytical information supplied by the laboratory with that required by the client to solve the social–economic problem

serious lack of communication between both parties (*e.g.* the client needs to know the concentration of metal traces and is supplied with that of hydrocarbons).

Also (Fig. 7.9, c), the information delivered by the laboratory may be inadequate to meet the client's information needs (*e.g.* speciation is required and only the total concentration is supplied or the quantitative composition is needed and only qualitative information is delivered).

Finally, the information supplied may be partly excessive and redundant (Fig. 7.9, d); this usually results in a waste of work and material resources (*e.g.* overall pollution figures are needed and the concentrations and uncertainties for an endless list of species are supplied; approximate concentrations are more than acceptable and data with very low uncertainties obtained under excellence conditions are delivered). This situation is relatively frequent: social-economic decisions rely on a small fraction of the analytical information received.

If the analytical problem is properly approached, the information sought and that supplied will be quite consistent. Many times, the client is unaware that the laboratory has the ability to provide special types of information that would allow the social-economic problem to be approached and solved with higher external quality. Other times, the client requests information with attributes (*e.g.* low cost, expeditiousness) that are technically unreachable. These situations never exist if the interface involved in the analytical problem-solving process works as it should.

_____ Box 7.10

Examples of Consistency and Divergence Between Required Information and Supplied Information

An analytical problem has been properly approached and solved when the information delivered by the laboratory is consistent with that requested by the client. The situations depicted in Fig. 7.9 are illustrated with several examples below.

Figure 7.9, a. Assessing the (external) quality of the output of one dairy factory entails determining the protein content in a 2-ton batch of Tetrabrik-packed milk that is located on a goods train. A mean content of $9.4 \pm 0.2\%$ is supplied that is the result of analysing 232 samples collected according to a statistical sampling plan. There is consistency between the global information required and that provided by the laboratory, which focussed on correct planning of the sampling operation.

Figure 7.9, b. There is the need to know whether the water supplied to an agricultural area is fit for consumption by humans (particularly infants) as nitrate ion (from fertilizers) is toxic above a given concentration level. Misunderstanding between the client and the analytical chemist has led to the latter providing the typical parameter values for water (pH, conductivity, COD, *etc.*) but not the concentration of the target analyte. The information supplied does not respond to the social problem, even though it could be used to assess the standard quality of the water.

Figure 7.9, c. With a view to determining whether water from a particular source has been made toxic by the potential presence of metal species, the total concentration of each metal will be inadequate as each may be present in different forms (oxidation states, complex species, chelates, organometallic compounds) of also different toxicity. Only a discriminate determination of each species will provide information consistent with the social-economic problem addressed.

Figure 7.9, d. For environmental legislation purposes, whether soil near an oil industry is contaminated with hydrocarbons must be determined. Legislation imposes a limit of 0.1 µg/kg to decide whether the soil is polluted. A global response encompassing all types of hydrocarbons (aliphatic, aromatic, nitrogen- and sulphur-containing, *etc.*) is thus needed. If the laboratory implements sophisticated CMPs based on hyphenated techniques such as gas chromatography-mass spectrometry, it will inevitably provide results of a high internal quality (*e.g.* a discriminate listing of the hydrocarbons present in the soil accompanied by their concentrations, in ppm or ppb, and uncertainties). However, such results will be inconsistent with the information required: ultimately, the client will not know whether the soil is contaminated as per the applicable legislation. Heavy human work and material expenditure will have led to useless results. Obviously, the sophisticated CMP used will have to be replaced with a more simple one providing a global figure encompassing all types of hydrocarbons rather than discriminate information.

7.7 The Analytical Problem in the Context of Quality

As shown in Chap. 8, the definition of quality encompasses the basic (properties, features) and applied approaches (meeting imposed or implicit needs), and their mutual relationships, for which analytical chemical counterparts exist, namely: analytical properties and metrological features (traceability, uncertainty), and the analytical problem-solving process (see Fig. 7.10). The relationship between the basic and applied sides of Analytical Chemistry is also self-obvious and materializes in the different steps of the analytical process.

As shown in Fig. 7.2, properly solving the analytical problem in order to ensure external quality (*i.e.* satisfying the client) is thus an essential element of analytical quality. It is interesting to note that, at present, analytical quality is

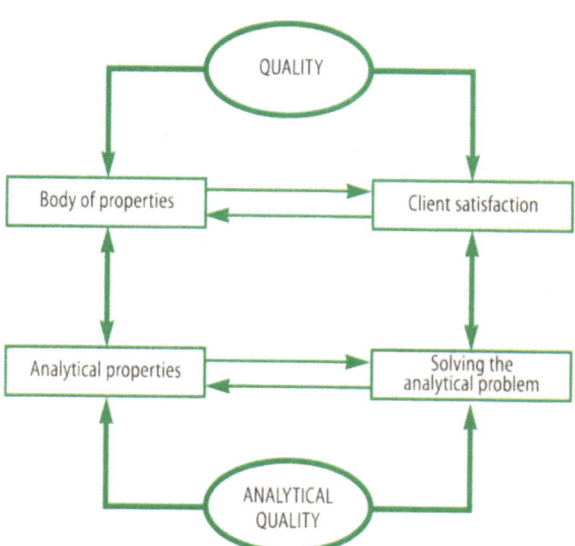

Fig. 7.10. Relationship between the basic and applied approaches to quality in general and analytical quality in particular

identified almost solely with its internal side (*viz.* the metrological quality of the results, laboratory quality control, *etc.*); few national or international bodies regard representativeness in the results and the analytical problem-solving process as essential elements of laboratory quality. This has confined Analytical Chemistry to a low-rank technical domain; in this situation, it will be unable to serve its true purpose unless its traditional boundaries are expanded, which inevitably entails considering these issues in dealing with analytical quality.

The countless definitions of quality reported to date further support this integral approach. Thus, it is widely assumed that "no quality is to be expected unless the objectives are properly defined"; in fact, the analytical problem provides the analytical chemist with such objectives. Also, "quality is the concern of us all rather than a few"; indeed, analytical quality only makes sense when integrated in a global concept of quality that lies outside the laboratory.

Questions

(1) What is "fitness for purpose" in Analytical Chemistry?
(2) Based on the contents of Box 7.1, name several social-economic problems calling for analytical information.
(3) Give some examples to materialize the different ways of defining "problem" in Analytical Chemistry (see Fig. 7.2).
(4) Why are the client's needs and opinion so significant to the applied side of Analytical Chemistry?
(5) Identify the ranking-problem-object-sample-analyte sequence in five specific examples. Use Box 1.11 as reference.
(6) Comment on the relationships materialized in the triangle below.

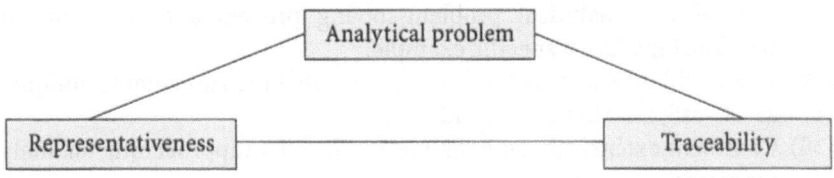

(7) Who is the client in the analytical context?
(8) Provide a definition for the intangible elements of an analytical problem.
(9) What theoretical and practical approaches are combined in the integral definition of an analytical problem?
(10) Why is "fitness for purpose" in analytical information encompassed in the capital analytical property representativeness?
(11) Which of the five steps of the analytical process involve planning? Which evaluation? Which correction?
(12) Provide a definition for the tangible elements of an analytical problem.
(13) Comment separately on the three interfaces that define the analytical problem and associate the client-related aspects to Analytical Chemistry (Fig. 7.3).
(15) Relate, via several examples, the analytical problem to the action of analysing, to reports and to external quality (Fig. 7.4).
(16) Does the analytical chemist fulfill his or her mission simply by delivering the results for the samples that reach the laboratory, with no other intervention?
(17) Relate the four aspects encompassed by the integral definition of analytical problem.
(18) Why is the top level of representativeness consistent with the analytical problem-solving process?
(19) Why is expanding the traditional boundaries of Analytical Chemistry essential to the "analytical problem" concept?
(20) Why does the analytical problem entail performing some actions before and after the samples are subjected to the CMP?
(21) Why does chemical metrology not consider the analytical problem in a systematic manner?

(22) Can an analytical problem result from a scientific or technical need? Give some examples.

(23) Can an analytical problem be approached without a deep knowledge of the information needs raised by the social-economic problem?

(24) Comment extensively on Example 1 in Box 7.4 (approaching an analytical problem).

(25) Comment extensively on the analytical problem derived from light foods (Box 7.5).

(26) In what types of analytical problem will the use of sensors be appropriate? Why? Give some examples.

(27) Comment extensively on Example 2 in Box 7.4 (approaching an analytical problem).

(28) Comment on the factors that influence the decision involved in the third step of the analytical problem-solving process and on their mutual relationships. Use a specific example.

(29) When will it be justified to use a hyphenated instrumental technique such as GC-MS, LC-MS or FTIR-MS?

(30) Comment extensively on Example 3 in Box 7.4 (approaching an analytical problem).

(31) Explain the constraints that may arise from the client's imposing a maximum cost per analysis.

(32) Give two examples of analytical problems involving the determination of a single or many analytes.

(33) In what way do relationships among analytical properties influence the manner an analytical problem is to be approached?

(34) Why do the basic properties sensitivity and selectivity of a CMP have a critical influence on the third step of the analytical problem-solving process?

(35) The first two steps of the analytical problem-solving process involve specifying information. What is the difference between them? In what way are they related?

(36) Give two examples of analytical problems where expeditiousness is the key analytical property.

(37) Comment extensively on the analytical problem faced by a vegetable and fruit growing industry as a result of its produce being contaminated with pesticides.

(38) What are the references to be used to validate the results in the fourth step of the analytical problem-solving process?

(39) Give some examples where the analytical information delivered is inadequate to solve the social-economic problem faced.

(40) Why is the analytical problem not properly solved when more analytical information than is required is supplied?

(41) Explain why the need for highly accurate, minimally uncertain information influences the choice of CMP.

(42) With the aid of an example, explain when a corrective action (fifth step of the analytical-problem solving process) must be implemented.

(43) Give some examples where excessive information was supplied with a view to solving the analytical problem.

(44) When can one state that an analytical problem has been properly approached and solved?

(45) Comment in detail on the analytical problem posed by the presence of cadmium in toy paint coatings.

(46) Comment, via some examples, on the impact of legislation on the planning and solving of analytical problems.

(47) What are the connections between quality in general and analytical quality in particular?

(48) Comment on the role of the analytical problem in the analytical quality context.

(49) What is one of the most valuable elements of the analytical approach when the object has a changing nature? What are the CMPs of choice like? Are any special tools required?

(50) Give some examples of analytical problems in the forensic domain.

Seminar 7.1

Solving an Analytical Problem in the Environmental Domain
The Polycyclic Aromatic Hydrocarbon Content in the Atmosphere

- **Learning objective**
  To understand how a specific analytical problem in the environmental domain is to be approached and solved.
- **Literature**
  "Airborne Particulate Matter", T. Kovimtzis and C. Samara, *Springer*, Berlin, 1995, Chap. 9: "Analysis of Organic Particulate Matter", by C. Samara (pp. 233–252).
- **Procedure: Aspects to be considered**
  - Describe PAHs (types, families).
  - The impact of PAHs on human and animal health.
  - Sources of atmospheric PAHs.
  - The organic fraction of suspended matter in the atmosphere.
  - Sampling: Distribution of PAHs between particulates and the vapour phase. Types of filters.
  - Concentration and clean-up from the sampling filter.
  - Systems for discriminate determination of PAHs in the clean concentrates:
    - GC with conventional detectors.
    - GC-MS.
    - LC with conventional detectors.
    - LC-MS.
  - Standards for calibration and global evaluation.
  - Validation of the procedure.
  - Discuss the features of the response to the information needs raised by the environmental problem.

Seminar 7.2

Solving an Analytical Problem in the Industrial Domain
Characterization of Soaps

- **Learning objective**
  To introduce an analytical problem related to industrial production.
- **Literature**
  "Official and Standardized Methods of Analysis" (3rd edn), C. Watson (Ed.), *Royal Society of Chemistry*, 1994, pp 427–445.
- **Procedure: Issues**
  - General description of soaps (commercially available types).
  - Definition of the key chemical features of soaps that dictate their quality.

- Determination of free fatty acids.
- Determination of the saponifiable and unsaponifiable fat fractions.
- Determination of the different types of alkalinity.
- Determination of rosin.
- Determination of phenols and cresols.
- Discuss the restrictions of the CMPs used for application to various types of soap. Justify them.
- Impact of the analytical results on soap industry and trade.

Seminar 7.3

Solving an Analytical Problem in the Nutritional Domain
The Aflatoxin Content in Cereals and Dried Fruits

■ **Learning objective**
To examine a typical example of major food toxicity problems and be able to choose an effective analytical approach in order to obtain pertinent, appropriate information.

■ **Literature**
"Official Methods of Analysis" (15th edn), Vol. 2, *AOAC*, Arlington, 1990, pp. 1184–1213.

■ **Procedure: Issues**
- Properties and toxicity of mycotoxins.
- Types and features of aflatoxins. Contaminated foods: types and sources.
- General cautions to be exercised in their analytical determination.
- Preparation of standards.
- General methods for the identification and quantitation of aflatoxins. Screening methods based on immunoassay.
- Description of the CMPs to be used to identify or determine aflatoxins in cereals and dried fruits.
- Critical comparison of available CMPs for the determination of aflatoxins in various types of food matrices.
- Impact of the presence of aflatoxins on various types of food.

Suggested Readings

(1) "The Analytical Problem", M. Valcárcel and A. Ríos, *Trends Anal. Chem.*, 16, 1997, 385
   This paper can be viewed as the "skeleton" for this chapter, which was inspired by it.
(2) "A Holistic Approach to Analytical Chemistry", H. Pardue and H. Woo, *J. Chem. Educ.*, 61, 1984, 409.
   This paper provides a broad view of the impact of Analytical Chemistry.

(3) "What Exactly is Fitness for Purpose in Analytical Measurement", M. Thompson and T. Fearn, *Analyst*, 121, 1996, 275.
  This paper relates metrological quality in the results to its impact on the information delivered with the aid of mathematical tools.

(4) "Analysis: What Analytical Chemists Do", J. Tyson, *Royal Society of Chemistry*, Cambridge, UK, 1988
  Chapter 1 of this interesting book discusses in a concise, entertaining manner, the impact of analytical information on various social and economic areas. Chapter 7 deals with the selection of analytical processes.

(5) "Analytical Chemistry", R. Kellner, J. M. Mermet, M. Otto and H. M. Widmer (Eds), *Wiley-VCH*, Weinheim, 1998.
  Chapters 1 and 2 discuss the role of the analytical chemist as the solver of clients' problems in the light of several examples.

Objectives

- To relate quality to the generic and specific goals of Analytical Chemistry.
- To introduce students to the planning and development of quality systems in the analytical laboratory.
- To describe the strategic and methodological tools needed to implement control and assessment actions in the analytical laboratory.

Table of Contents

8.1 Introduction

The word "quality" has been used systematically in the previous chapters of this book, devoted to the fundamentals of Analytical Chemistry, to refer to the degree of excellence of various analytical objects, systems and events. These are some of the more meaningful examples:

- In the definition of Analytical Chemistry (Sect. 1.1), quality is an attribute of the information produced.
- Analytical information encompasses three quality levels, *viz.* ideal, referential and real (see Fig. 1.6).
- Quality as applied to information, traceability and the analytical process, and as related to analytical properties, is one of the key words of Analytical Chemistry (see Fig. 1.10).
- Analytical quality (*viz.* quality of CMPs and of results) is directly related to analytical properties and their hierarchies (see Fig. 2.1).
- As shown in Box 3.15, standards contribute to quality of analytical laboratories.
- The principal aim of method validation is to assure analytical quality (see Sect. 4.6).
- Analytical properties and errors (false positives and false negatives) characterize the quality of the binary response in qualitative analysis (see Sects. 5.2.3 and 5.2.4).
- The analytical properties of a quantitative response somehow dictate its quality (Sect. 6.2).
- As shown in Sect. 7.7, the analytical problem is also related to quality.

This final chapter aims to provide an integral, comprehensive, though not exhaustive, picture of the Analytical Chemistry-quality binomial from two different points of view, namely:

(*a*) a basic one (most analytical chemical facets are related to quality in its broadest sense) and

(*b*) an applied one (the implementation of quality systems in the analytical laboratory).

Continuously improving analytical information in order to bring it as near as possible to the intrinsic information latent in the object or sample considered is an essential goal of Analytical Chemistry. Tests involving expert analytical laboratories that applied specific CMPs to identical samples in order to determine the same analytes have exposed differences in the results that can be taken as warnings for the absence of quality. Boxes 8.1 and 8.2 show two typical examples that more than justify the need for quality systems and for their inclusion in the basic curriculum taught to future analytical chemists.

_____ Box 8.1

Consistency Between the Results of the Determination of Metal Traces in the Same Lichen Specimen Produced by Fourteen Expert Laboratories

The BCR, a European Union body, organized an intercomparison exercise where fourteen laboratories selected on the grounds of experience, carried out the quintuplicate determination of element traces in a specimen of lichen (a bioaccumulator and a reliable indicator of environmental pollution). After the whole data collection was received by the BCR, the highest and lowest value among the fourteen for each element were chosen. The following table also includes the highest-to-lowest concentration ratio for each element.

Metal	Lowest value (µg/g)	Highest value (µg/g)	Ratio
Calcium	200 (XRF)	3155 (XRF)	15
Copper	0.08 (FAAS)	38.8 (ICP)	48
Mercury	0.008 (CVAAS)	0.551 (CVAAS)	110
Molybdenum	0.056 (ICPMS)	2.072 (ICP)	37
Zinc	9.7 (FAAS)	282.5 (ICP)	29

In brackets are the analytical techniques used in the different CMPs, namely: X-ray fluorescence (XRF), atomic emission spectroscopy using inductively coupled plasma (ICP), flame atomic absorption spectroscopy (FAAS), cold-vapour atomic absorption spectroscopy (CVAAS) and inductively coupled plasma-mass spectrometry (ICPMS).

As can be seen, the results provided by the different laboratories for the same analyte in the same type of sample were rather disparate in some cases. The implications on the underlying analytical problems are self-obvious. How many mistaken conclusions will have been drawn as a result of using poor-quality analytical information? How many preventive actions taken in response will have been wrong? What will the economic and social costs of the poor results have been?

This example exposes the need for an urgent institution of quality systems based on control, assessment and corrective actions in order to assure quality in the analytical results generated in this context.

_____ Box 8.2

Consistency Between the Results Produced by Ten Laboratories that Analysed the Same Sample Standard (a Blind Sample)

A water specimen was distributed between ten laboratories and included by their officials among the samples routinely analysed by each. The specimen was a "blind" sample because its reference nature was unknown by those implementing the CMP. It was artificial and its content known by the supervising body. The true value and the results obtained by the ten laboratories in the determination of element traces (mg/L) are given in the following table.

	As	Cd	Cr	Cu	Pb	Hg
True value	1	0.05	0.05	5	5	0.5
Laboratory						
1	**1200**	0.04	**0.011**	5.1	4.8	0.485
2	**< 0.01**	0.03	<0.05	4.78	**<0.05**	**0.05**
3	0.959	0.053	0.048	4.59	4.5	0.40
4	0.861	0.052	0.062	4.85	3.9	**0.05**
5	**0.052**	0.050	0.06	4.8	**3.6**	**0.05**
6	**<0.05**	0.05	0.05	5.2	5	**0.05**
7	0.86	0.057	0.05	5.2	**3.5**	0.41
8	0.87	0.05	0.057	5.0	5.2	0.5
9	0.90	0.051	0.051	4.98	4.9	**1.23**
10	0.86	0.067	0.067	4.9	4.6	0.49

The most significant discrepancies from the values held as true are boldfaced. The greatest divergences correspond to the determination of arsenic, the CMPs for which were more complicated and prone to error.

As in the example of Box 8.1, a need clearly exists to institute quality systems in these analytical laboratories.

8.2 A Generic Approach to Quality

8.2.1 Integral Definition of Quality

Most people know what quality is and use this word properly in a general context. Providing an accurate definition for quality, however, is made difficult by the many facets it encompasses, which call for a combined approach to its meaning.

In **basic terms**, quality can be defined as *the body of characteristics, properties, attributes or abilities of an entity that make it better or worse than, or equal to,*

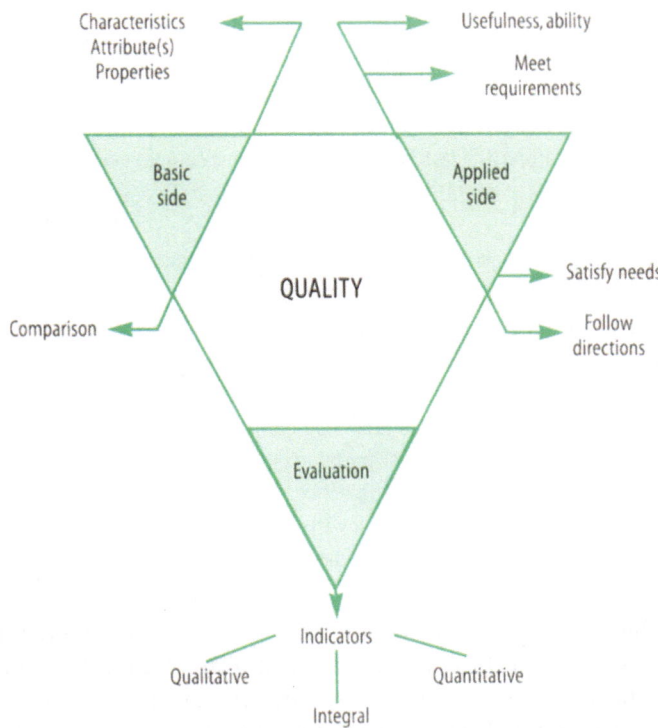

Fig. 8.1. Integral definition of quality that synthesizes its basic and applied aspects, and tools used to assess it. For details, see text

other entities of the same type. This approach, schematically depicted in Fig. 8.1, assumes that quality

(*a*) is attributed to an entity (an object, system or service);

(*b*) is unequivocally related to properties, attributes or abilities; and

(*c*) always involves comparing with a view to determining whether an entity is equal to, better or worse than another. The reference for comparison may either exist or be established through design or planning.

In **applied terms,** quality can be defined as *the totality of the characteristics of an entity that bear on its ability to satisfy stated and implicit needs imposed by a client or through legislation.* Consequently, quality in an entity involves usefulness, fulfilling requirements, satisfying needs and following clients' directions (see Fig. 8.1).

The comparisons with characteristics, requirements, needs, directions, *etc.*, involved in establishing quality call for the use of **quality indicators,** not only of the *quantitative* (numerical data) and *qualitative* (*e.g.* opinions), but also of the *integral* (qualitative and quantitative) type (see Fig. 8.1 and Box 8.3).

_____ Box 8.3

Quality Indicators

As noted in the text, quality indicators can be numerical data, opinions or a combination of both (integral indicators). The last are obviously the most comprehensive.

Defining quality via quantitative indicators alone is usually incorrect, especially in some contexts.

Thus, for a natural enclave to be properly characterized it does not suffice to ensure that the typical environmental parameters (including temperature, the air, water and soil pollutant concentrations, *etc.*) fall within the acceptable or legally established ranges; the human feeling of well-being often departs from the technical conception of "cleanliness" and "adequacy". The qualitative opinion is crucial with a view to characterizing environmental quality.

This is also the case in the agricultural food domain. It does not suffice that a given food meet specific regulations (*e.g.* lowest protein content, highest fat content, absence of additives). Its quality possesses a major subjective component that arises from its appearance, colour, flavour and other sensory properties. The judgements of so-called "sensory panels" can be considered a semi-quantitative complement of the results of food analyses.

Obviously, the three aspects that characterize quality are mutually related. The characteristics or properties compared materialize in indicators that should accurately reflect the ability, usefulness, requirements, needs or directions involved. The comparative facet is always present since, once characterized, the entity to be qualified must have its characteristics compared with the stated or implicit needs.

Quality thus defined materializes in so-called **quality systems**. A quality system can be defined in broad terms as the array of activities planned and conducted by a body in order that the generic entity may meet its clients' or users' requests. These systems possess some connotations that help one complete the generic approach to quality. Thus:

(*a*) Quality systems are not an end in themselves but a means to one: accomplishing the goals of the body that is to supply the entity.

(*b*) Quality systems make no sense unless the body concerned has clearly delineated its goals and defined the basic and applied attributes of the entity to be qualified. The comparison implicit in establishing quality is impossible in the absence of references.

(*c*) The human factor is the strongest support for quality systems; without it, they can hardly reach their goals. Quality is the concern of everybody, not only of the specific personnel in charge of its institution. Leadership, in the form of a firm commitment of executives (individual officials and boards), is also essential with a view to achieving it.

(*d*) Quality systems should never be a hindrance or an excuse for a body to avoid its generic goals (*e.g.* productivity, benefits).

(*e*) Quality systems must rely on compromises among the properties or characteristics of the entity; because some are contradictorily related, raising the levels of all at once is a chimeric goal.

(*f*) Quality systems are not constant in time; rather, they vary with the conception of quality of an entity (product, system, service) as a result of ceaseless social, economic, scientific and technical changes.

8.2.2 Types of Quality

The literature about quality is frequently confusing, partly as a result of the types of quality that arise in approaching it from various points of view not being clearly distinguished.

First, one should take into account that quality can be ascribed to the *output* (products or services) of a body or to its *organization*. The most immediate classification of quality, **external** and **internal**, is especially important with a view to avoiding so much misunderstanding in this context.

Second, quality can be assigned attributes that depend on its characteristics and on the qualifier. There are four types of quality according to this criterion, namely: (*a*) quality **designed** or **planned** by the body concerned; (*b*) quality **achieved** by it; (*c*) quality **required** and **expected** by the user or client, which should normally coincide with that planned by the body; and (*d*) quality **perceived** by the user or client.

The scheme of Fig. 8.2 synthesizes the previous two classifications. Internal quality is the quality of the body and can be designed or planned by it and achieved in relation to its product or service. External quality can be viewed in two different ways by clients; thus, it can be contractually agreed upon with the body or actually perceived by the clients themselves. The interfaces between quality types (Fig. 8.2) are very interesting as they involve the above-mentioned implicit comparisons. Thus, in pursuing internal quality (1), one seeks to match designed and achieved quality. As regards external quality (2), the client aims to obtain perceived quality of a level higher than that of expected quality; this, however, may involve increased costs for the supplier. The critical comparisons will always be those of achieved quality with expected quality (3) and perceived quality (4). In the ideal situation (total quality, 5), achieved, expected and perceived quality coincide.

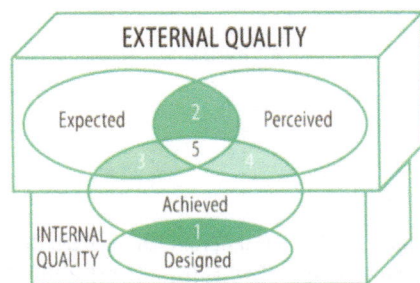

Fig. 8.2. Types of quality and interfaces between them (1 to 5). For details, see text

8.2.3 Quality Compromises

Quality should not regarded as an unreachable goal. The ideal situation is that where high levels of the intrinsic properties of the entity concerned are obtained in an expeditious, economical, safe manner. The actual situation is usually that intrinsic, economic, temporal and safety characteristics are usually mutually contradictory. These contradictory relationships are schematically depicted in the tetrahedron of Fig. 8.3. At each vertex is one of the previous performance characteristics; between each pair of vertices, the contradictions reflected on the tetrahedron edges arise. When one, two or three characteristics must be enhanced, it has to be at the expense of the rest and the tetrahedron is distorted as a result. A lower cost or more expeditiousness, for example, can never be achieved without lowering the levels of the other performance characteristics.

Quality compromises, which constitute an essential component of quality, must be clearly reflected in the different types of quality described in the previous section.

_____ Box 8.4

Quality Compromises

The mutual contradictions between the characteristics depicted in Fig. 8.3 require that those which are to be favoured at the expense of the rest be clearly established. In making the decision, one adopts a "quality compromise". Below are briefly described some typical situations.

If the quality of a product, system or service is to be maximized as regards intrinsic properties, then the client or end user will have to pay more, the process will be slower and the personnel will have to be more deeply involved in it.

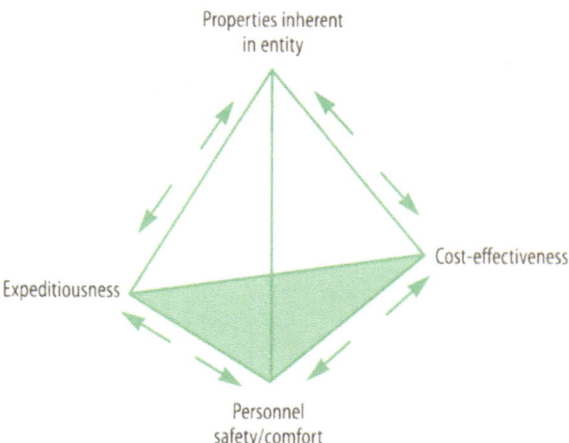

Fig. 8.3. This tetrahedron schematizes the contradictory relationships among the intrinsic characteristics of an entity, the expeditiousness with which they are achieved, the costs incurred and the environmental and health hazards involved. These relationships should be clearly reflected in quality processes. The shaded triangle represents the productivity of the body that supplies the entity

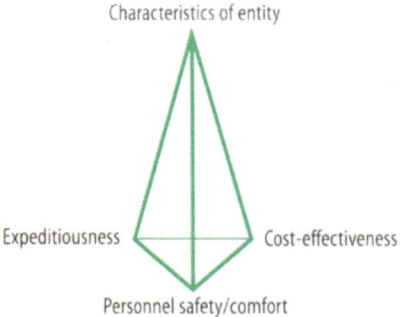

If the body concerned decides that expeditiousness should be the most substantial feature of its production or service, then some other characteristics of the entity will suffer, the cost will be higher and the personnel will be more deeply involved.

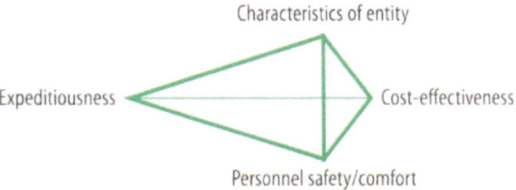

The case may also be that more than one characteristic (*e.g.* expeditiousness and some property of the entity) need be enhanced. As a result, a side rather than an edge of the tetrahedron of Fig. 8.3 will have to be expanded.

8.2.4 Structure of Quality

No doubt, a body that commits itself to achieving quality, whether internal or external, must organize itself in a structured manner. Figure 8.4 shows the cyclic cascade of decisions and actions that materialize in the policy-management-quality system-quality assurance sequence. Although a detailed description of this structure is beyond the scope of this textbook, its first and last elements warrant some comment.

The existence of a quality policy, reflected in a document passed by the body's top management, is the unavoidable starting point for the body's quality structure. Without a quality policy, the quality ship will sail courseless and any – occasionally heroic – swerve will be senseless. In practice, the quality policy translates into managerial elements and operating systems that materialize in so-called "Quality Assurance"; this can be defined as "the framework for the series of activities intended to assure the quality level planned for an entity". Assuring quality entails examining the entity (products, systems and/or services) and applying corrective actions if needed.

Quality assurance involves three well-defined types of activities that are often confused, namely: **quality control**, which is exerted by directly examining the entity via comparisons with essentially quantitative indicators; **quality assess-**

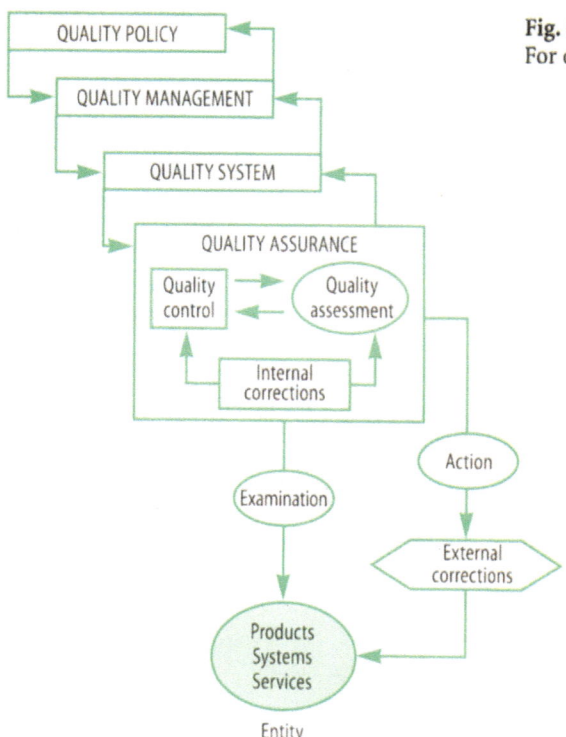

Fig. 8.4. Generic structure of quality. For details, see text

ment, which involves examining both control activities and the body's own activities, using integral indicators; and **external corrective actions**, which are an obvious consequence of the assessment of control activities.

8.2.5 Benefits of Quality

Applying quality systems to an entity yields both direct and indirect benefits.

Direct benefits are derived from client satisfaction, which in turn arises from improved quality characteristics of the product, system or service supplied. In addition, the body gains credit and prestige and generates confidence; these three elements are keys to the body's relations and future social and economic potential.

Indirect benefits of the institution of quality systems include the following:

- Planning (and documenting) of all activities.
- Rational performance.
- Minimized improvisation and indecision.
- Optimized human and material resources.
- Commitment to improvement and innovation.
- Motivated personnel.
- Job creation.

8.3 General Aspects of Quality in Analytical Chemistry

A combination of the analytical chemical principles dealt with in the preceding chapters and the generic approach to quality described in the previous section allows one to define and expand the facets of the Analytical Chemistry-Quality binomial. The aim is to answer the elementary questions of Fig. 8.5: why is the systematic institution of quality systems in the analytical domain justified?; how should quality systems be implemented?; where should they be applied?; when should they be launched?

8.3.1 Types of Quality in Analytical Chemistry

No approach to quality in Analytical Chemistry can be effective unless what is to be attributed quality is clearly established. Figure 8.6 schematizes the different facets of quality in Analytical Chemistry, ranked according to significance and scope.

One immediate distinction is that between **external quality**, which is quality as perceived by the client or end user, and as such external to the laboratory, and **analytical quality**, which helps achieve external quality by satisfying the client's or end user's information needs. These two concepts are frequently confused and a source of error in the specialized literature.

Analytical quality encompasses several elements that can be ranked according to significance and dependence. Thus, the **quality of the results** produced relies on the **quality of the CMPs** implemented, which in turn depends on the quality of the work and its organization, as well as on the **quality of the tools** used, both of the methodological (*e.g.* calibration methods) and of the material type (instruments, apparatus, reagents, standards).

This hierarchy allows one to deal systematically with the different facets of quality in Analytical Chemistry in a proper manner. Not only the significance sequence, but also the notions inherent in analytical quality, are important.

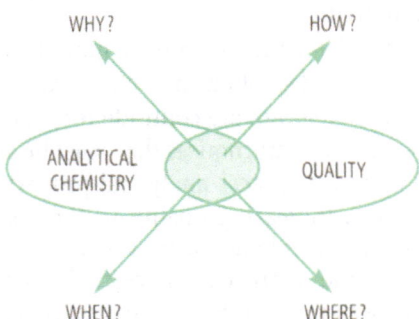

Fig. 8.5. General questions arising in approaching quality in Analytical Chemistry

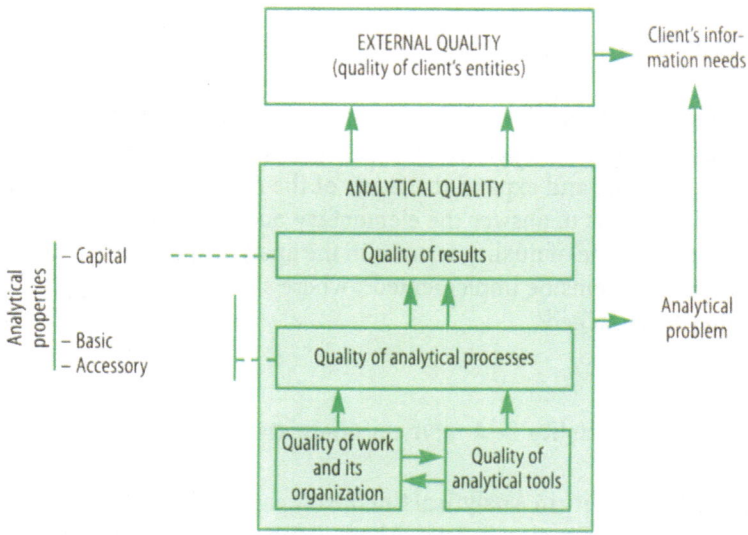

Fig. 8.6. Hierarchically ordered facets of the Analytical Chemistry-Quality binomial. The different types of analytical quality are ranked and their individual relationships to analytical properties, the analytical problem and the social or economic problem established

8.3.2 Quality and Analytical Properties

The basic approach to quality defines this term as a collection of characteristics or properties. As a result, the analytical properties described in Chap. 2 are directly related to analytical quality and allow its different facets to be characterized.

As shown in Figs. 2.1 and 8.6, quality of analytical results materializes in the two capital analytical properties, *viz.* accuracy (Sect. 2.4.1) and representativeness (Sect. 2.4.2), which must inevitably be integrated (see Fig. 2.7).

Quality of the analytical process (Figs. 2.1 and 8.6) materializes in two types of analytical properties, *viz.* basic (Sect. 2.5) and accessory (Sect. 2.6). Basic analytical properties (precision, sensitivity and proper sampling) provide support for capital analytical properties; on the other hand, accessory properties (expeditiousness, cost-effectiveness and personnel safety and comfort) characterize laboratory productivity.

The foundation, contradictory and complementary hierarchical relationships among analytical properties described in Sect. 2.7 are all essential aspects of analytical quality. Specially relevant are contradictory relationships (Sect. 2.7.3) as they give rise to the quality compromises that must be adopted in tackling the analytical problem. Figure 2.16 places analytical properties on two tetrahedra sharing a common vertex. Increasing quality in the results inevitably entails reducing the levels of the complementary properties of the CMP concerned – which will be slower and more expensive to implement, for example. If complementary properties are to be favoured, quality in the results cannot be expected to be too high.

8.3.3 Quality and the Analytical Problem

If analytical properties are related to the basic approach to quality, the analytical problem is the materialization, in the analytical domain, of the applied approach to quality, which identifies it with client satisfaction. Figure 7.10 realizes these correspondences, which are also reflected in Fig. 8.6.

Solving the analytical problem (Sect. 7.5) entails meeting the client's information needs and hence consistency between the analytical information required by the client and that supplied by the laboratory (see Sect. 7.6). Analytical quality, on which external quality – related to the client's social or economic problem – relies, will thus have been accomplished.

The basic and applied facets of analytical quality are obviously related; in fact, analytical properties are a substantial part in specifying the characteristics of the analytical information required (second step of the analytical problem-solving process) and an indispensable reference to assess the results produced (fourth step of the process).

8.4 Quality Systems in the Analytical Laboratory

The analytical laboratory can be viewed as a production body that receives raw materials (*e.g.* analytical problems, samples, reagents, standards) and delivers products (analytical information, results). Therefore, quality systems, which have been widely developed in the business realm over the past three decades, are also applicable here – following adaptation to the peculiarities of the analytical laboratory.

8.4.1 Elements of Quality Assurance

Quality Assurance as applied to a laboratory only makes sense when placed at the end of the hierarchical, cyclic chain of Fig. 8.4. Its components in the analytical chemical domain are depicted in Fig. 8.7.

Quality Assurance in Analytical Chemistry is the framework for the three components included in Fig. 8.4. Quality Control basically involves examining the laboratory and its results. Quality Assessment involves inspecting Quality Control, the laboratory, the results it supplies and their relationship to the solution to the analytical problem. Whenever needed, the outcome of Quality Assessment should include internal corrective actions within the framework of Quality Assurance.

Quality Assurance activities should lead to the implementation of corrective actions, which should initially focus on the laboratory (*viz.* on the quality of the analytical process in terms of performance and organization, and also on that of the analytical tools used); in any case, they should reflect in the quality of the analytical results and help improve the laboratory's ability to solve analytical problems.

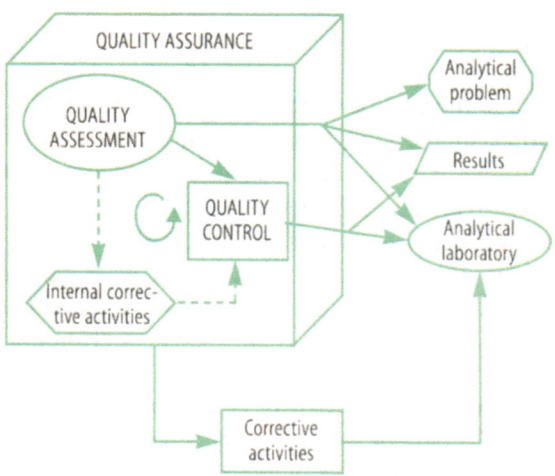

Fig. 8.7. General scheme of Quality Assurance activities in the context of Analytical Chemistry. For details, see text

Quality Assurance activities are cyclical in nature and have three time landmarks, namely: before, during and after the analytical process (see Fig. 8.8). Quality Control is performed before and during the analytical process, whereas Quality Assessment is done during and after it. Corrective actions are instituted *a posteriori* and lead to renewal or changes in control activities. All these activities are planned or designed at the beginning and subject to documenting and archiving operations; specific actions are also taken in between the two time extremes.

In practice, the quality systems applied to the analytical laboratory can be approached in different manners the most important of which are adherence to standards and compliance with legislation. Figure 8.9 shows the primary frame-

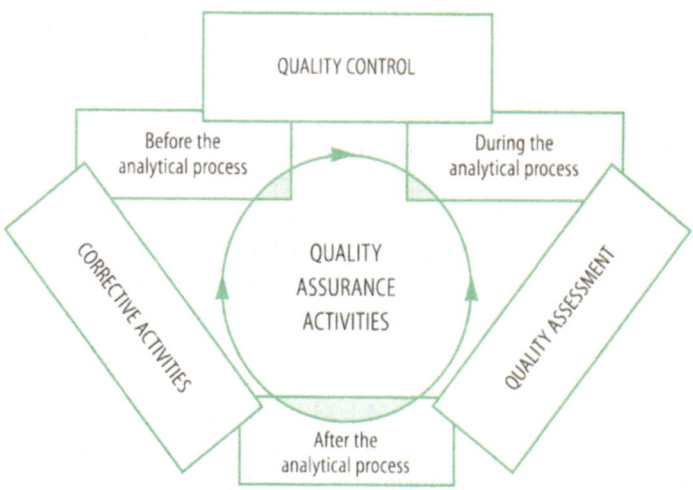

Fig. 8.8. Cyclic approach to the planning of Quality Assurance activities based on the course of the analytical process in the laboratory as the time framework

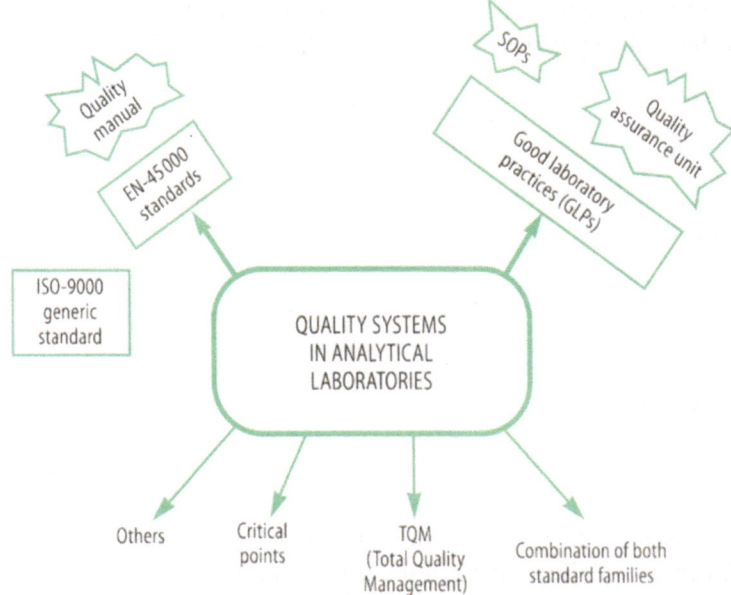

Fig. 8.9. Practical approaches to the institution of quality systems in the laboratory. The two most widely used are highlighted with bold arrows

works for development of quality systems as applied to the analytical laboratory. The most frequently used (ISO Guide 25 and Good Laboratory Practices) are discussed separately in the following sections. Others such as combinations of the principal families of standards, total quality systems and critical-point systems fall outside the scope of this textbook.

8.4.2 ISO Guide 25 (EN-45 000)

The International Standards Organization (ISO) has issued a variety of standards and guides that are directly or indirectly relevant to the implementation of quality in the analytical laboratory. The parent standards are those in the ISO 9000 series, which describe, in broad terms, the framework elements for the management of quality in the business world. At the laboratory level, these standards have materialized in ISO Guide 25 – currently being revised –, which is entitled "General Requirements for the Competence of Calibration and Testing Laboratories" and applicable to both physical and chemical metrology.

The European Centre for Standardization (CEN) has also developed a series of standards for the implementation of quality: the EN-29 000 and EN-45 000 series. Those in the EN-45 000 family, which is based on ISO 9000 and ISO Guide 25, are directly applicable to the laboratory. Each European country has had these standards translated and adapted by a national standardization body. In Spain, for example, the task was entrusted to AENOR (Spanish Association for Standardization) and the outcome was UNE-EN-ISO-25.

Box 8.5 lists the specific standards that constitute the EN-45000 family. As can be seen, they encompass every quality facet discussed in this chapter.

_____ Box 8.5

EN-45000

The European Centre for Standardization (CEN) has issued a series of standards based on ISO Guide 25 and, in broad terms, also on ISO 9000. The following are some of the more important standards in this family:

- EN-45001: General Criteria for the Operation of Testing Laboratories.
- EN-45002: General Criteria for the Assessment of Testing Laboratories.
- EN-45003: General Criteria for Laboratory Accreditation Bodies.
- EN-45004: General Criteria for the Performance of Various Types of Inspecting Bodies.
- EN-45012: General Criteria for Certification Bodies Operating Quality Systems Certification.
- EN-45013: General Criteria for Certification Bodies Operating Certification of Personnel.

The most relevant document in this quality system is the so-called **Quality Manual**, which contains a detailed description of all the characteristics of the laboratory; special emphasis is placed on Quality Assurance, which should be the reference, framework and guide for quality system activities, namely: documenting every aspect of analytical quality (organization, work, tools, materials, methods, results) and facilitating the implementation of external and internal quality assessment activities.

Compiling the Manual, which is usually based on specific standards, is a labour-intensive, expensive task; however, the Manual is indispensable with a view to the responsible, consistent institution of a quality system in the laboratory. ISO Guide 49 sets the guidelines to develop a quality manual for a testing laboratory.

The Quality Manual should be suited to the laboratory and its parent body – copying existing manuals is a frequent, but also dangerous and inadvisable temptation. However, the Manual is not for filing; rather, it should be a tool in permanent use to aid in the rigorous, consistent, efficient implementation of quality systems. It is thus an essential element of external evaluation (accreditation) of laboratories. The quality manual should be renewed whenever significant changes occur (*e.g.* when a CMP that was conducted in the classical manner at the time the Manual was compiled is automated). Box 8.6 describes the typical contents of the Quality Manual.

_____ Box 8.6

Structure of a Quality Manual

The Quality Manual is the reference and action document to be used in implementing quality systems in the laboratory within the framework of EN-45000. Its contents can vary slightly depending on the particular public or private body that endorses or supplies it. In any case, it contains the following items, all developed to an appropriate extent:

(1) The Quality Policy of the laboratory's parent body and the implementation of the hierarchical approach of Fig. 8.4.

(2) A detailed description of the laboratory.

(3) The organization and management systems.

(4) The staff, with specification of hierarchies, positions, experience, *etc.*

(5) A description of the available means (materials, instruments, apparatus).

(6) The analytical methods used, described in detail.

(7) The laboratory's ambient conditions and their maintenance.

(8) Documenting and archiving activities.

(9) Sample reception, monitoring and control systems.

(10) The systems used to verify the results, which involves checking both the analytical methods (*e. g.* with blind samples) and other aspects (*e. g.* sample-result correspondence).

(11) The systems used to diagnose generic errors and derive specific corrective actions.

(12) Other aspects such as participation in interlaboratory studies, audits, *etc.*

8.4.3 Good Laboratory Practices

Good Laboratory Practices (GLPs) constitute an alternative to the standards derived from ISO Guide 25. They have been established worldwide by institutions such as the Organization for Economic Cooperation and Development (OECD) and the European Union; also, national governments have adapted and published them in their official gazettes as imperative rules for laboratories engaged in the analysis and assessment of substances with potential social implications, which, as such, require regulation. Typical examples include pharmaceutical formulations, foods, cosmetics and environmentally aggressive products. GLPs can be defined as "a body of rules, operating procedures and practices established by a given organization (*e. g.* OECD, EU) that are considered to be mandatory with a view to ensuring quality and correctness in the results produced by a laboratory."

A detailed description of GLPs falls outside the scope of this book. Although they have a different structure, they share many practical connotations with ISO Guide 25. In addition, they contain two unique elements called Standard Operating Procedures (SOPs) and the Quality Assurance Unit (QAU).

Standard Operating Procedures are detailed descriptions of all the activities performed by a laboratory. SOPs must exist, for example, for sample handling; reagent usage; reference materials; method, instrument and apparatus supervision; archiving activities; *etc.* SOPs must be followed literally, with no alteration. If a change is pertinent, then it must be documented in written form and the actions involved be the subject of an SOP as well. Standard Operating Procedures are the equivalents of "procedures" in the Quality Manual.

One other essential element of GLPs is the so-called "Quality Assurance Unit" (QAU), which should be independent from the laboratory and answerable to the manager or president of the body with which the laboratory is affiliated. The QAU is directly responsible for instituting quality, controlling and assessing it, and proposing actions to enhance it. The unit provides an audit that its external to the laboratory but internal to its parent body (see Fig. 8.11).

8.4.4 Specific Quality Systems Used in Analytical Chemistry

Some aspects of Quality Assurance elements are directly related to analytical quality. Thus, the **methods** used by an analytical laboratory subject to (or benefiting from) a quality control system can be ranked according to accuracy, precision and institutor. Boxes 8.7 and 8.8 classify analytical methods according to quality. Only by proper, controlled application of a good method can quality in the results be achieved, which entails using the method within the framework of Quality Assurance.

_____   Box 8.7

A Primary Quantitation Method

Recently, the Consultative Committee for Quantity of Matter (CCQM) issued the following tentative definition of primary method:

A primary method is one that possesses the highest levels of metrological quality. The operations of the CMP involved must be accurately described and be understood and justified in such a way that a comprehensive, detailed expression of its uncertainty can be written in terms of SI base units. Therefore, its results should be acceptable without the need to use a reference to a standard of compositional quantity (an analytical standard containing the analyte).

A primary method must thus possess the following four essential features: (1) metrological quality, (2) an accurate description and understanding, (3) an uncertainty defined in terms of SI base standards and (4) the need for no analyte standard. Two essential requisites for a primary method in chemical metrology are specificity to an accurately defined analyte/sample couple and an accurate knowledge or the ability to calculate with a specified, acceptable uncertainty, every parameter and correction depending on other species and/or the sample matrix. According to CCQM, there are four types of chemical metrological methods with the potential to be considered primary methods, namely: gravimetry, titrimetry, coulometry and isotope dilution mass spectrometry (IDMS).

The following notes complement the previous definition of primary method:

(a) "Calculable" quantitation methods have the potential to be considered primary methods. Before they are categorized as such, however, they must be checked in other respects. Thus, whether they possess the required high metrological qualification should be confirmed via intercomparison studies of international scope at the highest quality levels. Also, a primary method requires not only monitoring of its procedure but also resting on an appropriate metrological infrastructure. For this reason, primary methods tend to be the patrimony of reference (*e. g.* national) laboratories rather than routine laboratories.

(b) No signal provided by a real sample standard containing the analyte (measurand) is required to express the result of a primary method. Coulometry and gravimetry use no analytical chemical standards; in titrimetry, the standard is the titrant, which is reacted with the analyte; in IDMS, the sample is supplied with a standard containing the pure analyte but in a different isotope ratio. This is also the case in physical metrology (*e. g.* in measuring length via the time light takes to travel the distance to be measured or in measuring temperature by means of a noise thermometer or from the Plank equation).

(c) Accomplishing an accurately defined, low uncertainty ($10^{-3} - 10^{-4}$ % in relative terms) is very difficult - and an arduous task as well. Such low uncertainty levels can be achieved if every potential source of error is carefully examined and efficiently corrected; this entails fulfilling the previous two requirements.

Box 8.8

Characterization of Analytical Quantitation Methods in the Context of Quality

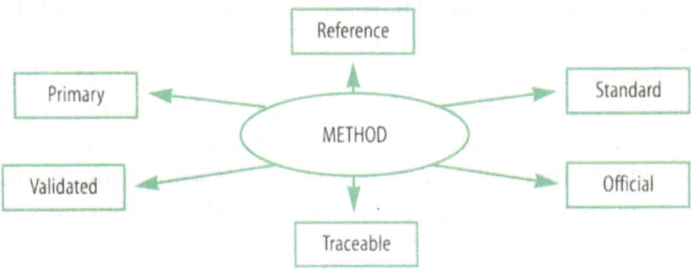

The figure above contains a series of adjectives that can be assigned to a method in order to characterize its quality. These attributes are not mutually exclusive. Also, there is a tendency to significantly reducing these choices in order to avoid the frequent confusion they raise in this context.

The definition of **primary method** is given in Box 8.7. This type of method can be placed at the top level of metrological quality based on its unequivocal relationship to base (SI) standards, its low uncertainty and the resulting proven traceability. Primary methods serve as "references" for all other methods, so they should be internationally validated via intercomparison exercises involving excellence laboratories (occasionally called "reference laboratories").

A **reference method** is one that is used to check the accuracy or uncertainty (traceability) of quantitation methods. It can be applied either by the same routine laboratory as a part of its quality control system or by another (national or international) laboratory.

A **standard method** is one that is developed, validated and published in detail by an institution engaged on standardization (*e.g.* ISO, CEN) or by an association created in support of Analytical Chemistry (*e.g.* the Association of Official Analytical Chemists, AOAC). An **official method** is one specified and issued by a governmental office (*e.g.* the US Environmental Protection Agency, EPA) for legal enforcement as the method to be used to endorse the results produced by laboratories. Many standard and official methods have no accurately defined uncertainty. In specific situations, they can be used as reference methods for quality control.

A **validated method** is one that, following detailed specification, has been subjected to a twofold study in order to establish (*a*) its internal characteristics (analytical properties), including accuracy and traceability (of its results), precision (uncertainty), sensitivity, concentration range, selectivity, expeditiousness, cost-effectiveness, *etc.*; and (*b*) its external characteristics, which materialize in its fitness for purpose (*i.e.* in its consistency with the analytical problem, Sect. 4.6).

A new designation has recently been coined in this context: **traceable method**, which is a standard, official or validated method that is clearly anchored (related) to a common reference which serves as the last link in the traceability chain. Such a reference can be a reference method (preferably a primary one), a certified reference material (CRM) or a base standard (an SI unit).

Like traceability, **validation** has lately become a buzzword even though it has always been an intrinsic, obvious concept with regard to the typical duties of the analytical laboratory. Validation is the formal demonstration that a system has performed as expected and continues to do so – or that an object has possessed given characteristics and retains them. In the analytical domain, validation is applicable to samples, data and methods. Validating a method entails formally demonstrating that it features given levels of the basic analytical parameters (accuracy, precision, sensitivity, selectivity, range, linearity, throughput, robustness) and is fit for the intended purpose within the framework of the analytical problem to be solved (see Sect. 4.6).

The **robustness** of an analytical method is a property of a high practical significance. It is related to constancy in the results in response to slight changes in the operating conditions, in such a way that the relative significance of each variable and the ranges over which it can fluctuate without altering the results in a statistically significant manner can be established. Robustness is related to precision (both change in the same direction but under different conditions), reliability (the ability to withstand slight alterations in the operating conditions with no change in the results) and transferability (the ability to lead to the same results when used by different laboratories). Robustness is essential for validation.

Quality of the results in terms of accuracy rests on two essential cornerstones of traceability, *viz.* correct, rigorous **calibration** and proper use of **reference materials**. Both are systematically discussed in Chap. 3. Chemometrics is of great help with the former. Reference materials, as primary standards, are a key to quality of results. While the availability of highly pure materials has grown dramatically over the last few years, certified real samples continue to be the greatest weakness of Analytical Chemistry as barely 5–10% of those needed are currently available. Much effort need still be devoted to obtaining certified reference materials to meet such a heavy demand.

8.5 Analytical Quality Control

Quality control activities are an essential element of Quality Assurance. They are planned, well-documented actions performed by laboratory staff and involve direct inspection of the laboratory organization, work or tools, as well as of the results produced. These activities are examined by quality assessment systems, which derive pertinent corrective actions. All should be properly recorded.

Quality Control possesses essentially quantitative connotations. As a rule, it is based on comparisons of data (preferably accompanied by their uncertainties) under the umbrella of Statistics (Chemometrics). They are thus required numerical references for quality control that are provided by data from reference materials (RMs) or certified reference materials (CRMs), the results of various methods applied to the same sample, *etc.*

The following are typical examples of quality control activities performed by analytical laboratories:

(*a*) Institution and utilization of **control charts** based on the use of RMs (see Box 3.15).

(*b*) Inspection and adjustment of **instruments** and **apparatus** in order to ensure correct performance. Equipment calibration with reference materials containing no analyte is one of the more important actions in this context (see Sect. 3.5.4.1) as it helps realize traceability.

(*c*) Examining the **purity** and **stability** of reagents and solutions used in CMPs.

(*d*) Examining the **ambient conditions** of the whole laboratory (temperature, moisture, presence of contaminants) or parts of it (*e. g.* "clean zones" for trace analysis).

(*e*) Supervising the **sample custody chain** to ensure correspondence between results and samples – and hence traceability of sample aliquots (Sect. 3.6.2).

(*f*) Using RMs and CRMs to conduct specific checking actions on a CMP. In some cases, it is advisable to use **blind samples** (*viz.* samples of unknown origin to the operator that are interspersed with those supplied by clients).

(*g*) Examining the results obtained by applying the same CMP to the same sample in order to determine the same analyte(s) following any **change in staff or tools** (*e. g.* when the chromatograph usually employed by the laboratory is replaced with one for routine analyses). In this way, the influence of a variety of factors and their implications on the analytical results are examined.

8.6 Assessment of Analytical Quality

Quality assessment activities are a substantial part of Quality Assurance and involve both direct inspection of the results and direct or indirect examination of quality control and laboratory activities. Such activities can be highly varied in nature. The most salient types and the actions they give rise to are commented on below.

8.6.1 Types of Assessment Systems

Figure 8.10 shows three classifications of Quality Assessment systems based on as many criteria (*viz.* the answers to the fundamental questions raised). The least useful of these classifications is that based on the point in time where actions are carried out (before, during or after the analytical process).

Depending on the nature of the target object and the manner it is examined, quality assessment activities can be of the qualitative, quantitative or integral type. **Qualitative assessment** involves the visual and documentary inspection of the laboratory and of the analytical processes and quality control systems it uses; the inspection comprises an *in situ* examination of the laboratory work, its

Fig. 8.10. Classifying criteria for Quality Assessment activities based on the questions what?, when? and who?

organization, the controls it performs, the documents it uses or produces (including laboratory notebooks, SOPs, primary data, results, reports, quality manual) and its archives. It additionally involves checking the results delivered against the information requested within the framework of the analytical problem. **Quantitative assessment** involves examining the metrological quality of the results obtained by subjecting certified reference materials (CRMs) to the analytical process and checking the results produced, and their uncertainties, against the certified values. This global assessment approach (see Sect. 3.9) is highly comprehensive since, if the outcome is favourable, the laboratory's work, organization and tools can be assumed to be of adequate quality. However, **integral assessment**, which is a combination of the previous two, is the most unequivocal and useful as it not only ensures quality of the results at the time of assessment, but also enables correction of defects that might lead to poor results in the future.

The classification of Quality Assessment systems based on the human factor (*i.e.* who implements them) is the most widely used. The attributes "external" and "internal" ascribed to assessment activities are a frequent source of confusion as they are used in at least two different ways (see Fig. 8.11). Quality

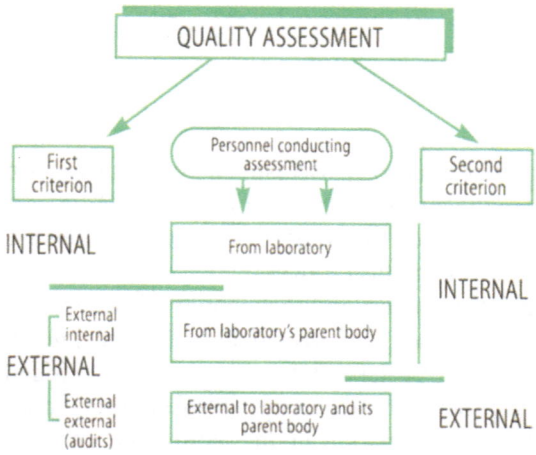

Fig. 8.11. General criteria for qualifying Quality Assessment activities as external or internal depending on the relationship of the personnel involved to the laboratory and its parent body

Assessment actions can be carried out by three different types of human groups, namely: laboratory staff, personnel from the body to which the laboratory belongs but not directly engaged in it (*e. g.* staff from the QAU established within the framework of GLPs), or individuals belonging to neither the laboratory nor its parent body. When the first criterion in Fig. 8.11 – the most widely used – is adopted, the laboratory staff sets the distinction; when the second criterion is assumed, the decision is made by the parent body. For simplicity, only the first criterion is considered here.

Internal assessment is conducted by laboratory staff concerned exclusively with laboratory quality alone or with routine tasks and assessment duties as well. Accurately, unequivocally distinguishing internal assessment and quality control activities is made difficult by the fact that both are conducted by the same personnel – no doubt, they share some characteristics. In any case, assessment activities involve more than quality control activities (*e. g.* qualitative aspects, global assessment, examining organization and management). Blind samples are a typical example of ill-defined interface between quality control and assessment – someone in the laboratory staff must ultimately introduce them among the samples to be analysed and thus act as an evaluator. Box 8.9 distinguishes these two types of activities in the context of Quality Assurance.

_____   Box 8.9

Difference Between Control and Assessment in the Context of Quality Assurance

Although quality control and quality assessment are clearly distinguished in the text and in Fig. 8.7, the boundary between these two activities is not always sharp. The figure below is intended to help distinguish them.

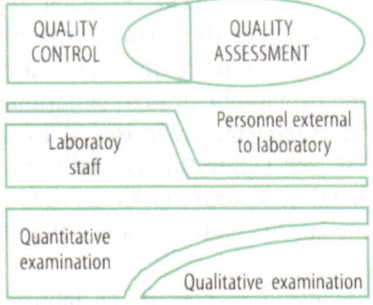

Some actions (*e. g.* the use of RMs or CMRs for various purposes) are performed both to control and to assess, so the two facets are undistinguishable.

A clearer distinction can be made on the basis of the type of examination involved and on the individuals that perform the activities. Thus, unlike Quality Assessment, Quality Control is largely carried out by laboratory staff. Also, Quality Control is of primarily quantitative nature whereas Quality Assessment has a qualitative slant (with quantitative connotations).

No doubt, **external assessment** (see first criterion in Fig. 8.11) is potentially more rational and objective. The ensuing activities are performed by individuals not attached to the laboratory or its technical management. There are two types of external assessment activities. **External-internal assessment** is carried out by staff affiliated with the body of which the laboratory is a part but not with the laboratory itself. Such staff may belong to the QAU established within the framework of GLPs when this is the system adopted or to the body's quality department otherwise.

The functions of this quality assessment group are the orientation, management and surveillance, not only of the quality control systems, but also of the rest of the laboratory. Using the established protocols and SOPs or the contents of the Quality Manual as guidelines, this personnel conducts a visual, documentary and, occasionally, also quantitative, inspection of the laboratory activities. Their work is similar to that involved in an internal audit; its institution raises major personal problems that eventually disappear as the laboratory staff gradually accept having their activities systematically examined on a routine basis. Incentives are very important in this context.

External-external assessment is conducted by experienced personnel from other bodies, so it is doubly external to the laboratory. In the quality domain, these activities are known as **audits**. Their targets can be systems (which are examined in a purely qualitative manner by visual and documentary inspection), performance (assessment of which is exclusively quantitative) or both (integrated assessment). There are two main types of analytical audits, namely: (*a*) direct audits, which materialize in laboratory accreditation; and (*b*) indirect audits, represented by proficiency testing. Both are commented on below.

8.6.2 Laboratory Accreditation

Laboratory accreditation is an external-external type of systems audit of laboratory quality. It is defined as the formal written acknowledgement that a laboratory is fit and competent to perform one or more given types of analysis. Accreditation is provided by a public or private national organization and relies on internationally established standards issued by institutions such as EU, OECD, ISO, ILAC, WELAC, *etc.* Laboratory accreditation is voluntary (conducted at the laboratory's request), temporary (it holds over a definite length of time) and partial (it refers to a specific activity rather than to the laboratory's whole work).

The accreditation process is started if the laboratory already possesses a quality system materialized in a Quality Manual. The auditor conducts a qualitative (visual and documentary) inspection and generates a report that can be appealed by the laboratory. If the report is eventually favourable, an Accreditation Certificate is issued that must be paid for by the laboratory and carries the twofold commitment of maintaining the existing quality systems and granting the auditor free access to the laboratory in order to conduct periodic controls over the period where accreditation is to hold. Renewing the accreditation (*e.g.* after the previous one has expired or the laboratory has undergone some change) theoretically entails reconducting the whole accreditation process; as a rule, however, the mutual knowledge of both parties relieves the task to some extent. The aspects to be considered during the accreditation process are very similar to the contents of the Quality Manual. Current accreditation bodies tend to perform purely qualitative audits; there is, however, a trend to instituting integral audits, which entails assessing results too and provides a more global picture of laboratory quality. Results can be audited by having the laboratory analyse certified reference materials provided by the auditor or requesting, on a regular basis, the results obtained by the laboratory in proficiency testing schemes.

8.6.3 Proficiency Testing

This is a somewhat special form of external-external quality assessment activity. The laboratory takes part in an interlaboratory study (see Boxes 8.10–8.12) where all participants analyse the same specimen for the same analyte(s) in order to have their proficiency evaluated. Through comparison, the results produced by each laboratory are assessed in a quantitative manner.

_____   Box 8.10

Interlaboratory Exercises

An interlaboratory exercise is a series of laboratory processes conducted on aliquots of the same specimen in order to have the same analytes determined independently by various laboratories. The main goal is to compare, in a critical, well-founded manner, the results produced and their uncertainties. The reference can be the mean of the results (with its uncertainty) or the reference value for the specimen (and its uncertainty) if this is an RM or CRM. For the exercise to be creditable, it should be designed, planned, executed and directed by an independent prestigious national or international body. The following figure shows two possible classifications of interlaboratory exercises.

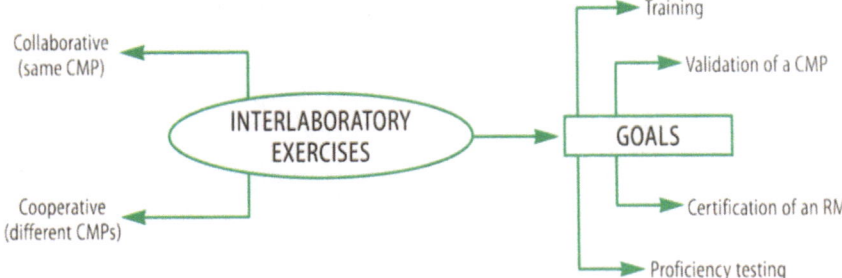

Interlaboratory exercises can be of the collaborative or cooperative type, depending on whether or not all the participating laboratories use the same method of analysis.

Interlaboratory exercises can have four basic goals, namely: (a) to train inexperienced laboratories in the implementation of CMPs; (b) to validate a given CMP developed in response to a new information request (see Section 4.6); (c) to certify data and uncertainties for a CRM (see example of Box 3.12); and (d) to assess the quality of the results produced by the participating laboratories (proficiency testing). The first goal is always present. If the sponsor creates a constructive atmosphere for discussions, where all participants should preferably be present, the quantitative information derived will be highly valuable as it could hardly be found in the analytical literature.

Participation in interlaboratory exercises is usually voluntary. However, it can be made compulsory by the Quality Manual or an accreditation scheme.

_____ Box 8.11

Improving the Quality of Results Via Interlaboratory Exercises

The benefits gained in participating in intercomparison exercises are undeniable. In addition to establishing a network of laboratories with which experience can be exchanged, the exercises provide results that improve substantially when several of them are carried out in sequence following discussion of any discrepancies and pinpointing of their sources. The European Union's BCR provides a number of applicable examples. One is described below.

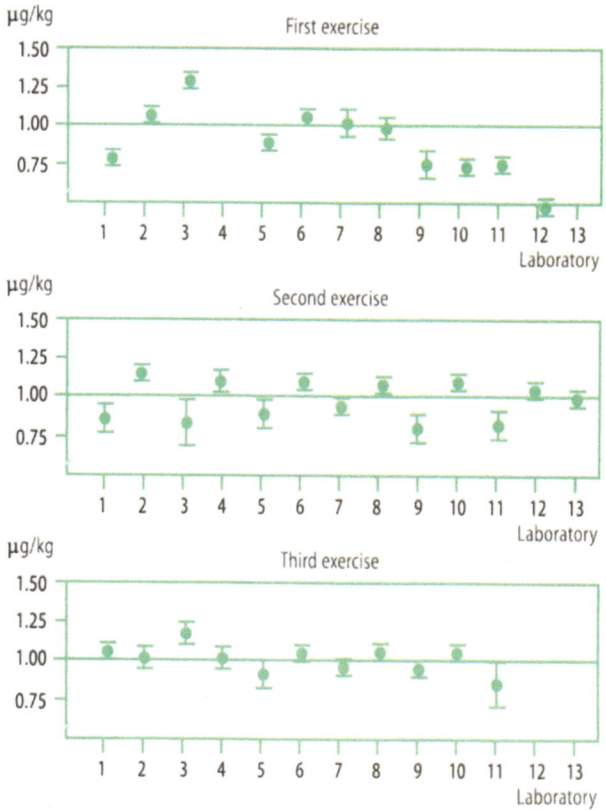

The determination of benzene retained on tenax is important because this sorbent is used to sample the atmosphere with a view to identifying contamination from aromatic organic substances. The BCR prepared an artificial sample with a benzene concentration of $1\,\mu g/kg$ and sponsored several intercomparison exercises involving thirteen laboratories. In the first round, the results were highly disperse; the underlying sources of errors were identified. The results from the second round were less disperse; however, those from several laboratories continued to be subject to a high uncertainty. Following discussion of the previous results, a third round was conducted the results from which were excellent except for two laboratories; data were close to the value held as true and subject to little uncertainty.

_____ Box 8.12

An Example of Proficiency Testing

A competent international body organized a proficiency testing scheme for laboratories concerned with the determination of cadmium traces in sea water. To this end, it previously prepared homogeneous aliquots from an RM or CRM containing an analyte concentration of 2.6 µg/L ($\hat{X}'$) after checking its stability. In requesting participation, the interested laboratories implicitly accepted the rules set by the exercise sponsor. The sponsor distributed aliquots between the laboratories, which, within the scheduled deadline, returned the individual results of i replicate determinations providing a mean $\bar{X}_i$. The z-score, defined mathematically as

$$z = \frac{\bar{X}_i - \hat{X}'}{\sigma}$$

where σ is the standard deviation of $\hat{X}'$, was calculated for each individual laboratory. z can be positive or negative, depending on the sign of the deviation of $\bar{X}_i$ from $\hat{X}'$. A graph was constructed to assess the quality of the results of each laboratory.

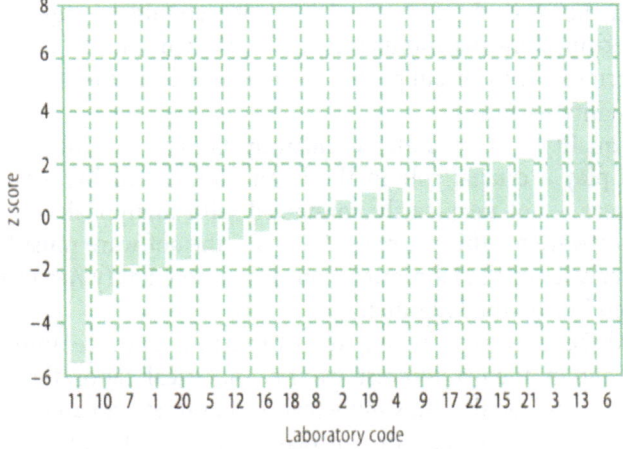

z values between $+2$ and -2 were acceptable; those greater than $+2$ or lower than -3 were questionable; and those greater than $+3$ or lower than -3 were out of control.

Unlike accreditation, proficiency testing is essentially quantitative in nature. However, the two can mutually complete their purpose. Thus, the decision to accredit a laboratory can be additionally supported by the results of the analyses of CRMs or the successful participation in proficiency testing schemes; these in turn will be more comprehensive if the potential sources of deviations are discussed in the light of information from the respective quality systems.

Other, complementary parameters provide information about the sustained performance of a laboratory through proficiency tests conducted on a regular basis.

Proficiency testing ensures quality of results as long as the sponsor is competent, creditable and completely independent of the participating laboratories. In addition, the interested laboratory should take part in the interlaboratory tests involved on a regular basis (one or twice a year), and preferably in the types of analysis it routinely performs.

Box 8.12 summarizes the performance of a proficiency test and the indicators used to assess its results.

Some authors and experts in Analytical Chemistry compare global assessment based on a CRM with participation in proficiency tests as the result (and its uncertainty) is compared with the certified value, which is also obtained in an interlaboratory study.

8.7 Supports of Analytical Quality Assurance

Quality Assurance activities in the laboratory make no sense unless they are supported by management and a consequence of the hierarchical cascade of Fig. 8.4. In addition, they should be wilfully accepted by the laboratory staff concerned, which should view quality systems not as an additional burden but as a means of working under excellence conditions. The mere willingness of those involved is not enough, however. In fact, the laboratory should be equipped with appropriate technical means and its staff trained to work in the new manner. This rests on various types of supports that are indispensable with a view to instituting Quality Assurance.

Thus, a quality system under exclusively human control is inconceivable. **Computers** play a crucial role in the institution of Quality Assurance. Some manufacturers – not necessarily connected with analytical equipment – sell computer software for this purpose. As a rule, the software must be adapted to the characteristics of the laboratory concerned, so Quality Assurance has to be implemented in a gradual manner.

The word **Qualimetrics** encompasses the triple interface among three subjects of a highly topical interest to the analytical laboratory: Computers, Chemometrics and Quality. In simple terms, Qualimetrics can be defined as the role that Chemometrics plays or can play in the establishment of Quality Assurance with the support of Computers. Note that this new concept affects two essential aspects, namely: (*a*) analytical information proper and (*b*) the optimization of analytical processes and quality systems.

All chemometric activities have analytical quality as their target. Qualimetrics focuses on those specific aspects of Chemometrics related to the institution of Quality Assurance in a laboratory that may enhance the quality of the information provided via suitable treatment of primary data and analytical results. Qualimetrics thus shares Chemometrics' primary aim of enhancing the analytical information obtained. One facet of Qualimetrics is the computer-assisted validation of analytical methods with a view to unequivocally establishing their accuracy, precision, sensitivity, selectivity, concentration range, robustness, *etc.* Validating data entails using a series of chemometric criteria to screen primary data for acceptance or rejection. As noted earlier, proper calibration is the basis for traceability of results; Chemometrics provides a firm support on which this analytical operation can be reliably and efficiently performed.

One other facet of Qualimetrics of a high interest in the context of analytical quality is **comparison of results and their respective uncertainties.** Systems for quality control and assessment rely on such comparisons. Comparing is thus a very frequent operation in this context. Chemometrics provides rigorous, reliable systems for comparing results; such systems produce not only primary information (whether results are comparable or not) but also secondary information that occasionally facilitates the identification of sources of divergence. In this way, data obtained from internal quality assessment, which are essentially quantitative in nature, are compared via control charts, changes in personnel or equipment, or the use of (usually validated) alternative methods. Chemometrics, however, also plays a prominent role in comparisons of results, the use of certified reference materials and interlaboratory studies.

Documenting and archiving activities pose the greatest hindrances to the institution and maintenance of a quality system in the analytical laboratory, and are the main source of personnel withdrawal from these processes, which involve time-consuming, tedious tasks. However, they constitute a key facet and the usual bottleneck of the Quality Assurance programme (Quality Manual). Any activity performed by an analytical laboratory under the umbrella of quality should be documented and recorded in written form. The laboratory should also record the sample custody chain, the performance of its instruments (*e.g.* gas chromatographs) and apparatuses (*e.g.* refrigerators, stoves) from the time they are installed; the monitoring of other available means; its operating procedures (SOPs), primary data, results and reports; and its self-documenting activities. This is an enormous task that calls for heavy dedication, strict discipline and rock-solid perseverance. Quality Assurance systems make no sense unless these activities are performed in a serious, efficient manner. Any document generated and archived should be ascribable to a specific individual. The system enforced should provide protection against loss, theft and tampering with. Documenting and archiving operations should encompass virtually all facets of the activities of an analytical laboratory, namely:

(*a*) primary data, results and reports;
(*b*) analytical process (sample handling, calibration and servicing of instruments and apparatus, calibration of methods);
(*c*) personnel and its characteristics; and
(*d*) Quality Assurance programme (Quality Manual, SOPs, *etc.*).

Documenting and archiving operations are completely consistent with the tracking facet of the integral concept of traceability discussed in Sect. 3.2. Establishing the traceability of results, standards, equipment (instruments and apparatus) and sample aliquots, which are mutually related in a hierarchical manner (see Fig. 3.16), also entails documenting and archiving.

8.8 Costs and Benefits of Analytical Quality Systems

Figure 8.12 illustrates the difficulties encountered in instituting quality systems in an analytical laboratory and the benefits derived from it. All are commented on below.

Although any action assuring quality could theoretically be beneficial and hence worthy of support, one should be aware that the institution and maintenance of Quality Assurance poses a series of **problems** (or risks) that should be known in order to assume the hazards involved and be able to adopt effective actions or attitudes in order to avoid mistakes. The following are some of the problems one can anticipate:

(a) The human factor is fundamental in the institution and subsequent implementation of a quality system. Those involved should be aware of the significance of this issue and address it willingly and enthusiastically. It is essential that the laboratory personnel be effectively motivated and stimulated to the extent that they become the strongest advocates of Quality Assurance. Imposition is advisable in the beginning but will be ineffective in the long run.

(b) If the laboratory's parent body has no clear-cut, well-defined goals (leadership), then any quality system instituted will be senseless and, as such, in-

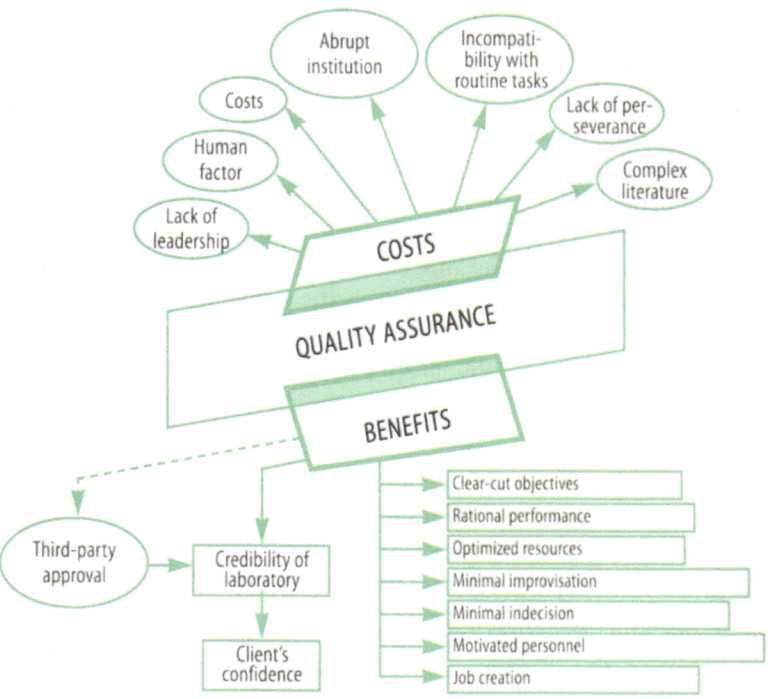

Fig. 8.12. Problems and benefits potentially arising from the institution of quality systems in the analytical laboratory

viable. Unless a quality compromise is adopted, the laboratory will be courseless. So will it if Quality Assurance does not rest on a Quality Policy. A quality system instituted merely for exhibition (*e.g.* in an accreditation process) rather than for strict development and fulfillment will be at a risk of failing.

(*c*) Implementing quality systems involves an initial investment and maintenance costs that should be considered in economic budgets. Effectively instituting Quality Assurance with a restrictive approach is virtually impossible. Some studies have estimated the additional costs to amount to 20–30% of the laboratory budget.

(*d*) Any attempt at establishing Quality Assurance in an abrupt manner is a serious mistake as it will raise opposition that will subsequently be difficult to overcome. Rather, Quality Assurance should be introduced in a gradual manner, beginning with training of staff at all levels and following with specific aspects such as the sample custody chain, SOPs, data validation, control charts, *etc.* Once these initial goals are reached, compilation of the Quality Manual can be tackled and the laboratory subjected to internal (creation of the Quality Assurance Unit within the framework of GLPs) and external audits (accreditation and intercomparison exercises).

(*e*) The implementation of Quality Assurance should never be an excuse for the analytical laboratory to avoid fulfilling its essential goal, *viz.* to generate quality analytical reports within the scheduled deadline and at the agreed cost. The duties implicit in the Quality Manual should be compatible with efficiency and expeditiousness in the laboratory's routine services and never a hindrance to them.

(*f*) A reliable quality system should be permanent and developed day by day, month by month and year after year. The initial implementation should not vanish with time. The personnel involved should be continuously assessed and encouraged so that the initial endeavour will not be worthless. Of great assistance in this context are audits (both internal and external). Special emphasis should be placed on documenting and archiving operations, which are the most costly and tedious.

(*g*) The literature on quality is atypical, abundant, occasionally contradictory and packed with abbreviations, which make it rather inaccessible to beginners. Once this initial barrier is overcome, however, one realizes that the topic is based on sensible, easily understandable premises.

Assuring the required level of quality, as a compromise between the characteristics of the results and laboratory productivity, provides a number of **benefits** that are equivalent to the (basic and complementary) objectives of Quality Assurance. The essential, most immediate benefit of Quality Assurance is that it provides well-grounded support for the credibility of the information supplied by the laboratory. Figure 8.13 illustrates the transparency and credibility gained by introducing an independent body in relations between organizations via the results produced by their respective laboratories. Indirect ternary relations, called in generic terms "third-party approval", involve mutual

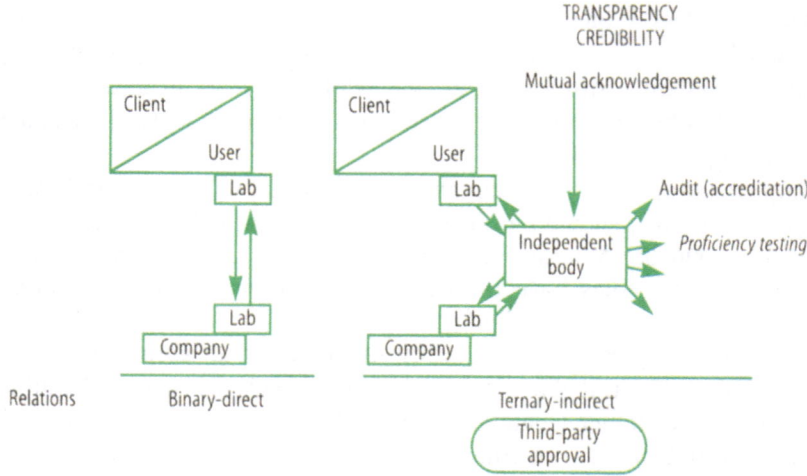

Fig. 8.13. Changes involved in instituting Quality Assurance in laboratory–client relations. For details, see text

acknowledgement and result in more fluent and extensive business relations than do direct binary relations. The former improve the external and internal image of the laboratory's parent body and inspire confidence from the client or end user.

In addition to this basic advantage, other, more specific aspects benefit from the institution of Quality Assurance. The following are some of the more salient:

(*a*) Work has to be rationalized by eliminating superfluous, redundant operations in response to the need to carefully plan the laboratory activities in compiling the Quality Manual. Any defects or mistakes arising during daily activities will thus be clearly exposed.

(*b*) Establishing and specifying goals is facilitated, hesitation is minimized and human relations are made more fluent.

(*c*) Extensive documentation is gathered that is highly valuable with a view to making educated decisions about the internal organization of the laboratory and facilitating achievement of the goals of its parent body.

(*d*) Training and encouragement for staff at all laboratory levels is provided.

(*e*) New jobs (*e. g.* members of the Quality Assurance Unit within the framework of GLPs) are created.

Questions

(1) How are the basic and applied sides of quality related?

(2) Why does accomplishing quality always involve some comparison?

(3) Comment in detail on the tabulated data in Boxes 8.1 and 8.2.

(4) What are the references used to establish, in broad terms, the quality of an entity?

(5) What tools are used to establish quality? Are they all the same?

(6) What constitutes "leadership" in the context of quality? How is its establishment influenced by the so-called "human factor"?

(7) Define some qualitative and quantitative indicators for the quality of a product (*e.g.* wine) and/or service (*e.g.* a restaurant).

(8) Why can a situation where the quality perceived is greater than that expected (by the client) be unfavourable to the body that supplies the product or service?

(9) How can "total quality" be defined?

(10) Distinguish external quality from internal quality and establish their mutual relationships via specific examples such as a fish farm, a governmental environmental agency, an analytical laboratory and various others.

(11) What are "quality compromises"? Give typical examples from various fields.

(12) What are the cornerstones of the quality structure of an entity?

(13) What activities are used to inspect Quality Control activities?

(14) Justify some of the benefits gained by a body as a result of implementing quality systems.

(15) Why are quality systems not an objective *per se*?

(16) Provide a generic answer to the basic questions that arise from the Analytical Chemistry-Quality binomial (Fig. 8.5).

(17) Which Quality Assurance elements examine consistency between the information produced and that expected in the context of the analytical problem?

(18) Distinguish between external and internal corrective actions within the framework of Quality Assurance.

(19) Explain the twofold examination of the results required in the context of Quality Assurance.

(20) Which Quality Assurance elements examine the analytical laboratory?

(21) Comment on the cyclic nature of Quality Assurance activities in the analytical chemical domain.

(22) On what types of quality does quality of analytical results rest? On what does quality of the analytical process?

(23) What are the elements of analytical quality?

(24) Systematically compare an analytical laboratory and a production factory in terms of raw materials, systems, intermediate products, finished products, clients, *etc.*

(25) Why do standards influence quality systems as applied to the analytical laboratory?

(26) What is a Quality Manual? What is its purpose?

(27) What are SOPs? In what context are they used?

(28) What is a primary method? How does it influence analytical quality?

(29) What is the difference between an official method and a standard method?

(30) Connect quality compromises to relationships among analytical properties.

(31) Relate the vertices of the following triangle:

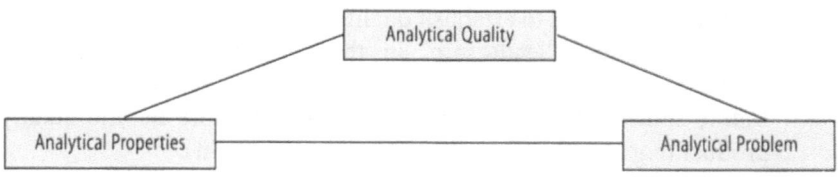

(32) Place quality of standards in the context of analytical quality.

(33) What are Quality Control activities?

(34) What is the difference between collaborative and cooperative interlaboratory exercises? How do they relate to their four objectives?

(35) What are EN-45000 and ISO 9000 standards?

(36) What are the essential differences between Quality Control and Quality Assessment?

(37) What are control charts? Where do they fall in the Quality Assurance context? What is their use?

(38) Relate traceability to the different facets of analytical quality.

(39) Are GLPs compulsory? Is accreditation?

(40) Why are the external and internal facets of Quality Assessment activities a source of confusion?

(41) What is the purpose of interlaboratory exercises? Where do they fall in the context of analytical quality?

(42) What is the meaning of "third-party approval"?

(43) What are "Good Laboratory Practices"? What are their most salient features as quality systems?

(44) Why can the quality of a laboratory be expected to rise through participation in an intercomparison exercise?

(45) Why are documenting and archiving activities a bottleneck in instituting quality schemes in a laboratory?

(46) What parameter is used to characterize the quality of the results of a laboratory in a proficiency test?

(47) What is external-external assessment? Give some examples and distinguish them from others of external-internal assessment.

(48) Where does the Quality Assurance Unit (QAU) fall in the context of Quality Assessment? Why?

(49) Of all the problems potentially encountered in instituting quality, which are the most relevant? Why?

(50) Give a specific example illustrating the different types of advantages gained by a laboratory and its personnel in working under the umbrella of quality.

Seminar 8.1

Quality Control Activities

- **Learning objective**
  To illustrate, via several examples, how analytical quality can be internally controlled by processing the quantitative data obtained in experiments conducted to this end.
- **Planning the Seminar**
  Compare the results of the different experiments conducted under the umbrella of Statistics. Use the parameters described in Chap. 2.
- **Exercises**
  (A) CONTROL CHARTS (see Box 3.15). An RM (or CRM) with a value $\bar{X} \pm U$ is subjected to the routinely applied CMP on a daily basis. Plot the results and identify the warning and action limits.
  (B) BLIND SAMPLES. Insert previously analysed samples or CRMs similar to the ordinary matrix in the sample set analysed by the laboratory each day. Compare the previously obtained results for the blind sample with those provided by the laboratory. Is it advisable that the analyst know what is being analysed?
  (C) PERIODIC GLOBAL VALIDATION OF THE CMP. Subject a CRM similar to the samples received by the laboratory to the CMP. Systematically compare the accuracy and precision achieved.
  (D) CHANGE THE OPERATOR OR INSTRUMENT. Using the same sample, compare the results obtained by having the CMP conducted by different analysts or using different analytical tools of identical characteristics. Discuss the conclusions reached if the results produced are disparate.

Seminar 8.2

Regulatory Frameworks for Implementation of Analytical Quality

- **Learning objective**
  To become acquainted with the use of the framework standards most widely used in connection with quality of analytical laboratories.
- **Material**
  - ISO 9000 standards.
  - ISO Guide 25. General Requirements of Competence for Testing and Calibration Laboratories.
  - EN-45000.
  - GLPs:
    - Decision of 13 May 1981 of the OECD Council about GLPs.
    - EU Directives 87/15/CEE and 88/320/CEE (European Communities Official Diaries of 18/12/87 and 6/11/88, respectively).

■ **Planning the Seminar**
- Definition of standard. Types of standards: generic and specific.
- Comment on the framework nature of ISO 9000 standards.
- Comment on EN-45000 in direct relation to laboratory quality.
- Comment on GLPs and the official publications through which they are issued.
- Discuss:
  - The significance of the regulatory framework.
  - Compulsoriness: voluntary and "forced" compliance.
  - Difficulties encountered in directly applying this regulatory framework to chemical metrology.
  - Advantages and disadvantages of the regulatory framework in relation to analytical quality.
  - Difference between "accreditation" and other processes such as "certification" and "homologation".

Seminar 8.3

Discussing the Results of a Proficiency Test

■ **Learning objective**
- To encourage students to take part in a simulated proficiency test.
- To emphasize the significance of interlaboratory exercises in general, and proficiency testing in particular, with a view to improving analytical quality.

■ **Literature**
- "Proficiency Testing in Analytical Chemistry", R.E. Lawn, M. Thompson and R.F. Walker, *Royal Society of Chemistry*, London, 1997.
- ISO Guide 43 (1984): "Development and Operation of Laboratory Proficiency Testing".

■ **General discussion**
Comment on the following aspects:
(1) Credibility of the sponsor.
(2) Preparation and characteristics of the material.
(3) Participation: instructions, submitting the material, using one or several methods, delivery of results, costs, *etc.*
(4) Processing the results: indicators used.
(5) Conclusions: qualification and ranking of laboratories.
(6) Continuity: performance indicators.
(7) Advantages and disadvantages of proficiency testing.

■ **Practical approach**
Thirty laboratories took part in a proficiency test organized by a non-profit excellence laboratory. Participants were supplied with a food sample with a fat

content of 6.65% (w/w) and asked to determine total fat matter in it. The two methods used (A and B) provided the following results:

- Method A: 5.10, 5.23, 5.24, 5.31, 5.32, 5.38, 5.38, 5.39, 5.49, 5.50, 5.59, 5.70, 5.71, 5.79, 5.91.
- Method B: 5.41, 5.61, 5.63, 5.63, 5.65, 5.67, 5.69, 5.69, 5.70, 5.70, 5.70, 5.71, 5.85, 5.95, 6.15.

■ **Discussion**
  - Calculate the z-score for each participating laboratory.
  - Compare the accuracy of the two methods.

Seminar 8.4

Accreditation of an Analytical Laboratory

■ **Learning objective**
To describe in detail the process followed to accredit an analytical chemical laboratory and discuss the impact of accreditation on analytical quality.

■ **Literature**
  - CGA-ENAC-LE (1997). General Criteria for Accreditation. Technical Qualification of Testing Laboratories.
  - PE-ENAC-LEC-01 (1997). General Procedure for Accreditation of Testing and Calibration Laboratories.
  - G-CGQ (1994). Guide to the Accreditation of Laboratories Conducting Chemical Tests.
  - ISO Guide 58. Calibration and Testing Laboratory Accreditation Systems. General Requirements for Operations and Recognition.
  - "Accreditation and Quality Assurance in Analytical Chemistry", H. Günzler (Ed.), *Springer*, Berlin, 1994.

■ **Topics for debate**
  - Characteristics of the accreditation body.
  - Documents pertinent to the process.
  - Requirements to be eligible for accreditation.
  - Accreditation process.
  - Auditors.
  - Rights and duties of an accredited laboratory.
  - Post-accreditation actions.
  - International bodies concerned with voluntary and "forced" accreditation.
  - Nature of audits; qualitative and/or quantitative.
  - Benefits of accreditation.
  - Accreditation in the context of analytical quality.

Suggested Readings

(1) "La Calidad en los Laboratorios Analíticos", M. Valcárcel and A. Ríos, *Reverté*, Barcelona (Spain), 1992.
   One of the earliest books in Spanish to provide a systematic view of the Analytical Chemistry-Quality binomial.

(2) "Analytical Chemistry and Quality", M. Valcárcel and A. Ríos, *Trends Anal. Chem.*, 1994, 13, 17.
   This paper can be viewed as the backbone for this chapter. It contains its definitions and considers the elements of analytical quality assurance.

(3) "Quality Assurance of Chemical Measurements", J.K. Taylor, *Lewis Pu.*, Michigan, 1987.
   This book is a landmark in the establishment of analytical quality. It deals with every aspect of it and continues to be a mandatory reference for the topics discussed in this chapter.

(4) "Las Buenas Prácticas de Laboratorio", J. Sabater and A. Vilumara, *Díaz de Santos*, Madrid (Spain), 1988.
   Discusses, in a systematic manner, the principles of GLPs and their practical implementation. Based on the authors' own experience and highly useful.

(5) "Quality Control in Analytical Chemistry", G. Kateman and L. Buydens, *Wiley*, New York, 1993.
   A comprehensive book that describes the chemometric principles on which quality control relies.

Glossary of Terms

Absolute error	S. "Error".
Absolute method	S. "Method".
Absolute trueness	An ideal analytical attribute of the intrinsic information latent in an object or sample. Related to the true value (Chap. 2, pp. 42, 44, 45; Fig. 2.5).
Accessory properties	S. "Analytical properties".
Accreditation	Formal written acknowledgement that a laboratory is fit and competent to perform one or more given types of analysis. Obtained by subjecting the laboratory to an audit conducted by personnel external to the laboratory and its parent body (Chap. 8, p. 322, 330–333, 337, 340, 342, 343).
Accuracy	The degree of consistency between a result, or the mean of a set of results, and the value held as true. A capital analytical property of a result or CMP (bias) (Chap. 2, pp. 40, 42, 44–46, 50–57, 65, 68, 69, 74, 76, 77, 79, 83–86, 89–91, 93).
Aliquot	A well-defined portion (mass, volume) of a sample.
Analyse, to	To subject a sample to an analytical process in order to extract information about measurands or analytes. To interpret analytical results with a view to producing a report. Top level in the information hierarchy of Fig. 1.15 (Chap. 1, pp. 1, 17, 23).
Analyser	An integrated system consisting of instruments, apparatuses and devices that performs virtually the whole analytical process (CMP). Top level in the hierarchy of Fig. 1.17 (Chap. 1, pp. 24–26).
Analysis	In a general sense, examination, study, acquisition of knowledge. Intended to provide information about objects, facts, systems, performance and attitudes (Chap. 1, pp. 2, 5, 6, 12, 17, 26, 28–31, 33). In the chemical field, the action of subjecting a sample to an analytical process in order to extract chemical information about it. Top level in the hierarchy of Fig. 1.18 (Chap. 1, pp. 28–30).
Analyte	A chemical or biochemical species in a sample about which qualitative or quantitative information is required (Chap. 1, pp. 4, 27, 31; Box 1.12).
Analytical chemical standard	S. "Standard".

Analytical Chemistry	A metrological science that develops, optimizes and applies measurement processes intended to derive quality (bio)chemical information about natural or artificial objects or systems with a view to solving analytical problems (Chap. 3, pp. 101–103, 133, 137, 141).
Analytical developments	Generic processes undertaken for various purposes in response to stimuli from inside and outside Analytical Chemistry (Chap. 1, pp. 15–17; Fig. 1.11).
Analytical error	S. "Error".
Analytical fundamentals	The cornerstones on which the theoretical and practical sides of Analytical Chemistry stand. Intrinsic to Analytical Chemistry or shared with other scientific and technical areas (Chap. 1, p. 13–14).
Analytical information	Chemical characteristics of an object or system, usually ascribed to its components (analytes) or the entity as a whole. The opposite of generic uncertainty and the primary goal of Analytical Chemistry (Chap. 1, pp. 6, 7, 9, 10, 21–26, 34, 35).

Information held as true. That obtained through special testing (*e.g.* an interlaboratory exercise) or that possessed by a CRM. Corresponds to referential quality.

Intrinsic information. That possessed by the object or system to be analysed. Corresponds to ideal quality.

Routine information. That ordinarily produced by laboratories. Corresponds to real quality.

Analytical method	S. "Method".
Analytical problem	An approach adopted to solve the client's information needs, via the design and planning of a CMP, and to interpret the ensuing results (Chap. 1, pp. 3, 12–16, 18, 22, 26, 28; Chap. 7, pp. 283–302, 303).
Analytical process	The body of operations that separate the uncollected, unmeasured, untreated sample from the analytical results. Second step in the hierarchy of Fig. 1.16 (Chap. 1, pp. 6, 9, 11–15, 22–24, 28, 31–33; Chap. 4, pp. 145, 146, 148, 149, 161, 165, 177, 182, 189, 194; Fig. 4.2).
Analytical properties	Attributes ascribed to results and/or a chemical measurement process (CMP). The quality indicators of Analytical Chemistry (Chap. 2).

Accessory properties. Those that can be ascribed to a CMP and define laboratory productivity (Chap. 2, pp. 40, 56, 73, 77, 81).

Basic properties. Those that can be ascribed to a CMP and support capital properties (Chap. 2, pp. 40, 41, 77, 81).

Capital properties. Those that can be ascribed to results (Chap. 2, pp. 40, 41, 52, 76, 77, 88).

Analytical quality   S. "Quality".

Analytical references   Landmarks used in the comparisons inherent in analytical measurements. Can be materials (standards) or methods (Chap. 1, pp. 10–11).

Analytical reports   The body of analytical results (data) and their interpretation in the light of the analytical problem concerned. Top level of the information hierarchy (Chap. 1, pp. 22).

Analytical results   Qualitative and quantitative data obtained by mathematical (chemometric) treatment of the primary data generated by an instrument in the analytical process. Second step in the information hierarchy (Chap. 1, pp. 10, 34; Fig. 1.15).

Analytical schemes   Sequential, orderly processes that use separation methods (*e. g.* precipitation) in Classical Qualitative Analysis to divide species into groups where each analyte can be reliably identified (Chap. 5, pp. 219, 221–223).

Analytical tools   Material, strategic and methodological means of varied nature used in chemical measurement processes (CMPs) (Chap. 1, pp. 12, 13, 15, 16, 24, 34).

Apparatus   A system consisting of devices that serves a specific function in a CMP but provides no analytical information. Produces secondary data. Third step in the hierarchy of Fig. 1.17 (Chap. 1, pp. 3, 4, 13, 24–26, 35).

Automation   Partial or total reduction of human participation in a CMP (Chap. 4, pp. 175, 176, 195, 200).

Avogadro's number   A chemical standard defined as the number (6.023×10^{23}) of atoms or molecules contained in one mole of any chemical substance (Chap. 3, pp. 110, 111).

Balance   An instrumental tool primarily used to measure the initial mass of the test sample to be subjected to a CMP or that of the weighed form in gravimetry (Chap. 4, pp. 147, 149, 161, 169, 170; Chap. 6, p. 249, 250, 258, 259, 261).

Base standard   S. "Standard".

Basic properties   S. "Analytical properties".

Bias   A systematic or determinate error ($\bar{X} - \hat{X}'$) that can be ascribed to an analytical method. Related to the accuracy of the method (Chap. 2, pp. 50, 52, 53, 91, 93).

Binary response	The result (YES or NO) of a qualitative analysis. The answer to various questions the most crucial of which are "is it the analyte?" and "is it in the sample?" (Chap. 5, pp. 201–210, 212, 213, 230, 240, 243, 244).
Black sample	S. "Sample".
Blank	A usually artificial sample containing no analyte. In theory, it should give no signal; if it does, it is called the "blank signal" (Chap. 2, pp. 56, 67, 68, 70, 81, 97–100).
Blind sample	S. "Sample".
Bulk sample	S. "Sample".
Burette	An instrumental tool used in titrations to measure the volume of titrant solution used. Manual or automatic in operation (Chap. 6, pp. 249, 250, 264, 265, 267–271).
Calculable method	S. "Method".
Calibration	The set of operations that establish the relationship between the signals provided by measurement standards and (a) the response corresponding to proper functioning of an instrument or (b) the concentration of the analyte.
	Equipment calibration. The process by which a standard containing no analyte is used to check that an instrument (or apparatus) operates as expected. Otherwise, corrections are introduced until the instrumental response (or the indication of an apparatus) reaches the value held as correct for the standard used in the operation (Chap. 3, pp. 123, 125, 126, 137, 139, 140; Fig. 3.10).
	Method calibration. The process by which an analytical standard is used to characterize the response of an instrument in terms of the properties of an analyte or analyte family. Relies on the establishment of an unequivocal relationship between the signal and the presence or concentration of the analyte (Chap. 3, pp. 120, 123–128, 130, 135, 137–140; Fig. 3.10).
Calibration curve	A two-dimensional graphical plot that shows the variation of the analytical signal with the concentration of analyte (standard) (Chap. 2, pp. 62, 66–70, 72, 81, 83, 88, 99).
Capillary electrophoresis	A separation process occurring within a capillary under a high electric field. Because the system is equipped with a detector, it can be considered an instrument (Chap. 5, pp. 214, 230, 236, 237; Box 5.17).
Capital properties	S. "Analytical properties".

Certified reference material
A reference material with certified values (specific uncertainties included) for one or more of its properties that are obtained by special procedures (*e.g.* interlaboratory exercises) under the supervision of a competent, independent organization. Should be accompanied by comprehensive documentation (Chap. 3, pp. 107, 114, 120, 130, 134, 142).

Characterize, to
To establish distinct features of an object or system from analytical results. Second level in the information hierarchy of Fig. 1.15 (Chap. 1, pp. 7, 23).

To establish the fitness for purpose, reactivity and macroscopic stability of the interaction or mobility of components at the microscopic level of the object of a chemical analysis (Chap. 6, p. 247; Fig. 6.1).

Chemical analysis
A process by which chemical measurement processes (CMPs) are used to extract information from objects or systems (Chap. 1, pp. 5, 12, 14, 34; Box 1.12).

Classical analysis. A type of qualitative and quantitative analysis based on chemical reactions in solution and involving the use of human senses for identification and a balance or burette for quantitation (Chap. 1, pp. 6, 30, 35).

Instrumental analysis. A type of qualitative, quantitative and structural analysis based on the use of instruments other than the balance, burette and human senses (Chap. 1, pp. 6, 30, 31, 35).

Macro analysis. A type of chemical analysis where the initial size of the sample aliquot subjected to the CMP is greater than 100 mg (Chap. 1, p. 32; Fig. 1.23).

Micro analysis. A type of chemical analysis where the initial size of the sample aliquot subjected to the CMP ranges from 10 to 1 mg (Chap. 1, p. 32; Fig. 1.23).

Qualitative analysis. A type of chemical analysis by which the analyte or analytes in a sample are identified. The result is a YES/NO binary response (Chap. 5, pp. 201–203, 208–210, 212–218, 220, 225, 229, 230, 234, 236, 240–242, 244, 245).

Quantitative analysis. A type of chemical analysis by which the proportion or amount of each analyte in a sample is determined. The result is a numerical response (Chap. 1, pp. 28–30; Chap. 6, p. 247–250, 253, 276–279, 282).

Structural analysis. A type of chemical analysis by which the structure of a sample (*viz.* the spatial distribution of its constituents) or a pure analyte is established (Chap. 1, p. 29).

Trace analysis. An analytical process specially suited to the identification or quantitation of analytes present in proportions below 0.01% (100 ppm) in the sample (Chap. 1, pp. 33, 35, Fig. 1.24).

Ultramicro analysis. A type of chemical analysis where the initial size of the sample aliquot subjected to the CMP is less than 1 mg (Chap. 1, pp. 32; Fig. 1.23).

Chemical measurement process (CMP) — S. "Analytical process".

Chemical metrology — The science of (bio)chemical measurements (Chap. 3, pp. 106, 109, 112, 115, 141, 142).

Chemical standard — S. "Standard".

Chemometrics — A multidisciplinary entity that aims to expand and enhance analytical chemical information by using less material, time and human resources (with the aid of computers).

Chromatograph — An analytical system that performs column chromatographic processes (gas or liquid chromatography) and possesses an on-line detector for continuous monitoring of the fluid that emerges from the separation column. Because it provides analytical information, a chromatograph is an instrument (Chap. 4, pp. 145, 148, 149, 151, 163–165, 167, 170, 171, 186, 187, 195).

Chromatographic separation — S. "Separation".

Chromatography — A word that describes a broad range of a highly efficient analytical separation techniques based on multiple mass transfer between a mobile phase and a stationary phase (Chap. 4, pp. 168, 188, 190).

Gas chromatography. A chromatographic technique where the mobile phase is a gas (into which the sample aliquot is inserted) and the stationary phase is a solid or a liquid supported on an inert solid that is placed in a column (Box 2.16).

Liquid chromatography. A chromatographic technique where the mobile phase is a liquid (into which the sample aliquot is inserted) and the stationary phase is a solid or a liquid that is either supported on an inert solid which is placed in a column or spread onto a thin layer of a support (in Thin Layer Chromatography) (Box 2.16).

Classical analysis — S. "Chemical analysis".

Clean-up — The process by which interferents in a sample are removed using a separation technique to indirectly enhance selectivity (Chap. 4, pp. 165, 194; Fig. 4.15.2).

Client	A generic designation applied to a body or individual requiring chemical information with a view to solving a social or economic problem.
Coefficient of variation	Relative standard deviation, expressed as a percentage (Chap. 2, pp. 48, 58, 82).
Comparative method	S. "Method".
Composite sample	S. "Sample".
Concentration	An expression of a relative quantitative result: the amount of analyte in a given volume or mass of sample (Chap. 6, pp. 248, 250, 252, 254–256, 259, 260, 262–269, 271–278, 280, 281). **Cut-off concentration.** The concentration chosen by the analyst in establishing a given probability level that the binary response will be correct (Chap. 5, pp. 206, 207, 211, 212, 240). **Threshold concentration** or **limiting concentration.** Highest or lowest level, established by the client or legislation, to be used in deciding whether a sample or object warrants assignation of a given attribute (Chap. 5, pp. 206–208, 240).
Convenience sample	S. "Sample".
Cut-off concentration	S. "Concentration".
Data processing	The body of mathematical calculations leading to the expression of the analytical result from tabulated data (chemical standards, constants, conversion factors) and test data produced by a CMP applied to the sample and standards (Chap. 4, pp. 144, 161, 171; Fig. 4.17).
Detect, to	To have an instrument produce a signal and transduce it into a readily measured physical quantity (a primary datum). Third step in the information hierarchy of Fig. 1.15 (Chap. 1, pp. 23, 28).
Detection	The action of detecting (Chap. 1, p. 21, 23, 28). The process of measuring for qualitative purposes (Chap. 5, pp. 202, 204, 205, 207–211, 213, 220, 230, 234, 236, 240; Box 5.1).
Determinate error	S. "Error".
Determination	The process by which the amount or concentration of an analyte (or analyte family) in a sample is established. The central level in the hierarchy of Fig. 1.18 (Chap. 1, pp. 5, 16, 21, 24, 26, 28, 31–35).
Deviation	The difference between an individual result in a set and the mean of the set (random error) (Chap. 2, pp. 47, 48, 50, 57–59, 63, 65, 67, 81, 82, 93, 94, 99).
Device	A part of an apparatus, instrument or analyser that can serve one or more of a wide variety of functions.

	Last step in the hierarchy of Fig. 1.17 (Chap. 1, pp. 23–26).
Dialysis	The process by which mass transfer between two miscible liquid phases separated by a membrane permeable to the analytes or their interferents takes place (Chap. 4, pp. 152, 164).
Disaggregation	A substep of the preliminary operations of a CMP involving the fusion of an insoluble solid sample mixed with a solid reagent (Chap. 4, pp. 146, 152, 164).
Dissolution	A substep of the preliminary operations of a CMP where a solid (or semi-solid) sample is completely dissolved by treatment with a solvent (Chap. 4, pp. 146, 151, 161–163).
Electrodeposition	A gravimetric method carried out by an electrochemical device in order to deposit the analyte quantitatively onto one of the electrodes (usually the cathode), which is weighed before and after the process (Chap. 6, p. 260).
End-point	In a titration, the volume of titrant added to the solution containing the analyte or standard by the time the indicator system produces the signal in response to which the titration should be stopped (Chap. 6, pp. 250, 263, 264, 266–271, 277).
End-point indicator	A system of the visual (Box 6.10) or instrumental type (Box 6.11) that exposes the end-point of a titration (Chap. 6, p. 269).
Equipment calibration	S. "Calibration".
Equivalence point	In a titration, the theoretical volume of titrant required to react in a quantitative, stoichiometric manner, with the analyte or standard (Chap. 6, p. 264).
Error	The difference between an observed or calculated value of a quantity and its assumed true value.
	Absolute error. The difference between the value to be qualified and its reference (Chap. 2, p. 45).
	Analytical error. In a generic sense, an alteration in the analytical information supplied. Of the random, systematic or gross type (Chap. 2, p. 45).
	Determinate error. S. "Systematic error".
	Gross error. A large systematic error (Chap. 2, pp. 50, 53; Box 2.1).
	Indeterminate error. S. "Random error".
	Negative error. A negative difference between the value to be qualified and the reference used to establish it (Chap. 2, pp. 50, 53, 70).

Percent error. The result of multiplying a relative error by 100.

Positive error. A positive difference between the value to be qualified and the reference used to establish it (Chap. 2, p. 45).

Random error. An error that can be ascribed to positive or negative (random) fluctuations typical of testing operations. The basis on which precision and specific uncertainty are established. Also called "indeterminate error" (Chap. 2, pp. 46, 50–53, 58, 59, 79, 93, 97; (Box 2.1).

Relative error. The ratio of an absolute error to the reference used to establish it (Chap. 2, p. 91).

Sampling error. An error that results from incorrect extrapolation of an individual result or the means of those for several aliquots to an object as a whole and detracts from representativeness in the collected samples. Accidental, systematic or random in nature (Chap. 4, pp. 153, 158, 160, 161).

Systematic error. An error ascribed to well-defined operational alterations in a CMP that has the true value or the value held as true as reference. Consistently positive or negative in sign. Can be assigned to a result or a CMP (Chap. 2, pp. 50–52, 57, 69, 70, 88, 91; Box 2.1).

Errors in qualitative analysis   S. "False positive" and "False negative".

Extraction   The process by which one or several substances are separated from a solid or liquid sample (Chap. 4, pp. 151, 152, 154, 163–165, 177, 180, 181, 184, 186–188, 199).

Liquid-liquid extraction. Treatment of a liquid sample with an immiscible solvent intended to separate the analytes or their interferents.

Liquid-solid extraction. Use of a solid sorbent to retain the analytes or interferents in a liquid sample. Usually called "Solid-Phase Extraction" (SPE).

Solid-liquid extraction. Treatment of a solid sample with a suitable solvent to dissolve the analytes of interest. Also called "leaching".

Supercritical fluid extraction. Treatment of a solid sample with a supercritical fluid to separate the soluble fraction.

False negative   An error in Qualitative Analysis that results when a NO response is obtained from a sample that should

	have yielded a YES response (Chap. 5, pp. 201, 210–212, 240).
False positive	An error in Qualitative Analysis that results when a YES response is obtained from a sample that should have yielded a NO response (Chap. 5, pp. 201, 210, 211, 240).
Faraday	A chemical standard defined as the amount of electricity (96487.3 coulombs) needed for one equivalent of a redox substance to be electrochemically transformed on an electrode (Chap. 3, pp. 110, 111).
Gas chromatography	S. "Chromatography".
Generic uncertainty	S. "Uncertainty".
Good Laboratory Practices	The body of rules and procedures that are held as mandatory with a view to assuring quality and correctness in the results produced by laboratories engaged in the analysis and evaluation of substances with direct social implications and as such necessitate regulation (Chap. 8, pp. 321, 323, 340).
Gravimetric factor	The ratio of the analyte's formula weight to the molecular weight of the weighed form in gravimetry. A dimensionless number by which the result of a gravimetric weighing must be multiplied in order to determine the analyte weight. A combination of chemical standards (atomic weights) (Chap. 6, pp. 260–262, 277).
Gravimetry	A type of calculable analytical quantitation method that uses no analytical standard and is based on measurement of the mass of an analyte or a chemical derivative of it (Chap. 6, pp. 250, 252, 257–259, 264, 277).
Grey sample	S. "Sample".
Gross error	S. "Error".
Group reagent	S. "Reagent".
Heterogeneity	A property of an object or sample in space or time that poses a problem which must be solved during sample collection if the results produced by the ensuing CMP are to be representative (Chap. 4, pp. 156, 160, 161, 194; Fig. 4.10).
Identification	A qualitative analytical process by which the presence of an analyte is ascertained on the basis of chemical or physico-chemical properties of the analyte itself or a reaction product (Chap. 5, pp. 202, 204, 207, 213–222, 224–231, 233–242, 244, 245).
Identification reagent	S. "Reagent".
Indeterminate error	S. "Error".

Instrument	A measuring system that produces analytical information (primary data) which can be related to the presence or concentration of analyte(s) in the sample. The realization of the analytical technique and the second level in the hierarchy of Fig. 1.17. A wide variety of instruments exist that are classified in Fig. 4.16 (Chap. 1, pp. 3, 4, 6, 10, 11, 13, 14, 19, 21, 23–27, 30, 31, 34, 35; Chap. 4, pp. 145, 147–149, 151, 153, 161, 163–166, 168–175, 179, 180, 194, 195, 199, 200).
Instrumental analysis	S. "Chemical analysis".
Interferences	Chemical or physical perturbations of various types that systematically alter one or more steps of a CMP and hence the analytical result in terms of selectivity (Chap. 2, pp. 69–71, 80, 81, 89, 96; Fig. 2.13).
Interlaboratory exercises	Series of CMPs carried out by different laboratories under the supervision of a competent body to analyse aliquots of the same sample with a view to determining the same analytes. Used to check the results produced (and their uncertainties) for a variety of purposes (Chap. 8, p. 331, 332, 340, 342).
Ion exchange	A process by which dissolved ionic species are separated using an active solid called an "ion-exchange resin" (Chap. 4, pp. 152).
Laboratory sample	S. "Sample".
Leaching	S. "Solid-liquid extraction".
Limit of confidence	The range of values about the mean where a result can be assumed to lie with a given probability when a CMP is applied to aliquots of the same sample (Chap. 2, p. 64).
Limit of detection	The analyte concentration providing an analytical signal that can be statistically distinguished from an analytical blank (Chap. 2, pp. 205, 207–211, 240).
Limit of quantitation	The analyte concentration that gives a signal taken to be the lower limit of the linear range of the calibration curve (Chap. 2, pp. 46, 67, 68, 98).
Limiting concentration	S. "Concentration".
Linear range	The linear portion of the calibration curve where the sensitivity (slope) remains constant (Chap. 2, pp. 67–69, 88).
Liquid chromatography	S. "Chromatography".
Liquid–liquid extraction	S. "Extraction".
Liquid–solid extraction	S. "Extraction".
Macro analysis	S. "Chemical analysis".
Macro components	Analytes the proportions of which in the sample exceed 1% of their masses (Chap. 1, p. 33; Fig. 1.24).

Masking	Use of a reagent intended to interact chemically in solution with interfering species in a sample in order to avoid their perturbation without the need to physically separate the reaction products from the medium (*pseudo*-separation) (Chap. 5, pp. 219–222, 242).
Masking reagent	S. "Reagent".
Maximum tolerated ratio	A parameter that describes the influence of an interfering species in the context of selectivity. The highest interferent-to-analyte concentration ratio that results in no perturbation to the CMP (Chap. 2, pp. 71, 88, 95).
Measurand	The quantity measured in a CMP, which may or may not be the analyte. Fourth level in the hierarchy of Fig. 1.18 (Chap. 1, pp. 26, 27, 35).
Measurement	The process by which a signal yielded by the analyte or a reaction product of it is compared with that produced by a standard. Lowest step in the hierarchy of Fig. 1.18 (Chap. 1, pp. 2–7, 10, 26–28, 30).
Method	The specific implementation of an analytical technique to determine one or more analytes in a given sample. The materialization of the analytical process.

Absolute method. A type of calculable quantitation method involving some (Chap. 6, pp. 252, 263, 277) or no analytical standard (Chap. 6, pp. 257, 266).

Analytical method. The body of specific operations used in the qualitative and quantitative characterization of an analyte (or analyte family) in a given sample. Entails the use of a technique (instrument) and is the materialization of a CMP (s. Fig. 1.16) (Chap. 1, p. 23).

Calculable method. One that provides results based on mathematical calculations involving both tabulated data and measurements made during the CMP. May or may not use analytical standards (Chap. 6, pp. 252, 256–258, 271, 276–279).

Comparative method. A type of relative quantitation method in which the final result is obtained by comparing the signal of the sample with that yielded by a sample standard (Chap. 6, pp. 252, 275–277).

Official method. A method adopted and issued by a governmental office for use in endorsing the results produced by laboratories (Chap. 8, pp. 325, 340).

Primary method. The type of method with the highest level of metrological quality (Chap. 8, pp. 324, 325, 340).

Reference method. A method that is used to compare the accuracy and uncertainty of routine methods (Chap. 8, p. 325).

Relative method. In Quantitative Analysis, a method based on comparisons between measurements of the sample and of a set of analytical standards. The outcome of such comparisons is the result (Chap. 6, pp. 252, 263, 271, 275, 277–279).

Relative extrapolation method. A relative quantitation method based on the establishment of a signal-concentration relation (the calibration curve) (Chap. 6, p. 271).

Relative interpolation method. S. "Relative extrapolation method".

Standard method. A method that is developed, validated and specified in detail by a competent body (Chap. 8, pp. 325, 340).

Traceable method. A method whose results (and uncertainties) are linked to a well-known standard (Chap. 8, p. 325).

Validated method. A method the properties of which have been thoroughly studied and specified (Chap. 8, p. 325).

Method calibration	S. "Calibration".
Metrology	The science of physical, chemical, biochemical and biological measurements (Chap. 1, pp. 2, 27; Chap. 3, pp. 102, 105–107, 109, 112, 115, 116, 120, 141, 142).
Micro analysis	S. "Chemical analysis".
Micro components	Analytes the proportions of which in the sample range from 0.01% to 0.1% of their masses (Chap. 1, p. 33; Fig. 1.24).
Miniaturization	A technological trend to dramatically reducing the size of analytical tools, integrating modules of a CMP or both (Chap. 4, pp. 175, 176).
Mole	A base standard and a base unit of the International System (SI) defined as the amount of substance that contains as many elemental units (atoms, molecules, ions, electrons or other individual particles or particle groups) as are in 0.012 kilograms of the isotope carbon-12 (Chap. 3, pp. 109–112, 119, 135, 137, 139, 140).
Multiple standard addition	An extrapolation procedure for method calibration used in relative quantitation methods (Chap. 6, p. 274).
Negative error	S. "Error".
Non-chromatographic separation	S. "Separation".

Object	A system from which chemical information is required and samples are collected for analysis. Second level in the hierarchy of Fig. 1.18 (Chap. 1, pp. 1–5, 7, 11, 15, 22, 23, 26, 27, 34, 35, 38).
Official method	S. "Method".
Operational information	Secondary data not directly related to analytical results that are derived from the performance of apparatuses, instruments and analysers (Chap. 1,, p. 22).
Outlier	A datum not belonging to a data set obtained under reproducible or repeatable conditions that exhibits a significantly greater or smaller difference from the mean of the set than the other data in it (Chap. 2, p. 63–65, 92).
Positive error	S. "Error".
Precision	The degree of mutual agreement of a set of results. Dispersion of such results around their mean, which is the reference used to calculate individual deviations (random errors) (Chap. 2, pp. 40, 44–46, 51–53, 62, 77, 79–86, 88–96).
Preconcentration	A process by which sensitivity is indirectly enhanced through the use of a separation technique. Involves reducing the original volume of a sample containing the analytes at low concentrations (Chap. 4, pp. 153, 156, 161, 165, 166, 177, 195, 196; Fig. 4.15.1).
Preconcentration factor	A dimensionless number greater than unity that results from dividing the original volume into the reduced volume obtained upon application of an analytical separation technique to a sample. Multiplied by the original analyte concentration, gives the final concentration of the aliquot that is subjected to the second step of the CMP (Chap. 4, p. 166).
Preliminary operations	The body of actions performed in the first step of an analytical process (CMP). The link between the uncollected, unmeasured, untreated sample and the principal measuring instrument. First step of a CMP (Chap. 4, pp. 143, 145, 147–151, 153–155, 161, 163, 168, 170, 175, 194, 197, 199).
Primary analytical standard	S. "Standard".
Primary data	Data produced by instruments in measurement processes. The most elementary form of information and the foundation of the results. Third step in the analytical information hierarchy. The result of detecting and sensing (Chap. 1, pp. 4, 14, 22, 23; Fig. 1.15).
Primary method	S. "Method".

Primary sample	S. "Sample, Bulk".
Procedure	A detailed specification of an analytical method. The lowest level of concreteness in the hierarchy of Fig. 1.16 (Chap. 1, pp. 12, 17, 18, 23, 24).
Productivity	A characteristic of a laboratory defined as the combination of its accessory analytical properties (expeditiousness, cost-effectiveness and personnel safety and comfort) (Chap. 2, pp. 73, 76, 77, 81, 89).
Proficiency testing	A form of external assessment of the quality of the results of an analytical laboratory that involves participation in a specially designed interlaboratory exercise (Chap. 8, pp. 330, 331, 333, 334, 342).
Qualimetrics	The triple interface that results from the convergence of Computers, Chemometrics and Quality on the laboratory (Chap. 8, pp. 334, 335).
Qualitative analysis	S. "Chemical analysis".
Quality	The totality of the characteristics or abilities of an entity that make it better, equal to or worse than others of the same kind. Identified with client satisfaction in practice (Chap. 8, pp. 307–344).
	Analytical quality. The degree of excellence in the chemical information supplied with a view to solving the analytical problem. Encompasses four components: quality of results, quality of CMPs, quality of the analytical tools used and quality of the work and its organization (Chap. 8, pp. 308, 317–319, 322, 324, 326, 327, 334–336, 339–344).
Quality assessment	Specific activities (audits) carried out by personnel from outside the laboratory to examine both the results produced and the laboratory as such and in relation to its quality control systems (Chap. 8, pp. 315, 319, 320, 322, 326–331, 335, 340).
Quality assurance	The body of activities performed in order to assure quality in the results produced by an analytical laboratory. Involves specific control, assessment and correction activities (Chap. 8, pp. 315, 319, 320, 322–324, 326, 327, 329, 334–340, 343, 344).
Quality control	The body of specific activities carried out by laboratory personnel in order to – basically – examine, in a direct manner, the results obtained and tools used by the laboratory (Chap. 8, pp. 315, 319, 320, 324–330, 335, 339, 340, 341, 344).
Quality indicator	A qualitative and quantitative aspect in which some characteristic or ability of an entity that meets the client's requirements materializes (Chap. 8, pp. 311, 312).

Quality manual	A detailed written description of a laboratory and all the activities it performs, with emphasis on quality control and assessment (Chap. 8, pp. 322, 323, 328, 330–332, 335, 337, 338, 340).
Quantitative analysis	S. "Chemical analysis".
Quantity	An attribute of an object that can be distinguished qualitatively and determined quantitatively. Encompasses the concepts of measurand and analyte (Chap. 1, pp. 23, 27).
Random error	S. "Error".
Random sample	S. "Sample".
Reagent	A chemical species that is added to a sample or standard in order to form a reaction product with the analyte(s).
	Group reagent. One that separates a small number of analytes from those present in the sample. Used within the framework of analytical schemes in Classical Qualitative Analysis (Chap. 5, pp. 217, 223, 224).
	Identification reagent. One that reacts with an analyte to produce an external effect that can be readily identified by human senses (*e.g.* in Classical Qualitative Analysis) (Chap. 5, pp. 219–222).
	Masking reagent. One that reacts in solution with species accompanying the analytes in the sample in order to cancel their interferences (Chap. 5, pp. 220–222, 242).
Reference material	A material or substance one or more properties of which are sufficiently uniform and well known for use in calibrating an instrument or apparatus, assigning values to materials and systems or assessing analytical methods (Chap. 3, pp. 103, 107, 114, 116, 120, 122, 129, 130, 134, 141, 142).
Reference method	S. "Method".
Relative error	S. "Error".
Relative method	S. "Method".
Relative extrapolation method	S. "Method".
Relative interpolation method	S. "Method".
Relative standard deviation	An expression of the standard deviation in relative terms (as a fraction of unity with respect to the mean of the set of results) (Chap. 2, pp. 48, 58, 59, 65, 94).
Relative trueness	A difference between values of the metrological scale (*e.g.* $\mu - \hat{X}$; $\hat{X}' - \hat{X}$; $\mu' - \hat{X}'$) as defined by different authors (Chap. 2, pp. 45, 50, 83; Fig. 2.5).

Reliability	A characteristic of a method (CMP) defined as its ability to preserve its accuracy and precision over time. Related to robustness and transferability (Chap. 2, p. 85). The proportion of correct identifications in individual qualitative tests performed on aliquots of the same sample. A capital property in Qualitative Analysis that combines accuracy and precision, and is used as an attribute of the binary response (Chap. 5, pp. 208, 210–213, 216, 217, 222, 230, 231, 240, 241).
Repeatability	A manner of expressing precision. Defined by ISO as "the dispersion of the results for mutually independent tests using the same method as applied to aliquots of the same sample, at the same laboratory, by the same operator, using the same equipment over a short interval of time." (Chap. 2, pp. 60–62, 88, 94).
Representative sample	S. "Sample".
Representativeness	A capital analytical property related to consistency between the results, the samples received, the object, the analytical problem and the social or economic problem (Chap. 2, pp. 40, 42, 46, 52, 54–56, 76, 77, 88, 89).
Reproducibility	A manner of expressing precision. Defined by ISO as "the dispersion of the results for mutually independent tests performed by applying the same method to aliquots of the same sample under different conditions: different operators, different equipment or different laboratories." (Chap. 2, pp. 60–62, 88, 94, 95).
Robustness	An analytical property of a CMP that reflects its resistance to slight changes in the experimental conditions under which it is performed (Chap. 2, pp. 84–86, 89).
Safety	An attribute of a laboratory or CMP related to hazards to human health and/or the environment (Chap. 2, pp. 40, 75, 85, 89).
Sample	A part (aliquot) of an object that contains the analyte. Third level in the hierarchy of Fig. 1.18. Possesses a broad range of variants (Chap. 1, pp. 5, 8, 11, 14, 21–24, 26, 28–33, 35, 36; Chap. 4, pp. 144–153, 155–163, 165–168, 170, 172, 174, 176, 177, 179–189, 191–194, 196–199; Fig. 4.11). **Black sample.** A sample the composition of which is completely unknown (Chap. 5, pp. 202, 217, 240). **Blind sample.** A sample of well-defined composition that is interspersed for quality control purposes with

those to be routinely analysed by a laboratory (Chap. 8, pp. 309, 323, 327, 329, 341).

Bulk sample or **primary sample.** The result of the first selection from the object. Usually of a large size.

Composite sample. The result of combining several portions of a bulk sample.

Convenience sample. One selected in terms of accessibility, cost-effectiveness, efficiency, *etc.*

Grey sample. A sample the composition of which is known only approximately (Chap. 5, pp. 202, 221, 222, 240).

Laboratory sample. A portion of the object that is submitted, in an appropriate container, for analysis by the laboratory.

Random sample. One selected in such a way that any portion of the object will have a specified probability (*e.g.* 95%) of being withdrawn.

Representative sample. A portion of the object that is selected by applying a sampling plan consistent with the analytical problem addressed.

Selective sample. A sample collected by following a selective sampling procedure.

Stratified sample. One withdrawn from a stratum or well-defined zone of the object.

Test sample (aliquot). The portion of a laboratory sample that is eventually subjected to the analytical process.

White sample. A sample with well-defined properties which, with some exceptions, remain virtually constant as a whole (Chap. 5, pp. 202, 240).

Sample collection	S. "Sampling".
Sample custody chain	The action or series of actions that ensures an unequivocal relationship between the sample aliquot subjected to a CMP and the result it produces (sample traceability) (Chap. 3, pp. 133, 138).
Sample standard	S. "Standard".
Sample throughput	A measure of expeditiousness of CMPs. The number of samples that can be processed per unit time (*e.g.* hour, day) (Chap. 2, p. 82).
Sample treatment	A generic term used to refer to the substeps of the preliminary operations of the CMP performed in order to condition the test sample or aliquot for measurement of the analytical signal (second step of the CMP) (Chap. 4, pp. 146, 150, 161, 168).
Sampling	An operation by which one or more portions (aliquots) of an object are chosen for individual or joint

	subjection (following size reduction) to a CMP. Of the intuitive, statistical, directed or protocol-based type (Chap. 4, pp. 146, 150, 153, 155–161, 173, 194, 195, 199).
Sampling error	S. "Error".
Sampling plan	The scheme used to ensure that the results of a CMP will be consistent with the analytical problem addressed (Chap. 4, pp. 156–158, 160, 194, 195).
Secondary analytical standard	S. "Standard".
Secondary data	Items of non-analytical information that characterize the performance of apparatuses and instruments in the analytical process. Lowest level in the information hierarchy (Chap. 1, p. 22; Fig. 1.15).
Selective sample	S. "Sample".
Selectivity	A basic analytical property of an analytical method defined as its ability to produce results exclusively dependent on the analyte for its identification or quantitation in the sample (Chap. 2, pp. 40, 56, 69, 70–72, 77, 79, 80, 81, 88, 89, 95).
Selectivity factor	A parameter describing the selectivity of a method with respect to another. Defined as the quotient of the tolerated interferent-to-analyte ratios obtained with the two methods in determining the same analyte (Chap. 2, p. 72).
Semi-micro analysis	A type of chemical analysis where the initial size of the sample aliquot subjected to the CMP ranges from 10 to 100 mg (Chap. 1, p. 32; Fig. 1.23).
Sense, to	To use a device responding to the presence or concentration of an analyte in a sample. Entails interacting with an instrument proper (Chap. 1, p. 23).
Sensor	A portable, easy to use miniature device or instrument that responds to the presence or concentration of an analyte (or analyte family) in a sample. Usually connected to or built into an instrument (Chap. 1, pp. 4, 16, 19, 23, 25, 26).
Separation	An operation involving mass transfer between two phases. A crucial element of the preliminary operations of a CMP. Discrete or continuous in nature (Chap. 1, pp. 30, 31, 35; Chap. 4, pp. 163, 165, 167, 170, 172, 194, 195, 199).
	Chromatographic separation. One where distribution between phases reaches equilibrium many times, thus significantly enhancing the separation efficiency.

Non-chromatographic separation. One where mass transfer between phases reaches equilibrium one or several times.

Simplification

A technological trend to reducing the number of steps traditionally involved in CMPs in order to increase expeditiousness and decrease costs (Chap. 4, pp. 175, 176, 195).

Social or economic problem

Client's issue that is to be solved or clarified and that, as such, raises a need for information. Top level in the hierarchy of Fig. 1.8 (Chap. 1, pp. 3, 4, 14).

Solid-liquid extraction

S. "Extraction".

Speciation

A type of analysis that provides qualitative and quantitative information about the different forms in which an analyte element may occur in a (usually environmental) sample (Chap. 1, pp. 21, 30, 34).

Specific uncertainty

S. "Uncertainty".

Standard

A tangible or intangible reference used to support or perform analytical chemical measurements (Chap. 3, pp. 101–116, 118–125, 127–131, 133–135, 137, 139–142).

Analytical chemical standard. Any of the standards used in ordinary practice. Of the primary or secondary type (Chap. 3, pp. 101, 107, 110, 112, 113, 116, 117, 139).

Base standard. A standard that coincides with one of the seven SI base units. Only the kilogram prototype is of the tangible type (Chap. 3, pp. 109, 110, 112, 116, 120, 130, 131, 134, 140).

Chemical standard. A standard that can be considered a (traceability) link between base (SI) standards and analytical chemical standards (Chap. 3, pp. 101, 107, 110–113, 116, 134, 137, 139; Fig. 3.6).

Primary analytical standard. A stable, homogeneous substance of well-known properties that can be directly related to a chemical standard via a traceability chain (Chap. 3, pp. 111, 113, 114, 118, 119, 130).

Sample standard. An artificial, naturally occurring or modified natural material intended to simulate as closely as possible an actual sample and possessing the properties of a reference material or certified reference material (Chap. 3, pp. 116, 118, 121, 128, 130, 134, 137, 140, 141).

Secondary analytical standard. A substance that lacks the properties required for direct use as an analytical chemical standard but is used anyway because an appropriate primary standard for the intended purpose either does not exist or is unavailable or expensive. Also known as "working standard" (Chap. 3, pp. 113, 119).

Standard deviation   A statistical parameter that reflects the precision of a set of results (Chap. 2, pp. 47, 48, 57–59, 63, 65, 67, 81, 82, 93, 94, 99).

Standard method   S. "Method".

Stratified sample   S. "Sample".

Structural analysis   S. "Chemical analysis".

Supercritical fluid extraction   S. "Extraction".

Systematic error   S. "Error".

Technique   A scientific principle used to obtain analytical information. Entails the use of an instrument. Top level in the hierarchy of Fig. 1.16 (Chap. 1, pp. 6, 19, 21, 23, 24, 30, 31, 35, 36).

Test sample   S. "Sample".

Threshold concentration   S. "Concentration".

Titration   The addition of a solution from a burette to a known volume of a second solution until a chemical reaction between the two is completed (Chap. 6, pp. 250, 264–269, 271, 277).

Titration curve   A logarithmic or linear plot of the monitored signal as a function of the titrant volume used in a titration (Chap. 6, pp. 268, 271, 277; Box 6.11).

Titrimetric factor   A dimensionless number by which the approximate concentration of a titrant solution prepared from a secondary standard must be multiplied in order to determine its actual concentration. Obtained through testing and calculation (Chap. 3, p. 113; Boxes 3.8 and 3.9).

Titrimetry   A classical quantitation technique involving an absolute method based on the use of analytical standards that relies on the accurate measurement of the volume of titrant solution required to react, in a quantitative manner, with the analyte present in a sample. Can be of the direct or indirect type and performed in a manual, semi-automatic, automatic or automated manner (Chap. 6, pp. 252, 257, 263, 264, 270).

Trace analysis   S. "Chemical analysis".

Traceability   An attribute that characterizes various analytical notions. An abstract concept that integrates two meanings: tracing (the history of production or performance) and relationship to standards (Chap. 3, pp. 101–114, 116, 119, 120, 122, 123, 130–141).

Traceability of an aliquot. Unequivocal relationship of a sample aliquot subjected to a CMP to both the social or economic problem (representativeness) and the result (sample custody chain). In this way, consistency between the problem and result (cyclic traceability) is assured (Chap. 3, pp. 133; Fig. 3.15).

Traceability of an instrument. Documented history of the performance of an instrument (installation, malfunctioning, repairs, servicing, calibration, correction, hours of use, samples processed, *etc.*). Through calibration, the relationship to standards implicit in the traceability concept is established (Chap. 3, pp. 134, 138).

Traceability of a method. A property of a method through which the results of its CMP are related to well-established references (Chap. 3, p. 135).

Traceability of a result. A property of a result or of the value of a standard through which the result or value is related to well-established national or international references via an unbroken chain of comparisons characterized by their respective uncertainties (Chap. 3, pp. 104, 130, 131, 133, 135, 136, 138).

Traceable method	S. "Method".
Traces	A word used to designate analytes present in proportions below 0.01% (100 ppm) in the sample (Chap. 1, p. 24, 31, 33).
Transfer weights	Objects of fixed mass used to calibrate balances. Available in various classes dependent on their uncertainty and issuer (Chap. 3, pp. 109, 114, 119, 125, 137; Box 3.5).
Transferability	An attribute of a CMP that reflects its ability to provide consistent results on application to the same samples in different laboratories. Related to robustness and reliability (Chap. 2, p. 85).
Ultramicro analysis	S. "Chemical analysis".
Unbroken chain of comparisons	The foundation of traceability of a result (Chap. 3, p. 130).
Uncertainty	The quality of being indeterminate in number, amount or extent.

Generic uncertainty. Dubiousness in the chemical composition of an object or sample. The opposite of information (Chap. 2, pp. 43, 44).

Specific uncertainty. The range of values where a result, a mean of such values and the value held as true may fall. Similar to, but not the same as, precision. Can be absolute, partial or zero (Chap. 2, pp. 42–44, 46, 51, 58, 59, 71, 84; Fig. 2.3).

Validated method S. "Method".

Validation The experimental, documented demonstration that a general process (CMP) or a particular step (*e.g.* sampling, data processing) has performed as expected and will continue to do so. Also, the experimental, documented demonstration that an object (*e.g.* an apparatus or instrument) possesses specified properties and will continue to do so (Chap. 4, pp. 173, 174, 194, 195; Fig. 4.19).

Value held as true A datum (accompanied by an uncertainty) derived by chemometric treatment of the results obtained by having many different laboratories process aliquots of the same sample (a CRM) in order to determine the same analyte. Related to referential quality (Chap. 2, pp. 44, 45, 48, 51, 53, 54, 69, 77, 91).

Volatilization A separation technique occasionally used for gravimetric purposes that relies on the difference between the mass of the sample prior to and after controlled heating (in the presence or absence of a reagent) (Chap. 6, p. 261).

White sample S. "Sample".

Symbols and Abbreviations used in the Text

Symbols

C_C	Cut-off concentration as used in the context of the binary response
C_L	Limiting concentration or threshold concentration, imposed by the client
C_{LOQ}	Analyte concentration at the limit of quantitation
C_{LOD}	Analyte concentration at the limit of detection
CV	Coefficient of variation
d_{x_i}	Deviation of a result with respect to the mean (random error)
δ	Sample throughput
e_{x_i}	Systematic error of a result
$e_{\bar{X}}$	Systematic error of a CMP when applied to $n < 30$ aliquots of the same sample. Ascribed to the mean $\bar{X}$ of the results (bias)
$e_{\mu'}$	Systematic error of a CMP when applied to $n > 30$ aliquots of the same sample. Ascribed to the mean μ' of the results (relative trueness)
F	Faraday (96 487.3 coulombs)
F_g	Gravimetric factor
g	Gram
I	Current intensity (amperes)
kg	Kilogram
L	Litre
µg	Microgram
µL	Microlitre
mg	Milligram
mL	Millilitre
MW	Molecular weight
μ	Arithmetic mean of ∞ analytical results obtained by processing n aliquots (an ideal) of the same sample in order to determine the same analyte
μ'	Arithmetic mean of $n > 30$ analytical results obtained by processing n aliquots of the same sample in order to determine the same analyte
N	Avogadro's number (6.023×10^{23} molecules or atoms in one mole)
n	Number of aliquots of the same sample and/or of analytical results
ng	Nanogram
Q	Amount of electricity (coulombs)
RSD (rsd)	Relative standard deviation
S	Sensitivity (variation of the analytical signal with concentration)
s	Standard deviation of a set of $n < 30$ results

σ	Standard deviation of a set of $n > 30$ results
s_B	Standard deviation of $n < 30$ analytical blank signals
σ_B	Standard deviation of $n > 30$ analytical blank signals
sdm	Standard deviation of the mean
t	Time
$U_{\mu'}$	Specific uncertainty of the mean of $n > 30$ results
$U_{\bar{X}}$	Specific uncertainty of the mean of $n < 30$ results
$U_{\hat{X}'}$	Specific uncertainty of the value held as true
U_{x_i}	Specific uncertainty of an individual result
v	Variance (standard deviation squared)
W_a	Weight of analyte, expressed as required
W_g	Gravimetric weighing
$\bar{X}$	Arithmetic mean of $n < 30$ analytical results obtained by processing n aliquots of the same sample in order to determine the same analyte
$\hat{X}$	True value (an unreachable ideal)
$\hat{X}'$	Value held as true, obtained through special testing (interlaboratory exercises)
$\bar{X}_B$	Arithmetic mean of n signals from analytical blanks
x_i	Individual analytical result
x_{LOQ}	Analytical signal at the limit of quantitation
x_{LOD}	Analytical signal at the limit of detection

Abbreviations

AOAC	Association of Official Analytical Chemists (USA)
BCR	Community Bureau of Reference. Former SMT
BIMP	International Bureau of Weights and Measures
CCF	Liquid chromatography as performed on a thin layer of stationary phase spread onto an planar inert support
CCQM	Consultative Committee for Quantity of Matter
CE	Capillary electrophoresis
CEN	European Centre for Standardization
CMP	Chemical measurement process. A synonym for "analytical process"
CRM	Certified reference material
CV	Coefficient of variation
ELISA	Enzyme Linked Immunosorbent Assay
EPA	Environmental Protection Agency (USA)
EU	European Union
EURACHEM	EURopean Analytical CHEMistry (a European network of associations working in support of chemical metrology)
FDA	Food and Drug Administration (USA)
GC	Gas chromatography
GLPs	Good Laboratory Practices
IDMS	Isotope Dilution Mass Spectrometry
IRM	Internal reference material
ISO	International Standards Organization
IUPAC	International Union of Pure and Applied Chemistry
LC (HPLC)	Liquid Chromatography (High Performance Liquid Chromatography)
LOD	Limit of detection
LOQ	Limit of quantitation
NIST	National Institute of Standards and Technology (USA)
OECD	Organization for Economic Cooperation and Development
ppb	Parts per billion: parts in 1000 million. Units of a relative quantitative result. Not endorsed by IUPAC
ppm	Parts per million. Units of a relative quantitative result. Not endorsed by IUPAC
ppt	Parts pert trillion: parts in 1000 (European) billion. Not endorsed by IUPAC
QAU	Quality Assurance Unit
RM	Reference material
RSD (rsd)	Relative standard deviation
SF	Selectivity Factor (of a method with respect to another)
SI	International System (of units)
SMT	Standards, Measurements and Testing (an EU body)
SOPs	Standard Operating Procedures (in the context of GLPs)